Biotechnology for Odor and Air Pollution Control

Zarook Shareefdeen • Ajay Singh (Eds.)

Biotechnology for Odor and Air Pollution Control

With 70 Figures

Springer

DR. ZAROOK SHAREEFDEEN

BIOREM Technologies Inc.
Research and Development
7496 Wellington Road 34
Guelph, ON N1H 6H9
Canada

e-mail:
zshareefdeen@bioremtech.com

DR. AJAY SINGH

Petrozyme Technologies Inc.
7496 Wellington Road 34
Guelph, ON N1H 6H9
Canada

Department of Biology
University of Waterloo
Waterloo, ON N2L 3G1
Canada

e-mail:
asingh@petrozyme.com
ajasingh@sciborg.uwaterloo.ca

Library of Congress Control Number: 2007932193

ISBN 978-3-540-74049-0 **Springer Berlin Heidelberg New York**

Springer is a part of Springer Science + Business Media
springer.com

5 4 3 2 1 0 - Printed on acid-free paper

Preface

Volatile organic and inorganic odorous compounds from various industries are emitted in large quantities and create hazards to the ecosystem and health effects to humans. The need for odor and air pollution control is driven by regulatory issues, generally enforced as a result of public complaints about poor local air quality and through emission monitoring by the enforcement agencies. With increasing populations, and new residential and industrial developments, the demand for air pollution control systems that provide nuisance-free, breathable air is growing.

Biotechnology offers one of the most economical and environmentally benign methods of air pollution control for industrial and municipal airstreams. Bioprocess is becoming an effective alternative to traditional physical and chemical odor removal methods used in wastewater treatment plants, rendering plants, chemical production facilities, and food and flavor manufacturing facilities.

This book, *Biotechnology for Odor and Air Pollution Control*, covers not only the basic microbiological and engineering aspects of various bioreactors, but also describes the design, modeling, and control of such bioreactors for their unique applications in odor and air pollution management, and control in industrial facilities and wastewater treatment plants. Specific topics include methods of odor and volatile organic compound (VOC) control, regulatory issues, microbiological aspects of bioreactors, selection of bioreactor media, description and design of various bioreactors for odor and air pollution control, such as biofilter, biotrickling filter, bioscrubber and membrane bioreactor, and applications and case studies related to these technologies for the treatment of air contaminants in municipal and industrial plants, and future prospects of biotechnology for odor and air pollution control.

The contributing authors are applied and industrial researchers from diverse backgrounds: universities, research institutes, and industries. They are experts in biological methods for odor and air pollution control. This book will be a valuable reference tool for graduate students, scientists, industrial consultants, biotechnologists, microbiologists, and chemical, biochemical, environmental and civil engineers who are interested in environmental sciences, and particularly, in innovative biological technologies for treatment and control of odor and air pollution. We hope that students, teachers, scientists and engineers, whether in academia, industry or government, will find the descriptive and practical contents of this book interesting and helpful.

We are grateful to all the authors for their excellent contributions. Several of our colleagues provided encouragement and help during the various stages of this editorial work. Continuous support and guidance provided by Dr. Jutta Lindenborn, Springer, during the preparation of this book is highly appreciated.

Guelph, Ontario, May 2004 *Zarook Shareefdeen and Ajay Singh*

Contents

Contributors

Aldrich, R.L., Sterling Environmental Engineering, P.C., 24 Wade Road, Latham, New York 12110, USA

Allen, D.G., Department of Chemical Engineering and Applied Chemistry, and Pulp & Paper Centre, University of Toronto, 200 College Street, Toronto, Ontario M5S 3E5, Canada

Andrès, Y., Ecole des Mines de Nantes, 4 rue Alfred Kastler, B.P. 20 722, 44307 Nantes Cedex 03, France

Bina, A., Department of Civil and Environmental Engineering, University of Southern California, Los Angeles, California 90089-2531, USA

Cudmore, R., Aurora Pacific Ltd., Christchurch, New Zealand

Datta, I., Department of Chemical Engineering and Applied Chemistry, and Pulp & Paper Centre, University of Toronto, 200 College Street, Toronto, Ontario M5S 3E5, Canada

Deshusses, M.A., Department of Chemical and Environmental Engineering, University of California, Riverside, California 92521, USA

Devinny, J.S., Department of Civil and Environmental Engineering, University of Southern California, Los Angeles, California 90089-2531, USA

Fitch, M.W., Department of Civil, Architectural and Environmental Engineering, University of Missouri-Rolla, Rolla, Missouri 65409, USA

Gabriel, D., Escola Tecnica Superior d'Enginyeria, Universitat Autonoma de Barcelona, Bellaterra, Barcelona 08193, Spain

Gérente, C., Ecole des Mines de Nantes, 4 rue Alfred Kastler, B.P. 20 722, 44307 Nantes Cedex 03, France

Gostomski, P., Department of Chemical & Process Engineering, University of Canterbury, Christchurch, New Zealand

Govind, R., Department of Chemical Engineering, Mail Location 171, University of Cincinnati, Cincinnati, Ohio 45221-0171, USA

Herner, B., BIOREM Technologies, 7496 Wellington Road 34, Guelph, Ontario N1H 6H9, Canada

Kraakman, B., BIOWAY bv, Rubensstraat 187, 6717 VE Ede, Postbus 361, 6710 BJ Ede, The Netherlands

Le Cloirec, P., Ecole des Mines de Nantes, 4 rue Alfred Kastler, B.P. 20 722, 44307 Nantes Cedex 03, France

Mohseni, M., Department of Chemical & Biological Engineering, University of British Columbia, 2216 Main Mall, Vancouver, British Columbia V6T 1Z4, Canada

Morgan-Sagastume, J.M., Environmental Bioprocesses Department, Engineering Institute, Universidad Nacional Autónoma de México (UNAM), Apdo. Postal 70-472, 04510 Mexico D.F., Mexico

Narayan, S., Department of Chemical Engineering, Mail Location 171, University of Cincinnati, Cincinnati, Ohio 45221-0171, USA

Ozis, F., Department of Civil and Environmental Engineering, University of Southern California, Los Angeles, California 90089-2531, USA

Popov, V., A.N. Bakh Institute of Biochemistry, Russian Academy of Sciences, Leninskiy pr. 33, 117071 Moscow, Russia

Pré, P., Ecole des Mines de Nantes, 4 rue Alfred Kastler, B.P. 20 722, 44307 Nantes Cedex 03, France

Revah, S., Department of Process Engineering, Universidad Autónoma Metropolitana-Iztapalapa (UAM-I), Apdo. Postal 55-534, 09340 Mexico D.F., Mexico

Shareefdeen, Z., BIOREM Technologies, 7496 Wellington Road 34, Guelph, Ontario N1H 6H9, Canada

Singh, A., Petrozyme Technologies, 7496 Wellington Road 34, Guelph, Ontario N1H 6H9, Canada

Ward, O., Department of Biology, University of Waterloo, Waterloo, Ontario N2L 3G1, Canada

Webster, T.S., Shaw Environmental, Inc., 1230 Columbia Street, Suite 1200, San Diego, California 92101-8517, USA

Zhukov, V., A.N. Bakh Institute of Biochemistry, Russian Academy of Sciences, Leninskiy pr. 33, 117071 Moscow, Russia

Part I

Introduction and Basic Theory

lid waste need to be land-filled or incinerated. In absorption or scrubbing, llutants from the air are absorbed into a scrubbing solution such as water or lvents. Often, chemical costs are high and liquid waste needs further treat-ent. Nozzle maintenance, complex chemical feed and control systems and gh operating costs are a few problems associated with this method (Devinny al. 1999). The incineration technology has been widely used due to its high iciency. To increase the efficiency of incineration and reduce fuel require-ent for combustion, several forms of this technology, such as recuperative, generative and catalytic oxidation, are practiced. Regardless, incineration is expensive method due to high energy requirements, and it is not economical concentration levels are low and large airflow volumes need treatment. This ocess also produces the highest amount of greenhouse gases (i.e., carbon oxide, or NO_X gases).

Biological methods are effective and economical for biodegradable odorants d VOC contaminants. Air-phase bioreactors used in industries for odor and OC removal include *biofilters*, *biotrickling filters*, and *bioscrubbers*. In gen-al, highly soluble and low molecular weight VOCs (i.e., methanol, ethanol, lehydes, acetates, ketones, and some aromatic hydrocarbons) and inorganic mpounds (i.e., hydrogen sulfide, ammonia) are easily biodegradable in these oreactors. Low molecular weight aliphatic hydrocarbons such as methane, ntane and some chlorinated compounds are difficult to biodegrade (Devinny al. 1999). Novel bioreactors for VOC and odor control include rotating drum ofilters, horizontal flow biofilters, foamed emulsion bioreactors, short contact ne biotrickling filters (Gabriel et al. 2002), higher plant-based biofilters (Guil-ult 2002), and microwave concentrator/biofilter integrated systems (Webster al. 2002). However, these bioreactors are in the early stages of development, d several pilot-scale demonstration studies are in progress. Chapter 3 gives detailed review on all physicochemical methods (i.e., condensation, mem-anes, UV oxidation, plasma, masking, and ozone oxidation) and biological ethods for odor and VOC control.

3
ological Reactors

.1
oreactor Media

edia selection for a bioreactor is, in general, based on the ability to support cterial growth. However, criteria vary depending on the reactor types. The rformance of a bioreactor for odor or VOC control depends on the nature the carrier or support media where the adhesion of microorganisms takes ace, resulting in the development of a biofilm due to contaminant degrada-n. Large surface area, pressure drop, cost, chemical reactivity, and void space e important factors considered in selecting a bioreactor medium. For biofil-s, in addition to biological properties, media must provide good absorption pacity, adsorption property, pH buffering capacity, good pore structure, and

1 Biotechnology for Air Pollution Control – an Overviev

Zarook Shareefdeen, Brian Herner, Ajay Singh

1.1 Introduction

Biotechnology offers the most economical and environmenta
for air pollution control when dealing with the removal of
contaminants from industrial and municipal airstreams.
large amounts, volatile organic compounds (VOCs) and i
compounds create hazards to the ecosystem and health e
Substances such as ammonia, amines, hydrogen sulfide, m
dimethyl sulfide, and dimethyl disulfide cause odor nuisan
ment. Increase in population density, new development o
dustrial facilities create a growing need for air pollution co
provide nuisance-free, breathable air. This chapter provid
various biotechnological methods used in odor and air poll

The need for the removal of odors and VOCs is driven by
generally enforced as a result of public complaints about po
and through emission monitoring by the enforcement age
1990s, it was not an easy task for an industry to select a biotec
reduce odor or VOC air emissions as a means of compliance
details on odor and VOC control laws, regulatory measures t
complaints, performance standards required for biological t
and review of regulations in several countries are discussed.

1.2 Methods of Odor and VOC Control

The treatment of off-gases has been practiced for years and i
on non-biological methods such as condensation, activate
tion, absorption/scrubbing, and incineration. In the cond
cooling and compression condense contaminant vapors fr
cess is economical for higher boiling point compounds and n
airstreams. In the adsorption process, pollutants are adsorbe
(i.e., activated carbon). This process is effective when the co
airstreams is low. Regeneration of the adsorbents is done u
air. However, recovery of compounds is costly, and spent ac

Shareefdeen/Singh (Eds.)
Biotechnology for Odor and

1 Biotechnology for Air Pollution Control – an Overview

Zarook Shareefdeen, Brian Herner, Ajay Singh

1.1 Introduction

Biotechnology offers the most economical and environmentally benign method for air pollution control when dealing with the removal of odorous and toxic contaminants from industrial and municipal airstreams. When emitted in large amounts, volatile organic compounds (VOCs) and inorganic odorous compounds create hazards to the ecosystem and health effects to humans. Substances such as ammonia, amines, hydrogen sulfide, methyl mercaptan, dimethyl sulfide, and dimethyl disulfide cause odor nuisance in the environment. Increase in population density, new development of housing and industrial facilities create a growing need for air pollution control systems that provide nuisance-free, breathable air. This chapter provides an overview of various biotechnological methods used in odor and air pollution control.

The need for the removal of odors and VOCs is driven by regulatory issues, generally enforced as a result of public complaints about poor local air quality and through emission monitoring by the enforcement agencies. In the early 1990s, it was not an easy task for an industry to select a biotechnology system to reduce odor or VOC air emissions as a means of compliance. In Chapter 2, the details on odor and VOC control laws, regulatory measures to handle citizens' complaints, performance standards required for biological treatment systems, and review of regulations in several countries are discussed.

1.2 Methods of Odor and VOC Control

The treatment of off-gases has been practiced for years and is primarily based on non-biological methods such as condensation, activated carbon adsorption, absorption/scrubbing, and incineration. In the condensation process, cooling and compression condense contaminant vapors from air. This process is economical for higher boiling point compounds and more concentrated airstreams. In the adsorption process, pollutants are adsorbed onto adsorbents (i.e., activated carbon). This process is effective when the concentration in the airstreams is low. Regeneration of the adsorbents is done using steam or hot air. However, recovery of compounds is costly, and spent adsorbents such as

Shareefdeen/Singh (Eds.)
Biotechnology for Odor and Air Pollution Control

solid waste need to be land-filled or incinerated. In absorption or scrubbing, pollutants from the air are absorbed into a scrubbing solution such as water or solvents. Often, chemical costs are high and liquid waste needs further treatment. Nozzle maintenance, complex chemical feed and control systems and high operating costs are a few problems associated with this method (Devinny et al. 1999). The incineration technology has been widely used due to its high efficiency. To increase the efficiency of incineration and reduce fuel requirement for combustion, several forms of this technology, such as recuperative, regenerative and catalytic oxidation, are practiced. Regardless, incineration is an expensive method due to high energy requirements, and it is not economical if concentration levels are low and large airflow volumes need treatment. This process also produces the highest amount of greenhouse gases (i.e., carbon dioxide, or NO_X gases).

Biological methods are effective and economical for biodegradable odorants and VOC contaminants. Air-phase bioreactors used in industries for odor and VOC removal include *biofilters*, *biotrickling filters*, and *bioscrubbers*. In general, highly soluble and low molecular weight VOCs (i.e., methanol, ethanol, aldehydes, acetates, ketones, and some aromatic hydrocarbons) and inorganic compounds (i.e., hydrogen sulfide, ammonia) are easily biodegradable in these bioreactors. Low molecular weight aliphatic hydrocarbons such as methane, pentane and some chlorinated compounds are difficult to biodegrade (Devinny et al. 1999). Novel bioreactors for VOC and odor control include rotating drum biofilters, horizontal flow biofilters, foamed emulsion bioreactors, short contact time biotrickling filters (Gabriel et al. 2002), higher plant-based biofilters (Guilbault 2002), and microwave concentrator/biofilter integrated systems (Webster et al. 2002). However, these bioreactors are in the early stages of development, and several pilot-scale demonstration studies are in progress. Chapter 3 gives a detailed review on all physicochemical methods (i.e., condensation, membranes, UV oxidation, plasma, masking, and ozone oxidation) and biological methods for odor and VOC control.

1.3 Biological Reactors

1.3.1 Bioreactor Media

Media selection for a bioreactor is, in general, based on the ability to support bacterial growth. However, criteria vary depending on the reactor types. The performance of a bioreactor for odor or VOC control depends on the nature of the carrier or support media where the adhesion of microorganisms takes place, resulting in the development of a biofilm due to contaminant degradation. Large surface area, pressure drop, cost, chemical reactivity, and void space are important factors considered in selecting a bioreactor medium. For biofilters, in addition to biological properties, media must provide good absorption capacity, adsorption property, pH buffering capacity, good pore structure, and very low compaction over time (Leson and Winer 1991; Williams and Miller 1992). Although the selection of biofilter media should be based on all these parameters, frequently only media with good biodegradation properties (i.e., peat, compost, soil, chicken manure) are selected, without giving consideration to structural, mass transfer and adsorption properties. Some of these issues have been resolved with the introduction of vendor-supplied manufactured media that are produced with definitive specifications. For example, Biorem's manufactured medium consists of hydrophilic mineral cores coated with hydrophobic sorption material to give a high specific surface area. The coating of the media includes nutrient-rich organic material for microbe hosting, and suitable binders to provide product strength, buffer capacity and stability. Through research and experience in field-scale demonstrations, bioreactor vendors have now put emphasis on media development and selection for more efficient odor and VOC removal performance that meets compliance. In Chapter 4, medium selection criteria, engineering specifications and limitations are discussed in detail.

1.3.2 Microbiology

Several groups of microorganisms, primarily bacterial species, are responsible for the degradation of the air pollutants in bioreactors. Naturally occurring biofilter media such as peat and compost contain about 1 billion microorganisms per gram, capable of degrading odor and VOC contaminants present in the air. Activated sludge suspensions from sewage treatment plants serve as inoculum for many compounds (Ottengraf and Diks 1990), but poorly biodegradable compounds such as chlorinated hydrocarbons and aromatics require inoculation with specially cultivated organisms (Ottengraf et al. 1986; Ottengraf and Diks 1990). In manufactured media supplied by some vendors, inoculum is added during the media-production stage. During treatment, the introduction of odorants or VOC contaminants into a bioreactor shifts the distribution of the existing microbial populations toward a strain that can metabolize the target odor or VOC pollutants.

Compared to pure cultures, the use of microbial consortia or mixed cultures is common in large-scale applications for odor or VOC contaminants. Wastewater sludge or a filter bed containing a diverse natural community are the most frequently used, non-defined inoculums for the startup of bioreactors. However, some studies with single strain or defined consortium inoculation at startup have reported high contaminant removal efficiencies (Cox et al. 1997; Veiga et al. 1999; Veiga and Kennes 2001). Biofilters using natural organic carriers are expected to contain a wide range of organisms including bacteria, fungi, yeasts, algae, and protozoa, as compared to synthetic or inert carriers. However, biofilter inoculation may speed up the startup period.

Various species of bacteria have been identified from the bioreactors treating waste gases. Some of the common species belong to the genera *Pseudomonas*, *Alcaligenes*, *Bacillus*, *Corynebacterium*, *Sphingomonas*, *Xanthomonas*, *Nocardia*, *Mycobacterium*, *Rhodococcus*, and *Xanthobacter*. Anaerobic bacterial species of *Clostridium* and *Enterobacter* have also been identified in biofilters (Arnold et al. 1997; Juneson et al. 2001; Kennes and Veiga 2001). Use of white-rot fungi such as *Phanerochaete chrysosporium*, *Tramtes versicolor*, *Pleurotus ostreatus* and *Bjerkandera adusta* has been reported in biofiltration studies for the elimination of compounds such as α-pinene, styrene and alcohols (Braun-Lüllemann et al. 1997). A fungal species of *Exophiala jeanselmei* has been reported to utilize pollutants as a sole carbon and energy source, and high styrene removal rates in biofilters containing this fungus as the main styrene-degrader have been reported (Cox et al. 1996, 1997). The aerial hyphae of fungi represent a large surface area for direct contact with pollutants in contaminated air, facilitating transfer of compounds to the cell surface without any phase transition issues. Fungi are generally resistant to extreme environmental conditions, but a potential problem is the accumulation of intermediate products. Fungi are suitable for removal of hydrophobic compounds, such as alkyl benzenes and styrene.

Although microbes can be isolated and identified by classical microbiological methods, only a limited number (approximately 20%) of the contaminant-degrading strains present in a bioreactor are cultivable, and can therefore be isolated and identified by classical microbiological methods. Advanced molecular tools have been recently developed for identifying "hard to culture" microorganisms to obtain useful information on microbial populations present in a particular ecosystem (van Elsas et al. 1998; Theron and Cloete 2000). Phospholipid fatty acid analysis (PLFA), fluorescent in-situ hybridization (FISH) and 16S rRNA characterization are some of the techniques used in such ecological studies. However, such investigations have not been popular in characterizing biofilter populations (Sakano and Kerkhof 1998).

Microorganisms require a range of nutrients for growth and activity to carry out biodegradation of pollutants. Nutrients may be naturally present in organic filter beds, but are added to the synthetic or inert beds. Microbial cells consist of carbon, nitrogen, oxygen and hydrogen, with typical cell compositions of $C_5H_{8.3}NO_{1.35}$ for bacteria and $C_4H_7N_{0.6}O_2$ for fungi. Trace elements such as P, K, Mg, Ca, S, Fe, and Mn may be required by some species. Yield of biomass depends on the nature of the nutrients. Oxygen is required for the aerobic degradation of pollutants, and could be limiting in thick biofilms or at high substrate loads. Microbial activity slows down considerably in a dry environment. Optimum water content (moisture) in polluted air treatment is crucial, as an excessively high water content may result in mass transfer limitation of contaminants and possible anaerobic conditions.

Although bacteria grow in a pH range of 5–9, while fungi can grow under the more acidic to neutral pH range of 2–7 for appreciable growth and biodegradation, microbial activity slows down below pH 4 or above pH 8. Most microorganisms do not tolerate pH fluctuations of more that 2–3 pH units. Although most waste gas treatment studies are made under mesophilic (15–40 °C) conditions, bioreactors operating efficiently under psychrophilic (< 10 °C) and thermophilic (> 45 °C) conditions have been described. The temperatures of contaminated air and waste gases are often higher than the mesophilic range.

Pressure drop and clogging are a well-known phenomenon in biofilters, due to the nature of the carrier, particle size and shape, moisture content, superficial gas velocity, and excessive microbial growth. Clogging due to extensive biomass formation is more common with inert material where a nutrient solution is fed at regular interval. Limiting the supply of some nutrients such as phosphate and potassium ion or nitrogen is a possible means to reduce biomass growth (Wübker and Friedrich 1996). In Chapter 5, the microbiology of bioreactors for waste gas treatment is discussed in detail.

1.3.3 Types of Bioreactors

In recent years, biofiltration has received increased acceptance for odor removal in the wastewater treatment industry. Industrial sectors including rendering, food processing, flavor manufacturers, and composting facilities are selecting biofilter systems for odor and VOC removal in their facilities. In biofilters, as the contaminated air is passed through a bed of media, the contaminants and oxygen are first transferred to the biofilms formed on the surface of the media particles, and then metabolized by bacteria. In order to sustain microbial growth on the media particles, moisture is provided by saturating the process air before it enters the biofilter unit. The moisture is also provided by intermittent, occasional spray irrigation of the media. The media within a biofilter are normally composed of material such as peat, wood bark, soil, compost, coated ceramic particles, synthetically manufactured media, or a combination of these products. If properly designed, biofilters can provide complete removal of the odor and VOC contaminants present in waste air. Chapter 6 covers the historical development, fundamental theory, process mechanisms, and important parameters of the biofiltration process.

The major difference between biotrickling filters and biofilters is the presence of continuous water flow in the reactor. The water phase carries nutrients for the microorganisms, and is usually neutralized before recirculation, for pH-control purposes. Microbial oxidation takes place in the water phase as well as in the immobilized biofilms attached on the media particles. Microorganisms in the biofilms degrade absorbed contaminants into harmless and odorless products. Excessive biomass growth and clogging are major problems encountered in biotrickling filters. Biotrickling biofilters are more complex to construct and operate than classical biofilters. For chlorinated VOCs and compounds that produce intermediate acidic by-products, however, biotrickling filters are very effective (Devinny et al. 1999). Chapter 7 covers recent advances in biotrickling filter technology.

While biofilters and biotrickling employ immobilized organisms, bioscrubbers utilize dispersed (suspended) cultures. Bioscrubbers consist of two units: the usual scrubber in which VOCs and odorous compounds are transported from the air to a water phase, and a liquid-phase bioreactor where the water exiting the scrubber is subjected to biological treatment in a liquid phase. The bioreactor, which contains suspended cultures, requires sufficient oxygen through aeration to maintain a high level of biodegradation. However, due to volatilization from bioreactor and scrubber operations, emissions may not be completely eliminated. Thus, a secondary polishing unit may be required in many cases. As in biotrickling filters, the water phase allows the addition of nutrients to the system. Bioscrubbers can handle varying inlet concentrations and flow rates, or shock-loading conditions more easily. Chapter 8 covers the basic principles, variations of designs, and main process parameters of bioscrubber technology.

Membrane bioreactor technology is becoming popular for odor and VOC removal. Membrane bioreactors are based on the utilization of microporous hydrophobic membranes that are made up of materials such as polypropylene and polyethylene. The membranes provide high gas permeabilities, while not allowing water transport across the membranes. Through diffusion, gases in the membranes are transferred into the surrounding liquid phase that contains essential nutrients and microorganisms capable of degrading contaminants. The arrangement is very much like a shell and tube heat exchanger where, in this case, the tubes are made of very small-diameter porous fibers. The membranes are manufactured to give a very high specific surface area. Ergas and McGrath (1997) reported several applications involving the use of membrane bioreactors for the removal of contaminants from airstreams. Chapter 9 gives a broad description of membrane bioreactors, membrane materials, and design aspects.

1.4 Modeling and Design of Bioreactors

1.4.1 Modeling of Bioreactors

In general, air-phase bioreactor models include terms for the accumulation of contaminants in all phases, dispersion effects in air, diffusional mass transfer in biofilms, mass transfer in liquid phase, consumption due to biological oxidation, adsorption of contaminants onto solid matrix, and biomass growth on the surface and pore structure of the media. Often, the resulting model equations are a complex set of partial differential equations with expressions for mass transfer parameters, kinetics of biological oxidation, adsorption, and absorption processes. The complexity of the model equations is substantially reduced with justifiable assumptions. In some cases, simplified assumptions make models unrealistic and often may lead to errors in design parameter estimations. For example, in a large-scale biofilter project, a 5-s overestimate can lead to design deficiencies and a significant difference in capital cost. Thus, predictive models play an important role in the scale-up and designing of bioreactors for odor and VOC control.

Laboratory and field pilot studies are often carried out to obtain kinetic and other parameters required for the model solutions. In the literature, these range from simple asymptotic (zero or first-order kinetics; Ottengraph and van den Oever 1983) to more realistic but complex kinetic expressions (Baltzis and Shareefdeen 1994) for biological oxidation under diffusion and reaction-limited regimes. In the case of mixtures, biodegradation kinetics may be more complex, as interactions such as competitive inhibition arise. Although detailed kinetic studies may be impractical (if at all possible) when dealing with multi-components, some understanding of the kinetics is needed for developing a predictive model.

Questions such as how well can a bioreactor respond to variations in volumetric flow rate, concentration, and compositions are of paramount importance for commercial applications. Transient models provide the predictive performance of bioreactors under shock-loading conditions. The availability of oxygen in the biofilm is also important for complete biological oxidation of contaminants under aerobic conditions. It was found that with water-soluble VOCs (e.g., methanol), oxygen (in the biofilm) depletes faster, leading to larger portions of inactive biofilm (Shareefdeen et al. 1993). Modeling of biomass growth and its effect on vapor-phase bioreactor (biofilter) performance are discussed by Ozis et al. (2002). Ozis et al. (2002) presented a discrete percolation model to describe biomass-clogging effects on biofilter performance. In Chapter 10, the principles of biofilter and biotrickling models are discussed in details, with case studies related to the application of modeling principles in full-scale designs.

1.4.2 Design of Bioreactors

In bioreactor design, characterization of contaminated airstream is the first and the most important step for selection of odor or VOC control methods. Airstream characterization includes identifying types of contaminants (odor and VOCs), concentration levels, fluctuations due to process conditions, temperature, humidity, and particulate levels (dust, aerosols, fat or oil) in the airstreams. These parameters are critical in determining the most suitable odor control method and design. In biofiltration, for example, air needs to be pretreated for the removal of particulates or fat aerosols, and then humidified in a packed tower or with pneumatic spray. Humidification is required to keep the biological activity within the solid medium of the biofilter. In some dual-stage treatment systems, for example, removal of contaminants through an adsorber or chemical scrubber is required prior to biological removal of contaminants. Non-biodegradable compounds that are odorous and toxic need to be removed by physical or chemical processes. Thus, the design of bioreactors requires knowledge of non-biological as well as biological treatment methods. The second step in the design is estimating the airflow volume. Generally, the number of air exchanges in an industrial plant, or simply the headspace of a wet well in a wastewater treatment facility, determines flow volume and concentration levels in the air that requires treatment.

One of the most commonly used terms in biofilter and biotrickling design is empty bed residence time (EBRT), which is defined as the medium volume to the volumetric airflow through the reactor. A very low EBRT leads to a high velocity, high pressure drop across the bed of media, and higher power consumption. On the other hand, a very high EBRT leads to a large footprint, large volume of media, and high capital cost. Thus, in designing these reactors, it is important to accurately estimate EBRT, which gives the specified efficiency of odor or VOC contaminant removal. EBRT can be determined based on literature data, laboratory and pilot-scale experiments, and through predictive modeling. Once the EBRT is determined, the bioreactor media volume can be easily estimated. The physical and structural properties of the media are also important in determining media depth and footprint of the bioreactors.

In evaluating reactor performance, elimination capacity (EC) or removal rate (RR), defined as the mass of contaminants removed per volume of media per time, are commonly used. Usually, EC has a unit of grams removed per cubic meter of media per hour. EC values are related to the biodegradability of the compounds, and also the types of media used. Percent removal efficiency (RE%), which is the percent reduction in odor or VOC concentration levels, is also used in evaluating reactor types. Removal efficiency is an important specification for the design. Accurate evaluation of bioreactor destruction efficiency should exclude removal due to physical processes such as adsorption/desorption and absorption in the overall process. Mass loading (mass load to the reactor per unit volume of media per unit time) is equal to EC when contaminants are completely (100% RE) degraded. Mass loading is related to the concentration of contaminants and flow rate, and is usually expressed in gram applied per cubic meter of media per hour. Variations in flow or concentrations lead to mass shock loadings, and therefore should be considered in full-scale designs of bioreactors. Chapter 11 covers details in design consideration, airstream characterization, pretreatment, air distribution, construction options, choice of bioactive medium, monitoring, operation and design failures.

1.5 Types of Contaminants

The literature shows that biotechnology for elimination of odor and VOC removal from industrial and municipal facilities is a viable option. The contaminants considered in the bioreactor-based studies include VOCs or inorganic compounds. Chapters 12 and 13 present in-depth coverage of biological treatment of volatile inorganic and organic compounds, respectively.

Examples of biofilter studies on VOCs are benzene and toluene mixtures (Baltzis and Shareefdeen 1994), butanol, butyl acetate, ethyl acetate (Ottengraf and van den Oever 1983), ethanol (Leson et al. 1993), methanol (Shareefdeen et al. 1993), methylformiate, methanol/isobutanol mixtures, styrene, vinylcyclohexene, butadiene (van Lith 1989), methyl-ethyl-ketone (MEK) and methyl-isobutyl-ketone (Deshusses and Hammer 1993), MEK and toluene (Llewellyn et al. 2002), phenol (Zilli et al. 1993), propane (Ebinger et al. 1987), tetrahydrofurane (Paul and Roos 1989), toluene, dichloroethane, and trichloroethene (Ergas et al. 1993).

Ammonia (Nicolai et al. 2002) and hydrogen sulfide (Aroca et al. 2002; Sologar et al. 2002; Shareefdeen et al. 2003) are the most studied inorganic compounds. Different reactor configurations and media have been used, but most studies were done on biofilters (Ottengraf and van den Oever 1983; Ebinger et al. 1987; Paul and Roos 1989; van Lith 1989; Deshusses and Hammer 1993; Ergas et al. 1993; Leson et al. 1993; Togna and Frisch 1993; Zilli et al. 1993). Compared to biofilter applications, fewer studies exist regarding VOC removal in biotrickling filters with inorganic support materials as supporting materials. Examples include dichloromethane removal in trickling filters with plastic support (Diks and Ottengraf 1989, 1991; Diks 1992), removal of gasoline fractions in biotrickling biofilters (Phipps and Ridgeway 1993), and methanol removal in a bioscrubber (Overcamp et al. 1991).

1.6 Case Studies

Application of biotechnology for elimination of odors and VOCs in industry and municipal wastewater treatment plants is discussed in Chapters 14 and 15, respectively.

Chapter 14 covers the history of industrial bioreactor development and applications involving food, tobacco, waste handling and leather processing. Selected bioreactor vendors from Europe, and case studies from Korea (trickling vs. open wastewater treatment), Russia (ethyl acetate and styrene), UK (formaldehyde at a chemical plant), Finland, and Italy are discussed. The chapter also covers carrier materials (peat, mussels, compost, sawdust, bark) used by European bioreactor vendors.

Chapter 15 presents eight different case studies from both designers and end users of biological odor control technologies at municipal wastewater treatment plants. This chapter, through the case studies, covers a wide range of bioreactors including a single-stage biofilter, stacked tray biofilters, manhole cover biofilters, biotrickling filters, and several biofilter applications related to sulfur compounds, the most odorous compound found at municipal wastewater treatment facilities. Readers should be aware that none of the authors of the book or chapters endorse any of the systems, bioreactor units, or their contents. However, the readers are encouraged to seek out more information with the individual designer, vendor, and/or end users of the system described.

An example of industrial odor emission application in the USA is the use of a biofilter for the elimination of odors at the Detroit, Michigan, plant (Fig. 1.1). The biofilter system was commissioned on 4 April 2001, and is successfully eliminating odorous emissions from 2,000 acfm (actual cubic foot per minute) air using synthetic media with a 20-s EBRT.

In Chapter 16, case studies in full-scale applications of bioscrubber and biotrickling filter technologies are presented. In this chapter, the robustness, implementation, and maintenance issues of these two bioreactors are discussed in detail. Through examples from the literature and case studies, recent de-

Fig. 1.1. A biofilter treating 2000-acfm air containing industrial odor emissions at Detroit, Michigan. (Courtesy of BIOREM Tech., ON, Canada)

velopments and suggestions for future developments in this technology are discussed.

1.7 Conclusion

Currently, the most widely used biotechnological processes for odor and VOC control are biofiltration, biotrickling, and bioscrubbing. In future, industries and municipal wastewater treatment plants around the globe are likely to be directed toward biotechnological processes over conventional methods for odor and VOC control needs. Furthermore, various forms or combinations of biotechnological processes are being developed to extend applications beyond industrial and municipal wastewater treatment markets. Chapter 17 elaborates more on the future of biotechnology and recommends key areas of research directions. The future of biotechnology for odor and air pollution control is promising.

References

Arnold M, Reittu A, von Wright A, Martikainen PJ, Sihko M-L (1997) Bacterial degradation of styrene in waste gases using a peat filter. Appl Microbiol Biotechnol 48:738–744

Aroca G, Oyarzun P, Nunez D (2002) Removal of hydrogen sulfide in a biotrickling filter. In: Proc USC-TRG Conf Biofiltration, 31 October–1 November 2002, Newport Beach, CA

Baltzis BC, Shareefdeen ZM (1994) Biofiltration of VOC mixtures: modeling and pilot scale experimental verification. In: Proc 87th Meet 10P Air and Waste Management Association, 19–24 June 1994, Cincinnati, OH, Pap no 94-WP260

Braun-Lüllemann A, Majcherczyk A, Hüttermann A (1997) Degradation of styrene by white-rot fungi. Appl Microbiol Biotechnol 47:150–155

Cox HHJ, Magielsen FJ, Doddema HJ, Harder W (1996) Influence of the water content and water activity on styrene degradation by *Exophiala jeanselmei* in biofilters. Appl Microbiol Biotechnol 45:851–856

Cox HHJ, Moerman RE, van Baalen S, van Heiningen WNM, Doddema HJ, Harder W (1997) Performance of a styrene-degrading biofilter containing the yeast *Exophiala jeanselmei*. Biotechnol Bioeng 53:259–266

Deshusses MA, Hammer G (1993) The removal of volatile ketone mixtures from air in biofilters. Bioproc Eng 9:141–146

Devinny JS, Deshusses MA, Webster TS (1999) Biofiltration for air pollution control. Lewis, New York

Diks RMM (1992) The removal of dichloromethane from waste gases in a biological trickling filter. PhD Thesis, Eindhoven University of Technology, The Netherlands

Diks RMM, Ottengraf SPP (1989) Process technological view on the elimination of chlorinated hydrocarbons from waste gases. In: Brasser LJ, Mulder WC (eds) Man and his ecosystem. Elsevier, Amsterdam, pp 405–410

Diks RMM, Ottengraf SPP (1991) Verification studies of a simplified model for removal of dichloromethane from waste gases using a biological trickling filter, part 1. Bioproc Eng 6:93–99

Ebinger MH, Bohn HL, Puls RW (1987) Propane removal from propane-air mixtures by soil beds. J Air Pollut Control Assoc 37:1486–1489

Ergas SJ, McGrath MS (1997) Membrane bioreactor for control of volatile organic compound emissions. J Environ Eng June:593–598

Ergas SJ, Schroeder ED, Chang DPY (1993) Control of air emissions of dichloromethane, tricholoroethene, and toluene by biofiltration. In: 86th Annu Meet Air and Waste Management Association, 13–18 June 1993, Denver, CO, Pap no 93-WA-52B

Gabriel D, Araya D, Brown J, Torress E, Deshusses MA (2002) Short contact time biotrickling filters for odor treatment: performance of a full-scale reactor at Orange County Sanitation District. In: Proc USC-TRG Conf Biofiltration, 31 October–1 November 2002, Newport Beach, CA

Guilbault MR (2002) Biofiltration and industry creating a climate change. In: Proc USC-TRG Conf Biofiltration, 31 October–1 November 2002, Newport Beach, CA

Juneson C, Ward OP, Singh A (2001) Microbial treatment of styrene contaminated air stream in a biofilter with high elimination capacities. J Ind Microbiol Biotechnol 26:196–202

Kennes C, Veiga C (2001) Conventional biofilters. In: Kennes C, Veiga C (eds) Bioreactors for waste gas treatment. Kluwer, Dordrecht, pp 47–98

Leson G, Winer AM (1991) Biofiltration: an innovative air pollution control technology for VOC emissions. J Air Waste Manage Assoc 41:1045–1054

Leson GDS, Hodge F, Tabatabai, Winer AM (1993) Biofilter demonstration projects for the control of ethanol emissions. In: 86th Annu Meet Air and Waste Management Association, 13–18 June 1993, Denver, CO, Pap no 93-WP-52C.04

Llewellyn D, Darlington A, Mallany J, Dixon M (2002) The influence of airflow on indoor air biofiltration: elimination of toluene and methyl ethyl ketone. In: Proc USC-TRG Conf Biofiltration, 31 October–1 November 2002, Newport Beach, CA

Nicolai RE, Clanton CJ, Janni KA (2002) Ammonia removal and nitrogen accumulation in biofilter media. In: Proc USC-TRG Conf Biofiltration, 31 October–1 November 2002, Newport Beach, CA

Ottengraf SPP, Diks RMM (1990) Biological purification of waste gases. Chimicaoggi May:41–45

Ottengraf SPP, van den Oever AHC (1983) Kinetics of organic compound removal from waste gases with a biological filter. Biotechnol Bioeng 25:3089–3102

Ottengraf SPP, van den Oever AHC, Kempenaars FJCM (1984) Waste gas purification in a biological filter bed. In: Houwink EH, van der Meer RR (eds) Innovations in biotechnology. Elsevier, Amsterdam

Ottengraf SPP, Meesters JJP, van den Oever AHC, Rozema HR (1986) Biological elimination of volatile xenobiotic compounds in biofilters. Bioproc Eng 1:61–69

Overcamp TJ, DeHollander GR, Chang HC, Grady CPL Jr (1991) A biologically-enhanced scrubber for volatile organic compounds. In: 84th Annu Meet Air and Waste Management Association, 16–21 June 1991, Vancouver, BC, Pap no 91-180.12

Ozis F, Yortsos Y, Devinny JS (2002) A discrete percolation model for describing biomass clogging in biofilters. In: Proc USC-TRG Conf Biofiltration, 31 October–1 November 2002, Newport Beach, CA

Paul PG, Roos C (1989) Biological waste air treatment with the Biobox. In: Brasser LJ, Mulder WC (eds) Man and his ecosystem. Proc 8th World Clean Air Congr. Elsevier, Amsterdam, pp 399–404

Phipps DW Jr, Ridgeway HF (1993) Modular bioreactor and computerized instrumentation package to identify critical design parameters for vapor phase bioreactors. In: Proc 2nd Int Symp In situ and On-site Bioreclamation, 5–8 April 1993, San Diego, CA

Sakano Y, Kerkhof L (1998) Assessment of changes in microbial community structure during operation of an ammonia biofilter with molecular tools. Appl Environ Microbiol 64:4877–4882

Shareefdeen Z, Baltzis BC, Oh YS, Bartha R (1993) Biofiltration of methanol vapor. Biotechnol Bioeng 41:533–538

Shareefdeen Z, Herner B, Webb D, Wilson S (2003) Hydrogen sulfide (H_2S) removal in synthetic media biofilters. Environ Prog 22:207–213

Sologar VS, Lu Z, Allen DG (2002) Modeling the biofiltration of air emissions containing reduced sulfur compounds (RSCs) and volatile organic compounds (VOCs). In: Proc USC-TRG Conf Biofiltration, 31 October–1 November 2002, Newport Beach, CA

Sorial GA, Smith FL, Smith PJ, Suidan MT, Biswas P (1993) Development of aerobic biofilter design criteria for treating VOCs. In: 86th Annu Meet Air and Waste Management Association, 13–18 June 1993, Denver, CO, Pap no 93-TP-52A.04

Theron J, Cloete TE (2000) Molecular techniques for determining microbial diversity and community structure in natural environments. Crit Rev Microbiol 26:37–57

Togna AP, Frisch S (1993) Field-pilot study of styrene biodegradation using biofiltration. In: 86th Annu Meet Air and Waste Management Association, 13–18 June 1993, Denver, CO, Pap no 93-WP-52C.03

Van Elsas JD, Duarte GF, Rosado AS, Smalla K (1998) Microbiological and molecular biological methods for monitoring microbial inoculants and their effects in the soil environment. J Microbiol Methods 32:133–154

Van Lith C (1989) Biofiltration, an essential technique in air pollution control. In: Brasser LJ, Mulder WC (eds) Man and his ecosystem. Elsevier, Amsterdam, pp 393–399

Veiga MC, Kennes C (2001) Parameters affecting performance and modeling of biofilters treating alkylbenzene-polluted air. Appl Microbiol Biotechnol 55:254–258

Veiga MC, Fraga M, Amor L, Kennes C (1999) Biofilter performance and characterization of a biocatalyst degrading alkylbenzene. Biodegradation 10:169–176

Webster TS, Togna AP, Guarini WJ, Albritton C, Carlisle C, Cha CY, Wander J (2002) The application of a microwave concentration/biofilter integrated system to treat spray paint booth emissions – final results. In: Proc 2002 USC-TRG Conf Biofiltration, 31 October–1 November 2002, Newport Beach, CA, pp 203–213

Williams TO, Miller FC (1992) Odor control using biofilters, part 1. Biocycle October:73–77

Wübker S-M, Friedrich CG (1996) Reduction of biomass in a bioscrubber for waste gas treatment by limited supply of phosphate and potassium ions. Appl Microbiol Biotechnol 46:475–480

Zilli MA, Converti AL, del Borghi M, Ferraiolo G (1993) Phenol removal from waste gases with a biological filter by *Pseudomonas putida*. Biotechnol Bioeng 41:693–699

2 Environmental Laws and Regulations Related to Odor and Waste Gas Contaminants

Rodney L. Aldrich

2.1 Introduction

Odor arouses some of the most vigorous reactions among those exposed. Odor perception is actually first processed by the limbic nervous system. The olfactory sensory system is connected, or in a sense "hard wired", to the amygdalae, which appear to be the center of emotions. This may be why odors can provoke such strong emotions (http://users.rcn.com/jkimball.ma.ultranet/BiologyPages/C/CNS.html).

Therefore, the fact that objectionable odor is one of the earliest types of pollution to be brought into the body of law is not surprising. Scholars acknowledge that odor entered common law as a possible trespass or nuisance. An odor wafting from one property to another can be a trespass. If sufficient evidence can be demonstrated, common law holds that the property owner who is the recipient of the odor has suffered a wrong. Courts have resolved such common law matters and their actions have depended on the costs for corrections (Rafson 2000). This type of person-to-person enforcement of odor control via the courts is still actionable today in any area of the world that has incorporated these concepts of common law into its legal system. However, beyond common law, many nations or sub-national governments have adopted a system of laws and regulations designed to control odor as a type of air pollution.

There are many substances that are perceptible in air as odors. The intensity of the perception varies by the compound, and by the individual smelling the odor. Considering the odor perception of one person, it is rather easily shown that each compound would have a different concentration in mass per volume of air at which this reference person could first perceive the odor. The concentration at which the average individual first can perceive an odor from a particular substance is called the odor threshold. Each particular volatile substance has its own unique odor threshold. Another complication regarding odor is that various individuals, assuming they are exposed to the same intensity of odor, apparently can have different reactions to the same concentration of odor.

This variability by compound and by individual reaction has made it challenging for a regulatory agency to develop an "odor scale" that could allow

Shareefdeen/Singh (Eds.)
Biotechnology for Odor and Air Pollution Control

consistent and quantitative measurement and enforcement across chemical species. As a result of these inherent difficulties, two systems of regulation that affect odor control have generally arisen. The first system involves the control of volatile organic compounds (VOCs). These controls come in the forms of control of total VOCs, and in the control of specific compounds that are VOCs.

Exposure to some VOCs can be harmful to health. The health effects of a VOC depend upon the type of compound, the concentration in the air, and the extent of exposure to the individuals. The initial restrictions developed by regulatory agencies on specific VOCs originally arose to protect the health of those potentially exposed. Over decades, the regulation of VOCs proceeded to strengthen because of the implication of VOCs involvement in ground-level ozone production, in depletion of protective stratospheric ozone, and in the exacerbation of global warming.

The second system of regulation that affects odor control involves the control of any odor-causing chemical that is the source of an offensive odor that causes interference in the use and enjoyment of the air by others, especially outside the originating property. These restrictions are similar to the nuisance provisions that are integrated into common law, but give the executive bodies within governments the authority to enforce these restrictions against nuisance odors without the direct, initial involvement of the courts. While the regulatory systems that regulate odorous substances to prevent nuisances have not necessarily garnered the global and national headlines that controls of VOCs have experienced, regulatory efforts regarding odorous substances probably have received more numerous and intense local media coverage, public speech in general, and comment to government regulatory authorities.

2.2 Control of VOCs

As discussed above, VOCs are generally regulated because they can affect people's health in several ways.

1. Individual VOCs have irritant or toxic effects. Regulatory programs have developed around controlling the volatilization and consequent human exposure of these individual volatile organic compounds.

   The classic dry-cleaning solvent, also referred to as tetrachloroethylene or perchloroethylene, is one example where regulation was originally based more on the compound's health effects than its other effects, such as its contribution to ozone.

2. VOCs help create ground-level ozone, which is a principal component of smog. Smog is formed by sunlight initiating the chemical reaction of VOCs and nitrogen oxides (NO_x) in the atmosphere (Fokkema 2003). Ozone in the troposphere can be harmful to health and can damage vegetation.

Ozone irritates the human respiratory system, causing coughing, irritation in the throat or a burning sensation in the airways. Ozone can reduce lung function, causing feelings of chest tightness, wheezing, or shortness of breath. Ozone can aggravate asthma and trigger asthma attacks. People at greater risk from ground-level ozone are people with lung diseases, such as asthma, and children and adults who are active outdoors (http://www.epa.gov/airnow/health.html).

In order to reduce ozone levels, the regulations have concentrated on limiting the man-made emissions of both VOCs and NO_X. In the United States, the limitations of VOCs to control ozone have concentrated on areas where the ozone level is considered to be excessive, and these are known as non-attainment areas.

Controls on VOCs, particularly on total VOCs, were implemented, especially in these urbanized non-attainment areas, in order to reduce the creation of ozone in the troposphere. The control of VOCs has indirectly caused the reduction in odors of these VOCs by limiting their concentration close to their source.

3. In the stratosphere of earth's atmosphere, some VOCs, particularly chlorinated or otherwise halogenated chemical compounds, were determined to cause the depletion of the protective, ultraviolet radiation-absorbing ozone layer in the earth's upper atmosphere. As excessive ultraviolet radiation exposure is associated with skin cancer, eye problems, and other negative effects, controls were placed on these specific ozone-depleting VOCs.

4. Finally, VOCs have been implicated in global warming, and many nations are attempting to limit the release of VOCs as a way of limiting global warming (Hansen et al. 2000).

VOCs gained so much attention for their negative environmental impacts that, in the United States, after a patchwork of regulations about individual VOCs, Congress amended the Clean Air Act in 1990 to force the United States Environmental Protection Agency (USEPA) to reduce the emissions of a list of 189 chemicals and chemical categories known as hazardous air pollutants or HAPs.

The USEPA is required to develop air emissions regulations that apply to "major sources" of HAP emissions, and such a major source is any facility that has the potential to emit, on an annual basis, 10 tons (t) or more of any single HAP or 25 t of all HAPs combined. Because these regulations are based on "maximum achievable control technology" (MACT), they are often referred to as the "MACT standards". The USEPA is developing the MACT standards on an industry-by-industry basis.

Again, by limiting the emission of these chemicals, government regulatory agencies are limiting the potential for the odors from these VOCs to be detected off the site of generation.

2.3 Control of Odor-Causing Chemicals

Control and regulation of odor-causing chemicals and mixtures, and their odors, has developed independent of VOC controls. These odor controls, which are independent of VOC controls, can be based on subjective or objective measurements, and can apply at different locations. These options are enumerated as follows:

1. Laws and regulations can forbid that an offensive odor reaches people off the property of generation. These controls are basically similar to the common law prohibitions of a nuisance odor.

   These "off-site receptor" controls have advantages and disadvantages. The advantages include:

   a) the offense is interpretable by the exposed party, thereby avoiding expensive analysis; and

   b) controls do not have to be met at an arbitrary location at which there is potentially no exposure, such as the source property boundary. Consequently, the concentration of odor at the source can be higher than would occur with a concentration limit enforced at the property boundary to the same level as must be met at an offsite receptor.

   The disadvantages of off-site receptor controls include:

   a) the exposed party may have other reasons for making a complaint than the odor, and the complaint may not be genuine;

   b) different exposed parties may have different odor thresholds;

   c) these controls often do not define the odorous substances chemically; and

   d) these controls often are subjective as to what constitutes an offensive odor.

2. Laws and regulations can control the concentration of a noxious chemical at the source property boundary. These controls can rely on subjective interpretations of odor, or may be based on measured concentrations of odors or chemicals. These controls seem to be stricter than the common law approach because they do not require an actual innocent recipient to be present when the offending chemical leaves the source property. When they are based on some form of measurement, these "source property boundary" controls offer the certainty of measurement and much objectivity in the judgment of an offense. Of course, such measurements have an associated expense, and usually cannot be measured continuously.

3. Laws and regulations can control the concentration of a noxious chemical at an emission source such as an air emission point or at a water discharge. These controls recognize that concentrations at this point may be more

quantifiable and, therefore, more easily measured and controlled. However, these types of controls have the disadvantage that they are not site- or situation-specific. Therefore, these controls can provide more protection than is actually needed, or they may not provide sufficient protection under certain unanticipated conditions.

2.4 Brief Overview of Odor Restrictions Around the World

As described in the preceding sections, odor control has developed in two parallel regulatory control systems:

1. control of volatile organic compounds or VOCs that coincidentally have odors;
2. control of odor-causing chemicals that have offensive odors.

Because the later is more relevant to the topic of this book, the regulation of odor-causing chemicals has been examined in a number of countries, and within certain states of the United States of America. Each government can be seen to rely on the types of odor regulation discussed in the preceding section.

2.4.1 The United States of America

While the federal government of the United States of America has directed the USEPA to study composting, biosolids, wastes from concentrated agricultural feeding operations (CAFOs), septic solids, and sewage sludge, it has left the actual enforcement of odors as nuisances to the State and local courts, and the State and local government executive agencies. The following sections describe the manner in which nuisance odors are regulated in certain states.

2.4.1.1 New York State

Nuisance odors are dealt with in New York State by the New York Code of Rules and Regulations (NYCRR) 211.2 (http://www.dec.state.ny.us/website/regs/211.htm). The effective requirement is that "No person shall cause or allow emissions of air contaminants to the outdoor atmosphere of such quantity, characteristic or duration which are injurious to human, plant or animal life or to property, or which unreasonably interfere with the comfortable enjoyment of life or property."

2.4.1.2 Michigan

In Michigan, odors are controlled via Part 9 of the Air Pollution Control Rules. Specifically, R 336.1901 states that "Notwithstanding the provisions of any other department rule, a person shall not cause or permit the emission of an air contaminant... in quantities that cause, alone or in reaction with other air contaminants, ... unreasonable interference with the comfortable enjoyment of life and property."

2.4.1.3
Massachusetts

Massachusetts' regulation at 310 Code of Massachusetts Regulation (CMR) 7.00 defines air pollution to mean "the presence in the ambient air space of one or more air contaminants or combinations thereof in such concentrations and of such duration as to:

> cause a nuisance;
>
> unreasonably interfere with the comfortable enjoyment of life and property or the conduct of business."

Then 310 CMR 7.01 goes on to say that "No person owning, leasing, or controlling the operation of any air contamination source shall willfully, negligently, or through failure to provide necessary equipment or to take necessary precautions, permit any emission from said air contamination source or sources of such quantities of air contaminants which will cause, by themselves or in conjunction with other air contaminants, a condition of air pollution."

Massachusetts has published draft guidance and policy (dated 23 December 1996) for the evaluation of odors at composting facilities. The Massachusetts draft guidance and policy regarding odors at composting facilities is designed for both proposed and existing facilities.

For all proposed facilities, air pollution control must be provided for all emissions from active composting operations, and consider non-active areas needing controls. The level of control is specified to be best available control technology (BACT), and shall include all reasonable practices to reduce/minimize odors and add-on controls as determined by BACT analysis. Also, air dispersion modeling must show that emitted odors will not result in a predicted off-site nuisance odor condition. Such facilities must also prepare and submit an odor management plan that must commit to best management practices (BMPs) that include specific operational procedures to minimize odor generation, a contingency plan for facility upset, a complaint response program, and a proposal for a community outreach/involvement program.

The Massachusetts draft guidance requires that proposed sources should be evaluated for an impact of five (5) dilutions to thresholds (D/T), as predicted by the Department-approved air dispersion modeling. The impacts are to be evaluated at the property boundary, or at the point of maximum impact beyond the property boundary. The guidance mentions that the most sensitive receptor might be allowed at the point of maximum impact beyond the property boundary, even though this may result in less stringent limitations than would result from use of the property boundary. This requirement does not relieve the odor-generating party of the requirement that nuisance conditions are not allowed beyond the property boundary.

To obtain these D/T measurements, the diluted sample is presented to odor panels of six to eight individuals. Each diluted sample is presented to the panel members, with two odor-free blanks for statistical validation.

The Massachusetts draft guidance document recognizes that there appears to be adequate testing data to demonstrate that biofilters can effectively remove odors from the emissions from sewage sludge (biosolids) composting facilities when the biofilters are operated and maintained correctly. The Massachusetts draft guidance document goes on to require that biofilters be designed to be operated at a loading rate not to exceed 3 $ft.^3$ min^{-1} $ft.^{-2}$ (3 CFM/SF), and that an irrigation/humidification system that is adequate to prevent drying out of the biofilter must be designed and installed. Also, pre-scrubbing to eliminate excessive ammonia and particulate loading is important for some facilities, and must be addressed by the project consultant. Also, biofilters must be designed with an empty bed retention time of 45–60 s, and should be 3–4 ft. in depth.

One permit that was issued with some recognition of this draft guidance was the one issued in the Southeast Regional Office of the Department of Environmental Protection of the Commonwealth of Massachusetts in November 2002. This permit uses 5.0 D/T at the property line for the municipal solid waste (MSW)/biosolids co-composting facility, and 20 D/T for two biofilter building stacks. Also, the biofilter section requires the following performance criteria be met:

media moisture content – 40–60%

maximum back pressure – 6 in.

maximum airflow rate – 3 acfm/$ft.^2$ (acfm, actual cubic foot per minute)

minimum odor control efficiency – 98%

maximum exhaust air temperature – 113 °F

2.4.1.4 Arkansas

Chapter 2 of the Arkansas Pollution Control and Ecology Commission Regulation 18, Arkansas Air Pollution Control Code (effective date: 15 February 1999) defines air pollution as "the presence in the outdoor atmosphere of one or more air contaminants in quantities, or characteristics, and or a duration ... which unreasonably interfere with enjoyment of life or use of property throughout the state or throughout the area of the state as shall be affected thereby." Further, Section 18.801 of Chapter 8, entitled "Emission of Air Contaminants such as to Constitute Air Pollution", states that:

> "No person shall cause or permit the emission of air contaminants, including odors or water vapor and including an air contaminant whose

emission is not otherwise prohibited by this Code, if the emission of the air contaminant constitutes air pollution."

Section 18.1201 of Chapter 12 permits the treatment of odors by dilution.

2.4.2
Japan

Japan's odor control law is entitled "Offensive Odor Control Law" and was Law No. 91 of 1971. The latest amendment was made by Law No. 71 of 1995.

Law 91 of 1971 applies to the odors generated in the course of business activities at factories or at other places of business. Law 91 of 1971 defines odor-causing substances such as ammonia, methyl mercaptan and other substances likely to cause unpleasant odors and disrupt the living environment, and applies in densely populated areas defined by the local governments. Maximum permissible concentrations are established for the odor-causing substances in air and water discharges of these substances, and a maximum odor index in air within the source property must be maintained, where the odor index is a multiple of the concentration at which the odor threshold occurs (see the discussion of dilution to threshold or D/T that occurs in the section entitled "Brief Overview of Odor Restrictions of Massachusetts" above). Emission standards may be established where site-specific compliance has not been effective. Stack height may be accounted for where the release is from a stack. Details are provided on the manner by which the compounds that generate odors are to be selected, the manner in which inspections are to be performed, which compliance measures are to be self-reported, and the manner by which the maximum permissible concentrations are to be established. Also, compliance and enforcement procedures and guidance are provided.

2.4.3
China

China requires the control of odors in its Law of The People's Republic of China on the Prevention and Control of Atmospheric Pollution that was adopted on 5 September 1987, amended on 29 August 1995. Article 32 of Chapter IV, entitled "Prevention and Control of Pollution by Waste Gases, Dust and Malodorous Substances", states that "Units that discharge malodorous gases into the atmosphere must take measures to prevent the pollution of neighboring residential areas." Article 34 bans the burning of certain materials (asphalt, asphalt felt, rubber, plastics, leather and other materials) that may produce malodorous gases, except under special circumstances with prior government approval. Also, compliance and enforcement procedures and guidance are provided.

2.4.4
The United Kingdom

Odors and other statutory nuisances are dealt with in British law by Part III of the Environmental Protection Act 1990 (http://www.hmso.gov.uk/acts/acts1990/

Ukpga_19900043_en_4.htm#mdiv79). In general, the law makes local authorities responsible to enforce odor nuisances. Requirements for compliance and enforcement are provided.

If a local authority fails to carry out its obligation to investigate complaints of odor nuisance properly, a Local Government Ombudsman can be asked to investigate the local authority. Alternatively, the person making the complaint can go directly to the local magistrates' court, as outlined in Section 82 of the Environmental Protection Act 1990. However, this route to the local magistrate court does require a financial commitment.

The central government has recently sought public comment on whether or not odors from sewage treatment works should be treated as a statutory nuisance under the Environmental Protection Act 1990. The results are still pending.

2.4.5 Canada

2.4.5.1 Ontario

The Ontario Environmental Protection Act (EPA) Section 14(1) states:

> Despite any other provision of this Act or the regulations, no person shall discharge a contaminant or cause or permit the discharge of a contaminant into the natural environment that causes or is likely to cause an adverse effect.

An adverse effect is defined as the loss of enjoyment of property, injury or damage to human, plant or animal life, and impairment of any person's safety or health. Thus, an adverse effect may range from actual physical symptoms or irregularities to simple, nuisance odor complaints. The Ontario EPA provides for large personal and corporate enforcement penalties.

With regards to specific chemical air contaminants, the Ontario Ministry of the Environment has published a document entitled "Point of Impingement Standards, Point of Impingement Guidelines, and Ambient Air Quality Criteria (AAQCs)", which provides emission standards for specific contaminants based on their odor characteristics. Also, Ontario takes special measures to control odors at sewage treatment facilities and landfills.

2.4.5.2 Quebec

In the R.S.Q., Chapter Q-2 of the Environment Quality Act, Quebec defines a contaminant to include "an odour ... likely to alter the quality of the environment in any way". The act then goes on to state that:

> The same prohibition applies to the emission, deposit, issuance or discharge of any contaminant the presence of which in the environment is

prohibited by regulation of the Government or is likely to affect the life, health, safety, welfare or comfort of human beings, or to cause damage to or otherwise impair the quality of the soil, vegetation, wildlife or property.

With regards to agricultural odors, Quebec has a regulation designated c. P-41.1, r.1.1 that establishes minimum separation distances for agricultural activities from surrounding activities to prevent odors from becoming a nuisance.

With regards to the regulation of industrial odors, the Quebec Ministry of Environment regulation c. Q-2, r.20, Division VI, requires certain industries to limit their odor emissions to certain multiples of the odor threshold.

With regards to all odor-producing industries, the Quebec Ministry of Environment's suggested guideline is one odor unit (10-min average) at the most impacted sensitive receptor. This level is achieved when 50% of the "trained noses" on a qualified Odour Panel detect an odor. Unfortunately, this is not equivalent to an adverse effect or nuisance. The Quebec Ministry of Environment has utilized dispersion modeling to predict what the emission concentration of an odor can be, in order to protect this most impacted sensitive receptor, and is attempting to refine the allowable models.

2.5 Conclusions

A survey of the types of odor controls or restrictions that were found in eight countries, or states within the United States of America (see Table 2.1) found that regulation of odors as a nuisance is universal. However, the types of controls that governments impose vary. Control of odors to off-site receptors is universal as well. However, some governments limit odors at the property boundary of the source. Also, some governments remove the subjectivity of the odor nuisance by requiring that the odor be below a certain number of dilutions to the odor threshold, or that certain identified compounds be below certain mass per volume concentrations.

The advantage of using the property boundary of the source is that, if someone happens to wander or travel close to the property boundary, they should not receive an offensive dose of the odor. A disadvantage is, especially in rural areas, that the use of the source property boundary may provide protection that is generally felt to be unneeded, especially by many rural residents.

The advantage of using limitations on odor to be below a certain number of dilutions to threshold is that the judgment of nuisance is moved away from subjective interpretation toward a more objective standard. This is a likely direction that governments will be attracted to, as it takes away the subjective interpretation of nuisance. This method is particularly useful with respect to mixtures of offensive substances that occur at concentrated agricultural feeding operations, biosolids composting facilities, and sewage treatment facilities.

The advantage of using limitations on the mass per unit volume concentration for certain identified compounds is, once again, that the judgment of

Table 2.1. Direct restrictions of odor

	Nuisance	Site of required compliance		D/T (dilutions to threshold)[a]	Chemical-specific[b]
		Off-site receptor	Property boundary		
Japan	Yes	Yes	Yes	Yes	Yes
Peoples Republic of China	Yes	Yes			
United Kingdom	Yes	Yes			
Canada					
Ontario	Yes	Yes			Yes
Quebec	Yes	Yes		Yes	
United States of America					
New York	Yes	Yes			
Michigan	Yes	Yes			
Massachusetts	Yes	Yes	Yes	Yes	Yes
Arkansas	Yes	Yes			

[a] Compliance level is measured in concentration of mixed chemicals whose odor first meets the odor threshold at a specific number of dilutions of an air sample of the odorous air
[b] For instance, may include ammonia, amines, hydrogen sulfide, mercaptans, dimethyl disulfide, dimethyl sulfide, fatty acids, limonene, other terpene compounds

nuisance is moved away from subjective determination toward a more objective standard. This advantage is shared with the use of limitations on odor to be below a certain number of dilutions to threshold, and, similarly, this is a likely direction that governments will be attracted to, as it eliminates the subjective interpretation of nuisance, and the difficulty of attempting to prove such a subjective interpretation in enforcement proceedings. However, because this type of limitation is specific to certain compounds, this method would seem to be most suitable to regulation of processes at which the odor is entirely or mostly due to a limited set of chemical compounds.

During the survey of government regulations, regulators reported that the subjectivity of nuisance odor enforcement was a common source of frustration. Often these regulators reported that they have taken steps to attempt to limit the subjectivity of monitoring for compliance. In some instances, the regulators had sent government inspectors to verify that malodorous conditions were as reported by complaints. In other instances, they had required complainants to keep logs of their odor detection, so that the logs could be compared to the generating source process information and to meteorological conditions.

Those regulators that have moved to the more objective forms of regulation, such as limitations on odor to be below a certain number of dilutions to threshold or limitations on the mass per volume concentration for certain identified compounds, reported a marked easing in the disputes and deliberations that existed under more subjective monitoring and regulation.

Acknowledgements. The author wishes to thank John K. Winkler of the Massachusetts DEP, Mark Reed of the Michigan DEQ, and Cissi Li of the ERM Corporation.

References

Fokkema B (2003) The solvents emission directive and product stewardship. European Solvents Industry Group (ESIG). In: 17th Congr Federation of Scandinavian Paint and Varnish Technologists, 7–9 September 2003, Stockholm, Sweden

Hansen J, Sato M, Ruedy R, Lacis A, Oinas V (2000) Global warming in the twenty-first century: an alternative scenario. Proc Natl Acad Sci USA 97:9875–9880

Rafson H (2000) About odors (http://www.urban-forestry.com/citytrees/v36n4a13.asp)

3 Methods of Odor and VOC Control

SERGIO REVAH, JUAN M. MORGAN-SAGASTUME

3.1 VOCs and Odor Definition

Volatile organic compounds (VOCs) are emitted as gases from certain solids or liquids. Title I of the Clean Air Amendment (CAA) defines a VOC as any compound containing carbon, excluding carbon monoxide, carbon dioxide, carbonic acid, metallic carbides or carbonates, and ammonium carbonate, which participates in atmospheric photochemical reactions. A more clear definition is provided by the United States Environmental Protection Agency (US EPA), and typically regards VOCs as organic compounds having vapor pressure exceeding 0.1 millimeters of mercury (mm Hg) at standard conditions (20 °C and 760 mm Hg). VOCs include a variety of chemical compounds, some of which may have short- and long-term adverse health effects. VOCs have been identified as a major contributor to smog, which can cause respiratory problems such as eye irritation, headache, haze, and damage to plant and animal life. The inorganic gases are called volatile inorganic compounds (VICs), and include inorganic gases such as hydrogen sulfide (H_2S), sulfur dioxide (SO_2), carbon disulfide (CS_2) and ammonia (NH_3).

Odor can be defined as a stimulus of olfactory cells in the presence of specific compounds including VOCs and VICs. In general, the most common odor problems are caused by mixtures of highly volatile compounds with very low threshold detection limits that are in low concentration in air. Many branches of the industry produce gases that contain odorous VOCs or VICs. These gases are generated in process industries, synthetic flavoring and painting factories, paper mills, pharmaceutical industries, and refineries, etc. Odorous components of natural origin are mainly discharged by industries such as slaughterhouses, breweries, bio-industries, textile industries, coffee roasting plants, yeast and alcohol factories, sewage treatment works, solid waste composting works, etc. Several authors identified VOCs together with H_2S as being the major odor stimuli in sewer pipes and aerobic wastewater treatment plants (Smet and van Langenhove 1998). Other odorous molecules include organic sulfides, mercaptans, ammonia, inorganic and organic amines, and organic acids, aldehydes and ketones.

Shareefdeen/Singh (Eds.)
Biotechnology for Odor and Air Pollution Control

Table 3.1. Henry's coefficient for some common compounds at 25 °C

Compound	Henry's coefficient (non-dimensional)
Hexane	30.9
Oxygen	29.1
Hydrogen sulfide	0.92
Toluene	0.25
Benzene	0.22
MIBK (methyl iso-butyl ketone)	0.016
Ethanol	0.0012
Ammonia	0.0005

The gas–liquid phase equilibrium of the pollutant can be described by Henry's law:

$$C_{gi} = H_i C_{li} , \tag{3.1}$$

where C_{gi} is the concentration of pollutant i in the gas phase, H_i is Henry's coefficient, and C_{li} is the concentration of i in the liquid phase.

Henry's coefficient (constant of proportionality H_i in Eq. 3.1) is found in the literature in different units. Using a non-dimensional Henry's coefficient (mg gas l^{-1} over mg liquid l^{-1}), substances with values over 0.01 are considered volatile, and the higher the value, the less soluble the substance is in water. Henry's coefficient depends on temperature and the chemical potential of the liquid phase. Table 3.1 compares the H_i for different substances in water.

Gas–liquid phase equilibrium can be affected if the compound reacts in the liquid. For example, H_2S has been considered one of the most common compounds responsible of odor generation. H_2S is released to the atmosphere due to commercial and industrial activities, such as the petrochemical industry, the photographic processing industry, and the paper industry. H_2S dissociation depends on pH and, as can be seen in the following reactions (Card 1998), when pH is under 5, practically all the sulfide is as H_2S and in physical equilibrium with the gas phase. At pH 10, sulfides are mainly as HS^-, which is a very soluble compound, and the apparent Henry coefficient can be three orders of magnitude lower than at pH 5.

$$H_2S_{(g)} \leftrightarrow H_2S_{(l)}$$

$$H_2S_{(l)} \leftrightarrow HS^- + H^+ \quad Ka_1 = 1 \times 10^{-7} , \quad pK_a = 7.0$$

$$HS^- \leftrightarrow S^{-2} + H^+ \quad Ka_2 = 1.3 \times 10^{-13} , \quad pK_a = 12.9 .$$

3.2 Methods for VOCs and Odor Control

To select an appropriate control method for VOCs and odors, it is important to consider the physical, thermodynamic and reaction properties. The difference in the properties of the target compounds and the conditions of the stream determine the control method. Table 3.2 lists the relevant properties that are commonly used to characterize VOCs and to select a control method.

A classification of common technologies applied for VOC, H_2S and odor control is shown in Fig. 3.1. The classification is based on the nature of each control technology, physical, chemical or biological. Generally, the physical processes are applied for waste gas streams where the flow and pollutant concentration are high. Important parameters for biological treatment are the solubility and the biodegradability of the compounds. The most important advantage of biological treatment methods over physical and chemical technologies is that the biological processes can be conducted at normal temperatures (10–40 °C) and atmospheric pressure. Biological methods are inexpensive, simple to operate, and ecologically clean as compared with the physicochemical treatments. However, to accomplish a good selection of a treatment method, it is recommended to take into account flow rate and pollutant concentration (Fig. 3.2) as well as the category of the compound that is considered to be removed. There are other important factors that determine the selection, such as temperature, oxygen content of waste gas, stream composition, solubility, operating schedule, utility and maintenance requirements. The investment and maintenance costs and the possible secondary environmental impacts must be evaluated, too.

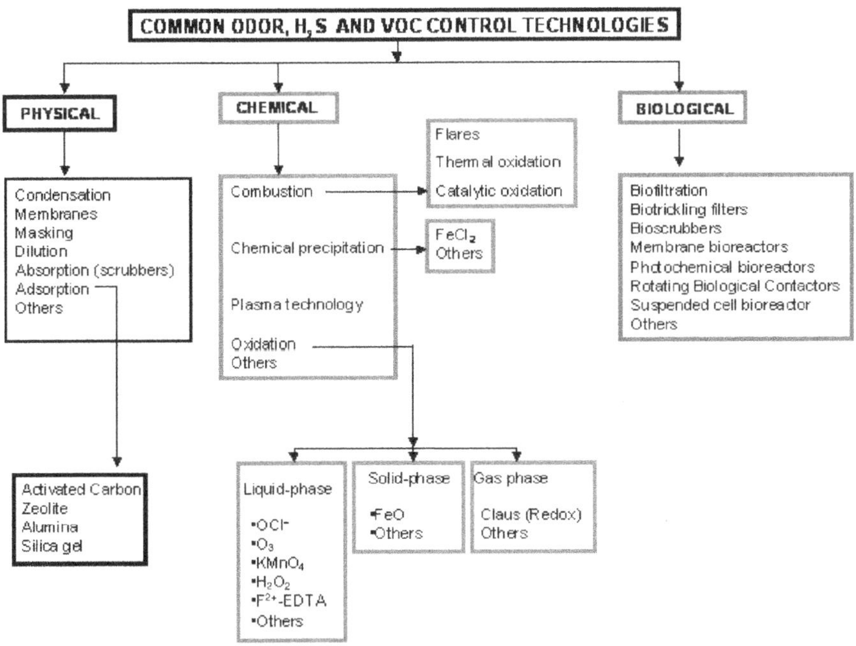

Fig. 3.1. Classification of common technologies for VOC and odor control

Table 3.2. Selection of odor and air pollution control methods

Property	Description	Comments	Implications for treatment method
Phase	Gases or liquids	Gas-phase properties are mostly independent of the particular mixture of compounds. Liquid-phase properties are highly dependent on composition	The knowledge of the phase of the VOC allows to determine their physical, transport and thermodynamic properties. This information is essential for the design and model development
Compounds with covalent or polar bonds	Covalent chemical bonds produce compounds that have a uniform electric charge throughout the compound. On the other hand, the polar compounds show an electric charge	Compounds that only have carbon and hydrogen (such as methane, benzene, butane) are usually fully covalent. Some organic compounds, such as acids or amines, have functional polar groups. This charge difference increases the energy required to separate the molecules from the liquid to the gas phase. Therefore, polar compounds are less volatile, in general, than covalent compounds. Polar compounds normally have much higher aqueous solubility than non-polar compounds	Aqueous solubility enhances treatment in the aqueous phase. This is particularly important for treatments that require that the pollutant be dissolved in water, such as biological treatment and most of the chemical treatments that are performed after pollutant scrubbing

Table 3.2. (continued)

Property	Description	Comments	Implications for treatment method
Ionization	Ionization is a rapid reaction. All acids, bases and salts produce ions in solutions	The theory of ionization is used to explain many of the observed phenomena in aqueous solution. One of the most important equilibrium of concern is the dissociation of water into a hydrogen ion (or proton) and hydroxyl ion. pH definition stems from the ionization concept. Also important is the formation of complex ions to establish the solubility product and ion effect	Ionization greatly increases solubility, as only the non-ionized species can volatilize. If a compound is ionized in solution, the performance and economics of liquid scrubbing for chemical or biological control can generally be enhanced
Vapor pressure	The pressure exerted by the vapor under conditions where it is in equilibrium with its liquid, under a particular temperature, is known as vapor pressure	Vapor pressure is highly dependent on temperature, and can be estimated using the Antoine equation. In general, compounds that are non-polar have higher vapor pressures because they require lower energy levels for the molecules to escape the liquid phase	The knowledge of the volatility of a compound greatly affects the options for odor and volatile organic compound control. An example of this is the condensation that depends on the contaminant vapor pressure

Table 3.2. (continued)

Property	Description	Comments	Implications for treatment method
Solubility in water	The actual quantity of a gas that can be present in solution is determined by the solubility of the gas as defined by Henry's law (cf. Eq. 1)	The equilibrium or saturation concentration of gas dissolved in a liquid is a function of the type of gas and the partial pressure of the gas in contact with the liquid	The ability of a compound to dissolve in water is the critical factor in determining whether the compound is suitable for control by liquid scrubbing or not. Biological treatment methods depend strongly on liquid solubility
Reactivity	Reactivity can be defined as the ability of a compound to react with itself or with another compound. It is quantified by a reaction rate that is dependent. The ability of microbes to transform a compound is called biodegradation	Most organic volatile compounds can be chemically oxidized at high temperatures. The use of a catalyst or strong oxidant such as ozone, hydrogen peroxide and chlorine can lower the temperature. Oxidation can be performed biologically at ambient temperature for many organic, sulfur, and nitrogen compounds	The reactivity of a compound will determine whether a physical, chemical or biological method is selected

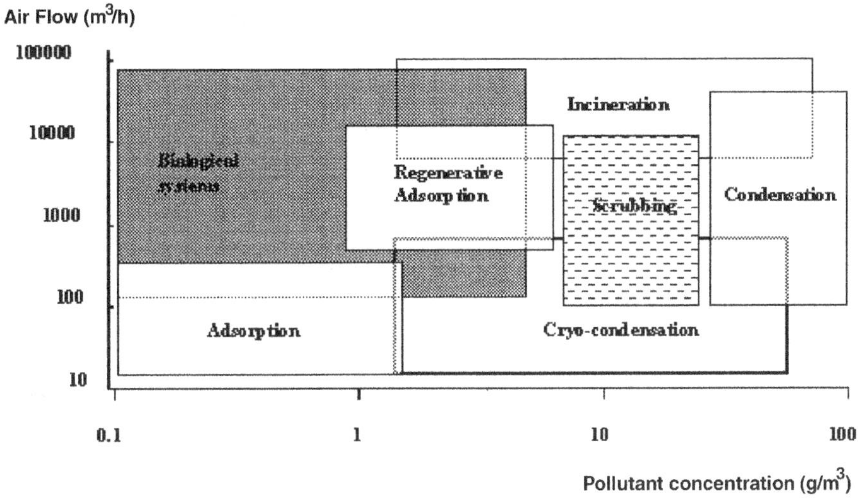

Fig. 3.2. Applicability of various air pollution control technologies based on airflow rates and concentrations to be treated

3.3 Physical-chemical Methods

3.3.1 Dilution

Dilution can occur by the addition of sufficient fresh air to reduce the odor concentration below the threshold level (Corbitt 1990). The method is useful in certain applications to reduce explosion risk.

Tall stacks and heated gas streams have been successfully used by the industry to dilute large volumes of gases by plume dispersion. The stack systems are useful for low concentrations of odor, and can be a second-stage system following another control method. Dilution cannot be considered as a treatment method, since pollution remains untreated.

3.3.2 Condensation

Condensation (Fig. 3.3) can occur by lowering the gas stream temperature at constant pressure or increasing the gas stream pressure at constant temperature (Bell 1988; Planker 1998; Waldrop 1998; Kennes et al. 2001).

It is generally applied to treat effluent streams consisting of a condensable pollutant vapor and a non-condensable gas. There are basically two types of condensers: surface and direct contact. Surface condensers are generally shell and tube heat exchangers where coolant flows inside the tubes. VOCs condense outside the tubes. Contact condensers operate by spraying a cool liquid directly into a gas stream to cool and condense the VOCs.

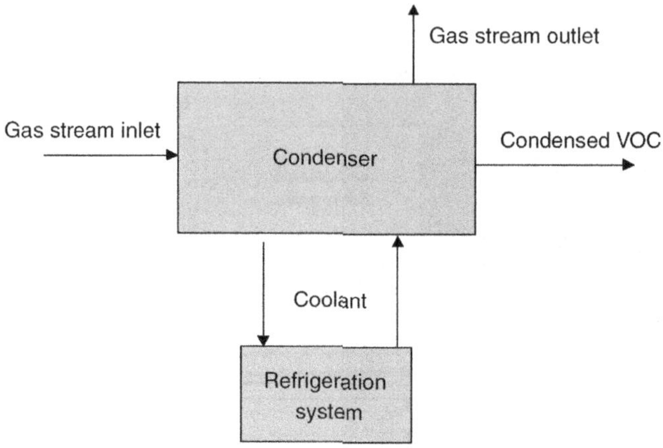

Fig. 3.3. Schematic diagram of condensation processes

3.3.3 Membranes

This refers to the use of a semi-permeable membrane to separate VOCs from a waste gas stream (see Fig. 3.4). A semi-permeable membrane is made of synthetic polymers wrapped around a perforated central collection pipe. The driving force for the airflow is the pressure drop across the membrane, using a vacuum pump. The membrane is permeable to VOCs but not air. Therefore, VOCs pass through the membrane while the purified air stream is released to the atmosphere (Mukhopadhyay and Moretti 1993; Kennes et al. 2001).

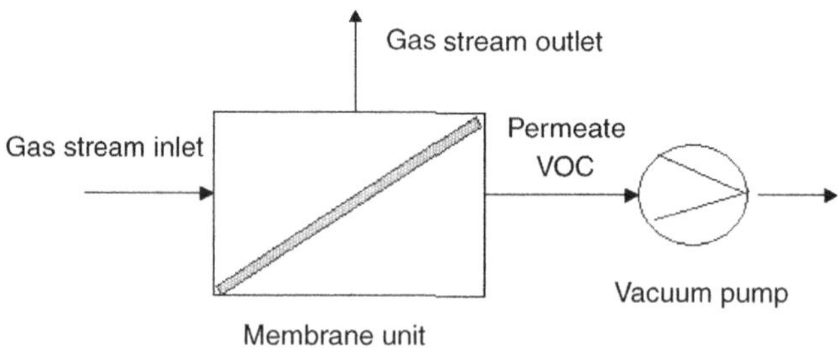

Fig. 3.4. Schematic diagram of membrane processes

3.3.4 UV Oxidation

Ultraviolet (UV) radiation is a physical process that is based on the transfer of electromagnetic energy from a source (lamp) to organic matter.

This is an emerging technology for the control of VOCs that uses oxygen-based oxidants like ozone, peroxide, OH, and O^- radicals to convert VOCs into CO_2 and H_2O in the presence of ultraviolet light (Mukhopadhyay and Moretti 1993; Qasim 1999; Kennes et al. 2001). UV light enhances the activity of oxygen-based oxidants. The primary source of UV energy is the low-pressure mercury lamp. It is almost universally accepted as the most efficient and effective source of UV radiation. The lamps are tubes, typically 0.75–1.5 m in length and 1.5–2.0 m in diameter. Approximately 35–40% of the energy is converted to light, and approximately 85% of the light has a wavelength of 253.7 nm.

3.3.5 Plasma

Plasma is a mixture of free-moving electrons and positively charged ions. Non-thermal plasmas are an excellent source of gas-phase free radicals (O, OH and H) and other active species useful for destroying pollutants. Reactive species in non-thermal plasma, such as OH radicals, O_3 molecules, and O and N atoms, can react with odorous and toxic gases and convert them into nonodorous and nontoxic molecules (see Fig. 3.5).

Plasma can be generated in a medium between two electrodes where a high voltage AC (10–30 kV) is applied for a very short time (1–10 ms). The plasmas can be generated by dielectric barrier discharge (DBD) or an electron beam. Dielectric barrier discharge, also called corona discharge, utilizes a dielectric material between the discharge gap and one of the two discharge electrodes used. When the potential across the gap reaches breakdown voltage, the dielectric acts as a stabilizing material, leading to the formation of a large number of microdischarges of short pulses that are spread over the discharge gap. On the other hand, the electron-beam technology is based on an electron gun that shoots high-energetic electrons to a target object. By interaction and collision

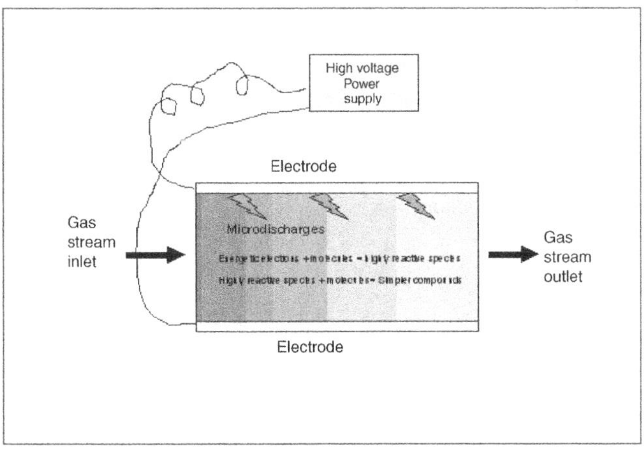

Fig. 3.5. Schematic diagram of plasma technology

with matter in their pathway, the high-energetic electrons generate a multitude of exited molecules that oxidize VOCs and inorganic volatile compounds. The use of nonthermal plasmas for gas pollution treatment is promising, but is still in its early stages of research and development (van Groenestijn 2001b). Plasma technology can be classified as an electrochemical process.

3.3.6 Adsorption

Adsorption (Fig. 3.6) refers to the process where gaseous VOC molecules contact the surface of a solid adsorbent and bond via weak intermolecular forces.

Carbon is the most common adsorbent in use today for VOC treatment. Others include silica gel, alumina and zeolite. They are "activated" by heat under controlled conditions at very high temperatures to remove the volatile non-carbon constituents and increase the surface area. Several types of carbon adsorption units are commercially available, but the most common is the fixed regenerative bed. It has of two or more beds of activated carbon working in parallel. Continuous system operation is possible by the concurrent adsorption by at least one bed and desorption by the other beds. Adsorbent regeneration is by volatilization of the adsorbed compounds, either by increased temperature with steam or by lowering the bed pressure. Other regeneration methods such as chemical or solvent washing, induction heating (Joule effect, electromagnetic induction) and biological oxidation have been developed and/or are currently being investigated (Mukhopadhyay and Moretti 1993; Smet and van Langenhove 1998; Kennes et al. 2001).

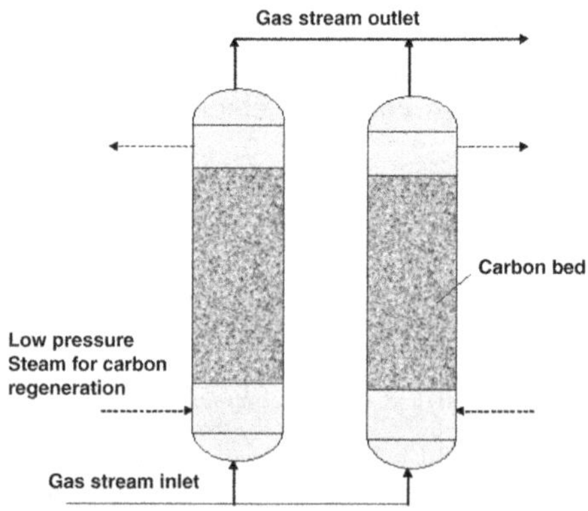

Fig. 3.6. Schematic diagram of adsorption processes

3.3.7 Combustion (Flares, Thermal and Catalytic Incinerators)

VOC is captured by a ventilation system, preheated and oxidized at high temperatures to form CO_2, SO_2, and H_2O (Corbitt 1990; Mukhopadhyay and Moretti 1993; Anguil 1998; Kennes et al. 2001). Primary applications are for control of VOCs, including hazardous wastes and odorous compounds (see Fig. 3.7).

This method is sensitive to sources that do not have low gas flow rates and low particle concentrations. There are three basic types of incinerators used as air pollution control devices: flares, thermal and catalytic incinerators. Flares are combustion devices in which an open flame and the open air assist in VOC destruction. They are used basically for emergency control but they may be used for VOC control. In petroleum refineries and chemical plants, flares function primarily as a VOC abatement technology. Thermal incinerators (afterburners) operate at high temperatures (650–800 °C) and are widely used.

Catalytic incinerators are an alternative to thermal incinerators; they operate by passing a preheated gas stream through a catalyst bed to oxidize the combustible emissions. The catalyst allows combustion at much lower temperature (250–500 °C) than those required for thermal incineration. Normally,

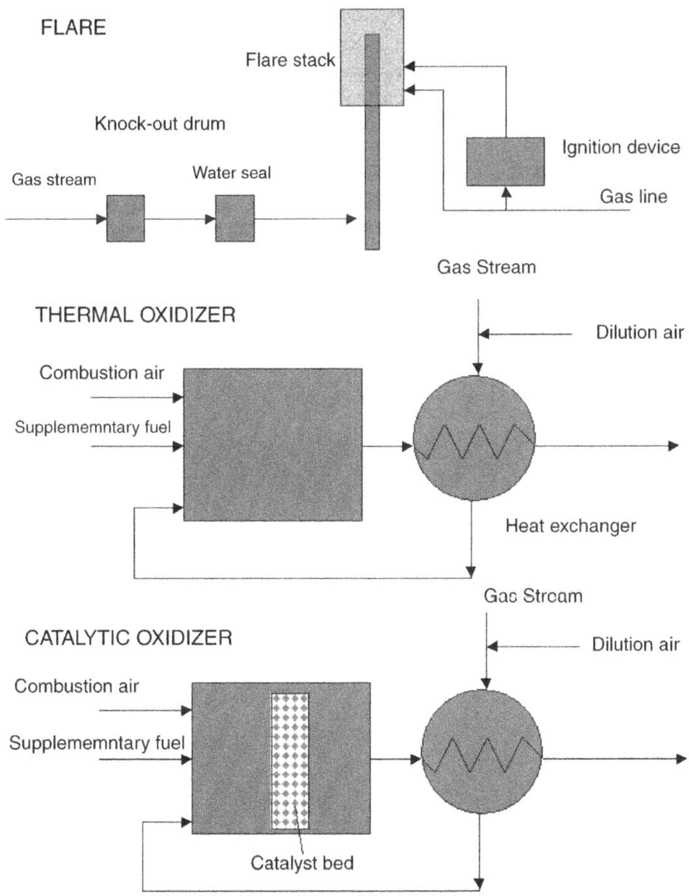

Fig. 3.7. Schematic diagram of combustion processes

catalysts are metals of platinum or palladium families that are thinly coated on an inert support material (ceramic, metal mesh mat, honeycomb, etc.). The flue gas from the incinerators is at a high temperature and contains recoverable heat energy used to preheat the gas emission stream before it enters the combustion chamber.

3.3.8 Masking

The process of masking involves overpowering the odorous molecules with a stronger molecule (Corbitt 1990; Planker 1998; Smet and van Langenhove 1998). Masking compounds frequently utilize only one essential oil, such as vanilla, citrus, pine or floral or as terpenes, to permeate the area with its aroma. Masking usually does not modify or neutralize the odorous molecules because there is only one essential oil. The application of these products becomes dangerous if they mask the odor of high and/or toxic concentrations of VOCs. In addition to masking agents, other deodorization products are commercially available. Some of these act as chemicals (acid/base reactions) and can be effective against H_2S and MeSH (methyl mercaptan), while others have enzymatic properties.

3.3.9 Caustic Scrubbing

Absorption is favored by highly alkaline conditions (Mansfield et al. 1992). A gas stream containing the pollutant is fed to an absorption tower with high alkalinity (i.e., NaOH 50% by weight, pH > 12). The absorbent is not regenerated in this process, which implies high reagent consumption (see Fig. 3.8).

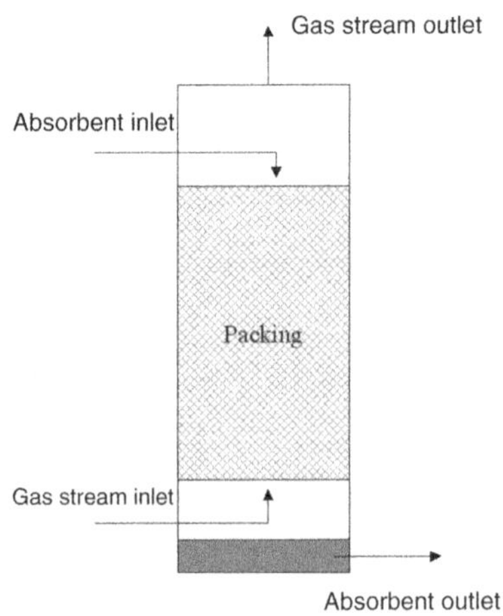

Fig. 3.8. Schematic diagram of caustic scrubbing processes

3.3.10
Regenerative Gas Scrubbing

Scrubbing aims to transfer the pollutant from the gas to the aqueous phase by intense contacting of the polluted air stream with the water phase (West 1983; Smet and van Langenhove 1998). As an example for H_2S treatment, it is possible to have the following reaction:

$$R_2NH + H_2S \leftrightarrow R_2NH_2HS + \text{heat} .$$

Mass transfer depends on the concentration, the air/water partitioning (Henry's law) coefficient, and the mass transfer resistance of the scrubber system. There are different types of absorbents that are acids or bases. Into the scrubbing process, the absorbent can be regenerated. During the H_2S scrubbing process, H_2S and CO_2 are discharged as exhaust products. The heat produced by the exothermic reaction during the absorption step is used to preheat the absorbent in the desorption step. Usually, the absorption tower works at low or ambient temperature to favor solubilization while desorption is favored at higher temperatures.

Absorbants commonly used include mono-ethanolamine, di-ethanolamine, di-glycolamine, methyl-di-ethanolamine, di-isopropanolamine, hot potassium carbonate, methanol, propylene carbonate, and *N*-methyl-2-pyrrolidone (see Fig. 3.9).

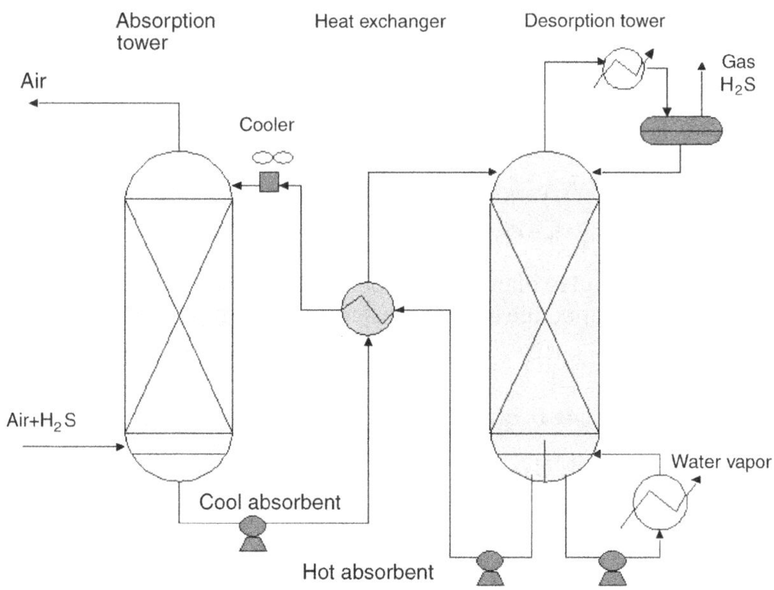

Fig. 3.9. Schematic diagram of regenerative gas scrubbing

3.3.11 Chemical Precipitation

The gas stream containing H_2S is treated in an absorption tower where, after the H_2S is dissolved; it precipitates as FeS when Fe^{2+} is added (Särner 1990).

$$\text{FeCl}_2 \text{ reaction:} \quad Fe^{2+} + H_2S \rightarrow FeS + 2H^+ .$$

The Fe^{2+} is not regenerated during the process, which means considerable reagent consumption.

3.3.12 Chlorine Oxidation

After H_2S is absorbed in a scrubbing tower, it is oxidized with sodium hypochlorite to produce either elemental sulfur or sulfate, depending on pH (Cadena and Peters 1988).

$$HS^- + OCl^- \rightarrow S^0 + OH^- + Cl^- \qquad \text{at pH} < 7.5$$
$$HS^- + 4OCl^- \rightarrow SO_4^{2-} + H^+ + 4Cl^- \qquad \text{at pH} > 7.5 .$$

The reagent is not regenerated in the process, and it is clear that a high cost associated to the reagent consumption is expected. In the presence of organic compounds, chlorination is not attractive, since the formation of undesirable organic chloride compounds is expected.

3.3.13 Ozone Oxidation

H_2S is dissolved in water with a scrubbing tower and then oxidized by ozone. The process can be applied to VOCs as well (Chen and Morris 1972).

$$HS^- + O_3 \rightarrow S^0 + OH^- + O_2$$
$$HS^- + 4O_3 \rightarrow SO_4^{2-} + 4O_2 + H^+ .$$

Ozone is a powerful oxidant but is expensive. The sulfur oxidation is practically instantaneous and, due to ozone instability, in situ generation is required.

3.3.14 Potassium Permanganate Oxidation

After H_2S is scrubbed, H_2S can be oxidized using potassium permanganate (Cadena and Peters 1988). This method is not attractive because of its high cost, and the manganese oxide must be adequately disposed to avoid a negative environmental impact. In addition, different sulfur compounds are produced depending on pH.

$$3H_2S + 2KMnO_4 \rightarrow 3S_0 + 2H_2O + 2MnO_2 + 2KOH \qquad (\text{pH} < 7.5)$$
$$3H_2S + 8KMnO_4 \rightarrow 8MnO_2 + 3K_2SO_4 + 2H_2O + 2KOH \qquad (\text{pH} > 7.5)$$

3.3.15 Catalytic Oxidation with Fe^{3+} (LO-CAT Process)

The ferric and ferrous ions are chelated with EDTA to avoid precipitation as FeOH or FeS (Thomson 1980). This allows the ferric ion regeneration using air (O_2). The H_2S oxidation and reagent regeneration are simultaneous.

$$H_2S + 2\left[Fe^{3+}\right] \rightarrow S^0 + 2\left[Fe^{2+}\right] + 2H^+$$

$$2\left[Fe^{2+}\right] + 0.5O_2 + H_2O \rightarrow 2\left[Fe^{3+}\right] + 2OH^-$$

3.3.16 Hydrogen Peroxide Oxidation

An H_2S scrubbing step is required. The oxidation rate of sulfide with hydrogen peroxide is relatively slow. Generally 20–30 min contact time is required for a complete reaction. The mechanisms of oxidation of H_2S by hydrogen peroxide in wastewater are not well understood. However, it is suggested that direct oxidation of sulfide by hydrogen peroxide depends on the direct reaction with the oxygen released during gradual decomposition of hydrogen peroxide (Cadena and Peters 1988).

$$H_2O_2 + H_2S \rightarrow S^0 + 2H_2O \quad \text{at pH} < 8.5$$

$$2H_2O_2 + S^{2-} \rightarrow SO_2^{2-} + 2H_2O \quad \text{at pH} > 8.5$$

3.3.17 Oxidation with FeO

This treatment is based on the interaction of H_2S with a dry, packed bed of Fe_2O_3 and FeO (Montalvo and Guerrero 2003). Some sulfur compounds are formed as reaction products. Regeneration is by contact with oxygen from an airstream.

$$Fe_2O_3 + 3H_2S \rightarrow Fe_2S_3 + 3H_2O \quad \text{Oxidation}$$

$$FeO + H_2S \rightarrow FeS + H_2O$$

$$Fe_2S_3 + 3/2O_2 \rightarrow Fe_2O_3 + 3S^0 \quad \text{Regeneration}$$

$$FeS + 0.5O_2 \rightarrow FeO + S^0$$

3.4 Biological Methods

Biological air treatment systems are based on the capability of microorganisms to transform certain organic and inorganic pollutants into less toxic and odorless compounds. As the pollutants are in air, the processes of microbial degradation are generally oxidative in nature and the end products are carbon dioxide, water, sulfate, and nitrate. The use of microorganisms to treat effluents has long been established for wastewater and solid waste. In the case of water treatment, it has been recognized that microbes can effectively degrade pollutants, especially when they are in low concentrations (less than 5,000 ppm) at mild temperature and pH and under aerobic or anaerobic conditions. In the case of solid waste treatment, microorganisms have been found effective to reduce bulk solids, as in the case of compost, or degrade hazardous pollutants from soil.

The first reports of biological air treatment systems were related to the treatment with soil beds of odors from sewage (Leson and Winer 1991). During the years 1960–1970, it was shown that the biological reaction was the main mechanism for the elimination of the pollutants, and the first commercial applications were established in The Netherlands, West Germany and the United States (Ottengraff 1986). The initial applications were generally open biofilters, and they were used to treat odors from a variety of sources such as sewage treatment plants, rendering facilities, compost, food processing, and farms (Leson and Winer 1991). Further developments included the use of better supports, which allowed improved performance, control and stability. Developments in biofilter configurations included closed systems and improved control. Since the 1980s (Ottengraff and van der Oever 1983), there has been very extensive research in the fundamentals of biological air treatment systems, and new systems have been successfully implemented (Bohn 1992). The technical improvements have allowed to increase biodegradation rates and efficiency, to reduce failures, to extend the number of applications, and to reduce investment and operating costs.

The elimination of volatile compounds from air requires first that the gaseous pollutant be transferred from the air stream to the biologically active aqueous phase. Then, the microorganisms utilize these molecules as a source of nutrients and energy for growth, producing more biomass and carbon dioxide, water, sulfate, nitrate, etc., as by-products. As the biological air treatment systems have generally a limited input of other nutrients, the biomass formed is partly recycled. The overall efficiency of the process is determined by the relative rates of the physical, chemical and biological processes.

Table 3.3. Classification of biological reactors

Biomass	Liquid phase	Reactor
Fixed on a support	Stationary	Biofilter
Fixed on a support	Flowing	Biotrickling filter Rotating contactors
Suspended	Flowing	Bioscrubber
Suspended or fixed	Stationary	Suspended growth
Fixed on a membrane	Flowing	Membrane

Table 3.4. Applications of biological air treatment systems

General	Specific	Specific (cont.)
Industrial activities including production, transport and storage	Asphalt Chemical Food, feed and beverage Foundries Fragrance Leather	Petroleum and petrochemicals Pharmaceutical Pulp and paper Textile Viscose processing
Naturally generated odors	Compost Farms Food and feed Landfill gases	Sewage Slaughter and rendering plants Tobacco Wastewater treatment
Other trades	Paint shops Print shops	Soil remediation

Although the basic mechanisms are the same for all biological air treatment systems, there are different equipment configurations to achieve transfer and biodegradation. These are grouped in Table 3.3 and will be covered in detail in the next chapters of this book. Biological air treatment systems have been used for eliminating pollutants in a wide range of applications, as presented in Table 3.4.

3.4.1 Terminology

Since the first works published on biofiltration, a common terminology has been established in biological air treatment systems, facilitating communication and comparison among the various processes. This terminology, with the most common units used, is defined below.

3.4.1.1 Empty Bed Residence Time (EBRT)

This term, also called empty bed retention time (EBRT), is related to the time that the air spends in the reactor, considering that the reactor is empty. EBRT considers the volume of the reactor instead of a more real air residence time that would include the porosity of the packing. The difference may be substantial for some configurations. In some cases, the real or "true" residence time is used. EBRT and true retention time are defined below:

$$\mathrm{EBRT} = \frac{V_\mathrm{r}}{Q}\,, \tag{3.2}$$

$$\tau = \frac{V_\mathrm{r} P}{Q}\,, \tag{3.3}$$

where EBRT denotes the empty bed residence time (min), V_r is the bed volume of the reactor (m^3 reactor), Q is the inlet airflow ($m^3\,min^{-1}$), τ is the true air residence time (min), and P is the porosity (1 m^3 packing m^{-3} reactor).

The EBRT is a critical parameter as it has direct influence in the economy of the process. It may go from a few seconds to few minutes, depending on the characteristics of the inlet stream, but generally falls between 0.5 and 2 min.

3.4.1.2
Volumetric Load and Superficial Velocity

Volumetric load and superficial velocity are terms that normalize the amount of air fed to the system. They are defined as:

$$\mathrm{VL} = \frac{Q}{V_r} \tag{3.4}$$

$$v_s = \frac{Q}{A}\,, \tag{3.5}$$

where VL is the volumetric load (m^3 flow m^{-3} reactor min^{-1}), v_s is the superficial velocity ($m\,min^{-1}$), and A is the surface area perpendicular to the flow (m^2).

3.4.1.3
Inlet Mass Load

Inlet mass load refers to the quantity of pollutant that enters the system. The inlet load is generally normalized to the reactor volume and expressed as:

$$\mathrm{ML} = \frac{QC_{gi}}{V_r}\,, \tag{3.6}$$

where ML is the mass load (g pollutant m^{-3} reactor min^{-1}), and C_{gi} is the inlet concentration of pollutant (g pollutant m^{-3}).

3.4.1.4
Elimination Capacity and Removal Efficiency

Both terms are used in conjunction to describe the performance of the system. They are defined as:

$$\mathrm{EC} = \frac{Q\left(C_{gi} - C_{go}\right)}{V_r}\,, \tag{3.7}$$

$$\%R = 100\frac{\left(C_{gi} - C_{go}\right)}{C_{gi}}\,, \tag{3.8}$$

where EC is the elimination capacity (g pollutant m^{-3} reactor min^{-1}), C_{go} is the outlet pollutant concentration (g pollutant m^{-3}), and $\%R$ is the removal efficiency (%).

3.4.2 Mechanisms

In biological air treatment systems, the pollutants cannot be degraded directly in the gaseous phase, so the compounds, including oxygen, have first to be transferred to the liquid phase where the microbes, either growing suspended or attached to a support, will attack them. The overall process includes complex physical, chemical and biological interactions that result in the transformation of the pollutants to other substances with less health and environmental impact.

3.4.2.1 *Gas–Liquid Transfer*

The pollutants and oxygen are first solubilized from a flowing gaseous phase to the liquid phase containing the microorganisms. The liquid phase may be static, as in the biofilm formed in a biofilter, flowing over the biofilm, as in a biotrickling filter, rotating contactor and membrane bioreactor, or contain the suspended microorganisms, as in a bioscrubber or suspended growth bioreactor. The transfer mechanism is dependant on the gradient, the transfer surface, and an overall mass transfer coefficient. The gradient is the difference between the equilibrium and the actual concentration of the pollutant in the liquid phase, as described by Henry's law (Eq. 3.1). In general, values of H_i in water are considered as a reference, but it has been found that microorganisms in the aqueous phase, biofilms and the supports generally used for biofiltration favor the solubilization of hydrophobic substances (Acuña et al. 1999; Davison et al. 2000). For example, Davison et al. (2000) reported that propane solubility was an order of magnitude greater in a microbial suspension than in water.

Mass transfer can be described by Eq. (3.9), which shows that transfer can be improved by increased surface, better coefficients, determined by both process and chemical properties of the components involved, and a larger gradient. In the membrane bioreactor, the overall transfer step includes the resistance caused by the membrane, which is proportional to the wall width and the effective diffusion coefficient of the pollutant in the membrane material.

$$\mathrm{d}C_\mathrm{li}/\mathrm{d}t = k_\mathrm{l}a\left(C^*_\mathrm{li} - C_\mathrm{li}\right) = k_\mathrm{l}a\left(C_\mathrm{gi}/H_\mathrm{i} - C_\mathrm{li}\right)\,, \tag{3.9}$$

where k_l is the overall mass transfer coefficient, a is the transfer area, and C^*_li is the equilibrium concentration of i in the liquid phase.

3.4.2.2 *The Liquid Phase: Physical and Chemical Processes*

Once the pollutants and oxygen are solubilized in the moist solid phase, they interact with the water, the insoluble and soluble compounds, and the microorganisms. Among the physical interactions, there are equilibrium-driven processes such as sorption (absorption and adsorption) in the insoluble matter, such as the support in biofilters and the microbial debris. In these equilibrium processes, there may be desorption if the pollutant concentration decreases in the gas phase. Sorption tends to substantially increase the amount of pollutant in the aqueous phase, especially in the case of a biofilter, but the substances are not necessarily available to the microorganisms, as there may be strong adsorptive interactions. On the other hand, sorption provides a buffer capacity for systems with variable incoming concentrations. Sorption capacity of the biofilter is easily detected at startup.

Certain pollutants undergo chemical reactions in the liquid phase. The most common reactions take place when charged molecules establish ionic equilibrium depending on the pH of the aqueous phase. This ionic interaction is relevant as it may displace gas–liquid equilibrium – for example, H_2S may be ten times more soluble at pH 8 than at pH 4, as previously shown. The composition of soluble minerals and the influence of the carbon dioxide formed by the oxidation reactions determine the buffer characteristics of the aqueous phase. Absorption of volatile inorganic acids such as H_2S, NO_X, or organic acids such as formic or acetic acids tends to acidify the medium while bases, such as ammonia and amines, tend to alkaline. For certain compounds, there are chemical reactions that occur simultaneously with the biological reaction. This is the case for H_2S where spontaneous oxidation occurs, although the rates are much lower than those promoted by microbes under the conditions used in biological processes.

3.4.3 The Biological Phase

The microorganisms present in the biological air treatment systems are the biocatalysts in the degradation process. Through their metabolic capacities, they transform the pollutants into biomass and byproducts. They may be suspended in the liquid, as in bioscrubber and suspended growth bioreactors, forming a biofilm within a moving liquid (biotrickling filter, or membrane bioreactor), or colonizing a moist support, as in a biofilter. The initial mixed microbial population, or consortium, is continually evolving in response to changes in pollutant feed, environmental conditions (temperature, pH, nutrient availability, moisture), and the microorganisms continually entering the reactors with the incoming air.

3.4.3.1 Type of Microorganisms

Biological air treatment systems are based on the extensive and efficient capacity of microorganisms to degrade air pollutants. Microbial diversity ranges from bacteria, actinomycetes, and yeast to fungi. The active microbial population in a determined system will be dependent on the incoming pollutant and the prevailing environmental conditions.

The microbial population will grow from the energy (as ATP) derived from the transformation of the pollutant. For the case of VOCs, heterotrophic or-

ganisms derive energy from the oxidation of the organic molecule, and part of the carbon is incorporated into biomass. When the pollutant contains sulfur, nitrogen or halogens, some of these elements accumulate in the system. Reduced inorganic pollutants such as H_2S and NH_3 are treated with autotrophic microorganisms that derive energy from the oxidation of the molecule and use CO_2 as a carbon source.

The selection of the microbial population is determined by the characteristics of the pollutant to be treated (Devinny et al. 1999; Deshusses and Johnson 2000). Naturally produced substances, such as those found in foul odors (water treatment plants, composting facilities, food and feed factories, etc.), are in general highly biodegradable at low concentrations. They can be treated using the natural population found in biofilter media, or can be inoculated in other systems with a diverse mixed population, such as that found in activated sludge. Appropriate elimination capacity and efficiency require that the initial biomass attains the necessary density. The startup period depends on the amount of initial microbial population that can degrade the pollutants, so inoculation from an adapted population is desirable. The same startup approach can be used for industrial applications when the pollutant has been shown to be biodegradable, in which case a less diverse population is established. There are extreme cases of substances that are difficult to degrade and require that the reactor be inoculated with specialized strains or consortia to initiate degradation (van Groenestijn and Hesselink 1993; Devinny et al. 1999). For example, from Table 3.5, it can be seen that ethers are generally easily degraded; this is the case for diethyl or dimethyl ether, and nevertheless microorganisms that degrade methyl ter-butyl ether (MTBE) are very rare.

For most applications, biological air treatment systems have traditionally relied on microorganisms, generally bacteria, found under ambient environmental conditions. Considering that in certain applications the effluents are at high temperature, the use of thermophilic microorganisms, those growing above 55 °C, has been shown to be feasible for various pollutants (Heslinga

Table 3.5. Biodegradability of various compounds

Biodegradability	Compounds
Rapid	Alcohols[a,b], aldehydes[a], ketones[a,b], esters[a,b], ethers[a], organic acids[a], terpenes, amines[a], thiols[a], sulfides, ammonia
Slow	Hydrocarbons[a,b], phenols[a]
Very slow	Halogenated hydrocarbons[a,c], polycyclic aromatic hydrocarbons

[a] Branched molecules are less biodegradable

[b] Removal in biofilters follows alcohols > esters > ketones > aromatics > alkanes

[c] Biodegradability decreases with higher number of halogens

and van Groenestijn 1997; Dhamwichukorn et al. 2001; Cox et al. 2001). In the case of highly hydrophobic hydrocarbons, it has recently been shown by various authors that fungal reactors offer some interesting advantages over flat bacterial biofilms. Fungi mycelia occupy the internal void space in the reactor, increasing substantially the transfer surface; EC values for toluene close to 300 g pollutant m^{-3} reactor min^{-1}, and 150 g pollutant m^{-3} reactor min^{-1} for hexane have been reported (García-Peña et al. 2001; Woertz et al. 2001a; Spingo et al. 2003), which are higher than those reported for bacterial films. Studies with fungal reactors have been done for α-pinene (van Groenestijn and Liu 2002) and nitric oxide (Woertz et al. 2001b).

The evaluation of the microbial concentration or activity in the bioreactors can be made through classical plate count methods (Aizpuru et al. 2003), by determining the pollutant uptake from the headspace of closed vessels (microcosms) containing samples of the support (Acuña et al. 1999), by measuring in-vitro enzymatic activity (Song and Kinney 2000), by the use of molecular biology tools (de Castro et al. 1997), or by in situ isotope labeling (Lipski and Altendorf 2002).

3.4.3.2
Biological Reactions

Microorganisms use pollutants to draw chemical energy for growth and the maintenance of biological activity. Microbial activity can be realized under aerobic or anaerobic conditions but, as the pollutants are in air, oxygen plays a pivotal role in the biotransformation. Several types of reactions take place in biological air treatment systems.

Microbial growth is a consequence of the uptake and metabolism of the pollutants. The derived chemical energy is used to synthesize new biological material, using available minerals in the aqueous phase. Minerals needed for growth include N, S, P, Ca, K, Na, Mg, Fe, Co, and Zn (Shuler and Kargi 1992). These nutrients can be found in the traditional supports in biofilters, and are available in the incoming water in certain suspended growth reactors but need to be added in the other systems to sustain growth. When nutrients are limited, new biomass is formed from the minerals derived from dead cells, and generally the transformation rates are low. On the other hand, excess growth can occur when nutrients are continually supplied, and undesirable plugging may come about.

Mineralization takes place when the pollutants are completely oxidized. These are the preferred reactions, as the end products are not toxic. Mineralization requires stoichiometric amounts of oxygen that may not be easy to provide when the pollutant is in high concentration in the liquid (from high solubility or high gas concentration). Partial oxidation may produce, in some cases, volatile intermediates that exit with the gas. This is the case for the transformation of carbon disulfide (CS_2) that produces the volatile intermediate carbonyl sulfide, COS (Alcántara et al. 1999); only under circumstances of high load, this intermediate is detected in the exit air (Table 3.6). In the absence of sufficient oxygen, the microorganisms may derive some energy

Table 3.6. Examples of biological reactions

Type	Example
Growth	$C + H + N + S + P$ + minerals → biomass
Mineralization with O_2	$C \rightarrow CO_2$
	$S \rightarrow SO_4$
	$N \rightarrow NO_3^-$
	$H \rightarrow H_2O$
	$R{-}Cl \rightarrow CO_2 + HCl$
Partial oxidation	Ethanol → CO_2 + acetaldehyde + acetic acid + ethyl acetate
	Isopropanol → acetone
	1-propanol → propianaldehyde → propionic acid
	Sulfide → elemental sulfur → sulfite
Cometabolism	MTBE in the presence of linear alkanes
	TCE in the presence of methane or phenol
	Benzene in the presence of toluene, ethylbenzene and xylenes

from the partial oxidation of certain pollutants, thus producing intermediates that may accumulate in the liquid or, if they are volatile, be carried away with the exiting air (Devinny and Hodge 1995; Kirchner et al. 1996; Christen et al. 2002).

Certain applications require that the pollutant be reductively transformed for further treatment. Highly oxidized chlorinated solvents, such as carbon tetrachloride, can be transformed to methane under anaerobic conditions, with a supply of electrons as hydrogen or from another carbon source (Lee et al. 1999). Another interesting example is the biological elimination of SO_2 from combustion gases by a three-step process that includes first the oxidation of the absorbed SO_2 to sulfate, then the reduction of the formed sulfate to H_2S and its partial re-oxidation to elemental sulfur under micro-aerophilic conditions (Philip and Deshusses 2003).

When there is a mixture of pollutants, the biological reactions can be competitive or noncompetitive, depending of the chemical structure of the pollutants and, in general, this is difficult to predict (Devinny et al. 1999; Aizpuru et al. 2001). In some cases, one pollutant is preferred by biodegradability or availability, and the elimination of the second is inhibited. For a biofilter treating butyl acetate and styrene, it was found that the ester inhibited styrene degradation. Nevertheless, the selection and inoculation of a butyl acetate-tolerant microorganism allowed its simultaneous removal (McGrath et al. 1999). In other cases, overall elimination is diminished by the presence of the mixture, as in BTEX (Lu et al. 2000). In some cases, proper conditions can be established to allow the simultaneous elimination of very different compounds, such as methanol and H_2S emissions found in the pulp and paper industry (Sologar et al. 2003). In other cases, it is preferred to treat the pollutants sequentially (Kraakman et al. 1996; Chitwood et al. 1999).

Cometabolism is the metabolic transformation of a substance while a second substance serves as primary energy or carbon source. In co-metabolism the transformation of the pollutant yields generally a byproduct that may be sometimes very toxic, as is the case for vinyl chloride production from trichloroethane transformation.

3.4.3.3 *Biological Kinetics*

The behavior of the microorganisms in biological air treatment systems is similar to that found in other bioreactors. First, increase in biomass can be described by a net or observed growth rate (μ) as:

$$\frac{dX}{dt} = \mu X = \left(\mu' - \alpha\right) X \,, \tag{3.10}$$

where X is the biomass concentration ($g l^{-1}$), is the observed (net) cell growth (h^{-1}), μ' is the specific cell growth rate (h^{-1}), and α is the specific cell death rate (h^{-1}).

In mixed populations, such as those generally found in biological air treatment systems, the different species are growing, or even dying, at different rates, and the structure of the consortium is evolving with time. Furthermore, new species arrive through the air and establish in the reactors, their fate being dependent to the adaptation to the prevailing conditions. The growth rate is particular to each microbial species, and depends on the concentration of the carbon and energy source, or other limiting substrates such as oxygen, and the surrounding micro-environmental conditions. These include temperature, pH, water content (described by the water activity, A_W), available nutrients to sustain growth (N, S, P, Mg, etc.), and the presence of inhibitors. The relation between growth and the concentration of the limiting substrate can be described by the Monod model:

$$\mu = \frac{\mu_{max} C_L}{K_s + C_L} \,, \tag{3.11}$$

where C_L is the concentration of limiting substrate ($g l^{-1}$), μ_{max} is the maximum cell growth (h^{-1}), and K_s is the half-saturation constant.

The Monod model predicts that at high concentration of the pollutant ($C_L \gg K_s$) growth attains μ_{max} but, at low concentration ($C_L \ll K_s$), growth is a first-order reaction. K_s values have been reported for various substances (Shuler and Kargi 1998). The Haldane variation includes in the denominator the term (C_L^2/K_i), where K_i is an inhibition constant to account for the possible effect of high pollutant concentration in the liquid, resulting from high concentration in the gas phase or low Henry's constant.

In biological air treatment systems, the interest is to sustain a high EC so that the pollutant degradation rate is more important than the growth rate.

Microbes degrade the pollutants when there is growth but it has also been shown that degradation occurs even when the net growth rate is zero. This can be attributed to the maintenance energy (m), the amount of substrate needed to obtain the required energy to sustain microbial life. The pollutant uptake rate can thus be defined as:

$$\frac{dC_L}{dt} = \frac{\mu X}{Y_{xs}} + mX \,, \qquad (3.12)$$

where Y_{xs} is the biomass yield from the pollutant (gxg pollutant^{-1}), and m is the maintenance coefficient (g pollutant g x^{-1} h^{-1}).

As can be seen, pollutant uptake is directly dependent on biomass concentration, provided that all the biomass is active. Experiments with different substrates in biofilters have shown that overcoming growth limitation can improve pollutant removal substantially (Cherry and Thompson 1997; Morales et al. 1998). Increasing biomass in biological air treatment systems allows improved performance to a point where other factors become limiting.

3.4.3.4 *Modeling*

The integration of the knowledge of the physical, chemical and biological phenomena taking place in biological air treatment systems has been carried out through the formulation of mathematical models intended to describe and predict performance, and serve as tools in dimensioning and designing. There is a strong challenge to model biological air treatment systems, as they are complex systems involving various phenomena. A good analysis of this complexity has been given by Devinny et al. (1999). One of the first models was published by Ottengraff and van der Oever (1983), describing the diffusion of the pollutant in the biofilm and Monod kinetics, and where the gas was considered to move as plug flow in the reactor. Several expressions were described for situations where reaction or diffusion are limiting, and for zero- or first-order kinetics. This basic model was improved by including double substrate and oxygen limitation and inhibition for high substrate concentration (Shareefdeen et al. 1993). Partial coverage of the support and adsorption phenomena was later described by Shareefdeen and Baltzis (1994). A different approach was used by Hodge and Devinny (1995), using a two-phase system (air and solids/water). Research in modeling has seen an increased interest in recent years, and some of the latest developments are presented in the following chapters of this book.

3.5 Types of Bioreactors

The elimination of pollutants from air with microorganisms has been accomplished in a diversity of reactors, as shown in Table 3.3. This section explores each configuration in detail.

3.5.1 Biofilter

In biofilters the contaminated air passes through a moist, packed bed that contains the microorganisms. Microbes grow on the surface and crevices of the support, forming a biofilm. The biofilm activity is determined by its microbial density and the environmental conditions, such as temperature, nutrient availability, pH, and humidity. The humidity of the biofilm is one of the critical steps to maintain a proper performance, as biological activity is highly dependent on water activity (A_W). The heat generated by the biological reaction and the humidity of the incoming air determines the rate of water loss and requirements for water restoration (Morales et al. 2003). Increased drying rates are attained with dry air and high elimination capacity. To maintain performance, air is generally pre-humidified, and biofilters have intermittent water spraying. To reduce ventilation costs, they generally have a high void fraction to limit pressure drop.

The supports can be either natural bioactive or inert (Fig. 3.10). The natural bioactive supports are soil, peat, compost, bark, etc., which can retain water and generally contain enough mineral nutrients to support an initial active microbial population (Cardenas-Gonzalez et al. 1999). They are relatively inexpensive and easy to obtain, and have been used for many applications. To obtain suitable structural characteristics, a mixture of materials is generally used, including a coarser fraction to prevent high pressure drop in the packing. Sometimes inert synthetic materials, such as plastics or ceramic, are included. Other additives that can be used are pH regulators and slow released nutrients

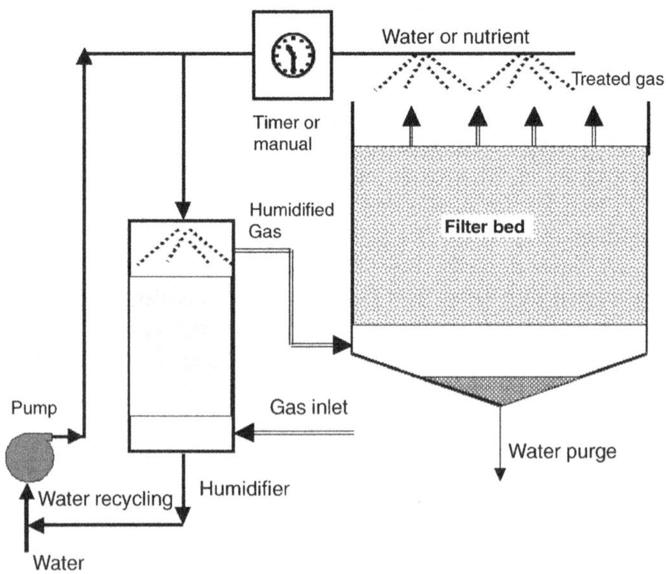

Fig. 3.10. Schematic diagram of an open biofilter

to increase biomass. The natural supports may degrade with time and loose the structure and water-retaining capacity, inducing channeling and the loss of performance (Morgan-Sagastume et al. 2003). In some cases, remixing the support with some fresh material and nutrients allows to recover the activity (Auria et al. 2000), but eventually it will need to be replaced. With proper maintenance, the support can be used for several years.

Inert natural or synthetic supports have been explored to be used in biofilters. These porous materials, such as activated carbon, ceramics, sintered glass, lava rock, polyurethane foam, vermiculite and perlite, do not contain the required nutrients to sustain microbial activity and, hence, it is necessary to provide them intermittently. On the other hand, they are not degraded and, in theory, could be engineered to have optimal properties such as controlled head loss, porosity, adsorptive capacity, etc. This remains an area of active research (Kennes and Veiga 2002). The high surface and low water content of biofilters make them appropriate for the treatment of the less hydrophilic pollutants ($H < 10$). The structural characteristics of the support determine the height of the packed bed, which is about 0.8–1.2 m for typical packings. This limitation necessitates having a large footprint for the biofilter, which may be a disadvantage for situations where space is limited.

3.5.2 Biotrickling Filters

In the biotrickling filters, the polluted air is passed through a packed column where liquid is continuously recirculated through the packing (Fig. 3.11). The pollutant is first solubilized in the falling liquid film, and then transferred to the microorganisms growing on the surface of these supports. The liquid provides moisture, nutrients, pH control to the biofilm, and allows the removal of

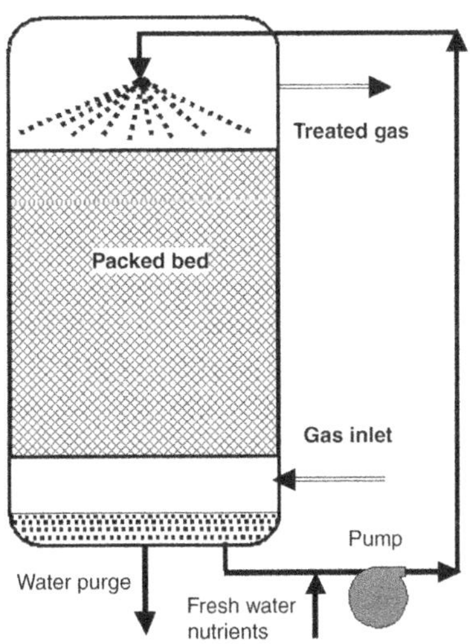

Fig. 3.11. Schematic diagram of a biotrickling filter

inhibiting products. Ideally, the excess biomass is sloughed off by the trickling liquid, and a stable operation can be achieved.

The supports are inert random or structured packings that are similar to those used in traditional scrubbers (plastic Raschig or Pall rings and saddles), although others such as lava rock or polyurethane foam have been tested. The air may be directed upflow or downflow, which is countercurrent or co-current with the liquid flow, respectively. It has been shown that generally there is no difference, except when there is stripping from the incoming liquid in the countercurrent configuration. To maintain low pressure drop and reduce clogging, the supports have low porosity and low specific surface ($100-400\,m^2\,m^{-3}$). EBRTs are normally around 30 s but systems with EBRTs as low as 1.2 s have been reported for low H_2S concentration (Gabriel and Deshusses 2003).

Overgrowth produces increased pressure drop, reduced real residence time, and a drop in performance. Several strategies (Diks and Ottengraff 1994) have been studied to control overgrowth while maintaining appropriate microbial activity. These include choosing a specific size and structure for the packing, limiting organic load, limiting nutrients, and adding inhibitors. The possibility of using microbial predators, such as protozoa and nematodes (Cox and Deshusses 1999), has been reported. In extreme cases the reactor has to be shut down and cleaned.

3.5.3 Rotating Biological Contactors

Rotating biological contactors were developed initially for water treatment. The polluted air passes through the headspace of the reactor, containing discs mounted on a rotating shaft that serve as support for a biofilm. The shaft is rotated (~ 2 rpm), and the discs are partially wetted in water containing nutrients and other additives. The movement of the discs favors mass transfer and the control of the fixed biomass. Air can be fed tangentially to the disks or through perforations in a hollow shaft (von Rohr and Ruediger 2001).

3.5.4 Bioscrubbers

In bioscrubbers, the pollutant in the gas phase is removed by absorption in the re-circulation liquid in a gas–liquid contactor (Fig. 3.12). Subsequently, this pollutant-laden liquid is regenerated by the microorganisms suspended in the liquid in a bioreactor with supplementary oxygen, to be returned to the contactor (van Groenestijn 2001a). Nutrient addition and pH are continually controlled to maintain microbial growth and high activity. The excess biomass and byproducts are continually purged from the system.

The gas–liquid contactors are designed to favor mass transfer from the air to the liquid phase, thus reducing equipment volume, while maintaining a low pressure drop (< 3 cm $H_2O\ m^{-1}$). The contactors can be packed towers, venture scrubbers, spray towers, etc. Typical gas velocities are $1-3\,m\,s^{-1}$ in packed

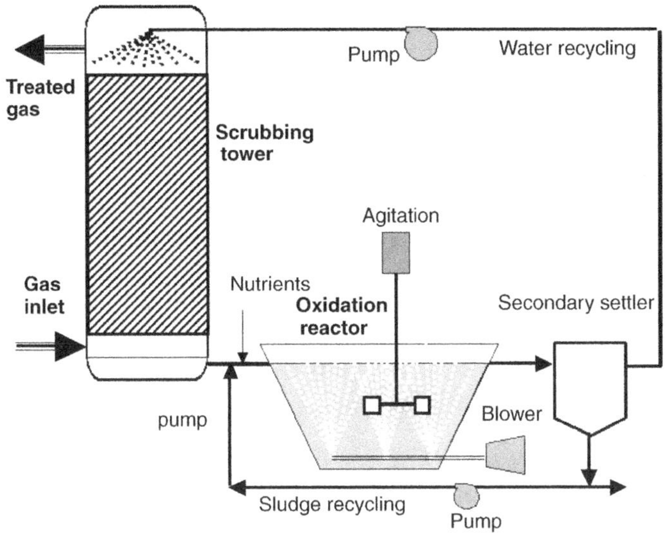

Fig. 3.12. Schematic diagram of a bioscrubber system

contactors. In the bioreactor, the liquid is regenerated by the suspended microorganisms, and CO_2, H_2O and other mineral products are produced. Most of the reactors are vessels where air is bubbled, and resemble activated sludge tanks. Water retention time in the reactors is high, and biomass concentration is expected to be around 5–8 $g\,l^{-1}$ (Ottengraff 1986), which is reasonable for high volumetric rates, while reducing clogging problems in the packing. The low specific gas/liquid area restricts the use of bioscrubbers to pollutants with low Henry's coefficients (< 5). A large absorber or high water flow rates are to be avoided (van Groenestijn and Hesselink 1993).

3.5.5 Membrane Bioreactors

In the membrane bioreactors, the pollutant in the gas phase is transferred through a membrane to the biofilm, attached to the other side and where nutrients and oxygen are provided. The basic configurations are hollow fibers and flat sheets. In hollow fibers, the gas is usually passed through the lumen of the tube, and the biomass is on the shell side. These reactors have been used for other waste treatment applications where the conditions of the stream exclude the possibility of direct contact with the biomass (van Groenestijn and Hesselink 1993; Ergas 2001).

Membranes can be constructed of very diverse materials, and have different chemical (solubility and selectivity) and physical properties (mechanical strength, pore size, thickness and porosity). In semi-permeable hydrophobic membranes, such as latex or silicon, the transfer rate is related to the solubility and diffusivity of the pollutant. In micro-porous membranes, the pollutant diffuses through the gaseous void volume and water, from the biofilm, is excluded by selecting hydrophobic materials. Micro-porous membranes are made of Teflon, polypropylene, polytetrafluorethylene (PTFE) and different composites. They have pore sizes of 0.1 – 1.0 µm, small diameters (200 – 400 µm ID), and are usually packed in bundles in reactors containing between 30 and $100\,cm^2\,cm^{-3}$.

A distinct characteristic of the membrane bioreactors is the fact that the polluted gaseous stream and the biomass are physically segregated, which allows the biological waste gas treatment to be used in certain applications such as indoor air and, in an extreme case, for spaceship air treatment.

3.5.6 Suspended Cell Bioreactor

In the suspended cell bioreactors, the polluted air is bubbled in the bulk of the liquid containing suspended microorganisms (Fig. 3.13). Several configurations have been proposed (Bielefeldt 2001). In the activated sludge process, the biological activity from the treatment of municipal wastewater is used to simultaneously treat the sparged polluted air. The characteristics of the reactor, such as biomass concentration, aeration and sparger design, are generally imposed by the requirements of the wastewater treatment. A review of the characteristics of several facilities is given by Bowker (1998).

In other cases, the reactor is conceived specifically for the treatment of the pollutant in air, and can be tanks or columns with different aeration configurations such as airlift, external loop, split cylinder, etc. (Shuler and Kargi 1992). These systems are designed to optimize mass transfer from the bubble to the bulk liquid where biodegradation occurs, and to allow the control of the conditions of the microbial population. An interesting example of this type of reactor is the cometabolic degradation of TCE with phenol and toluene (Ensley and Kurisko 1994).

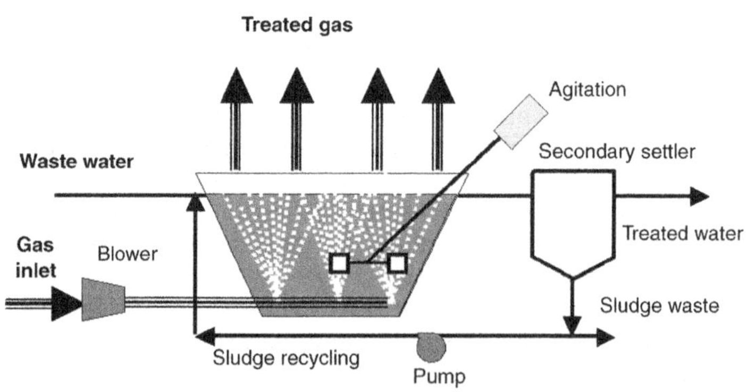

Fig. 3.13. Schematic diagram of a suspended growth bioreactor

3.6 Conclusions

The technologies that can be applied to VOCs and odor control are relatively wide. The choice of a particular technology should include technical, health, legal and economic considerations. From a technical standpoint, variables such as stream (flow, temperature, presence of particles, humidity, etc.) and pollutant characteristics (composition, concentration, reactivity, solubility, and biodegradability) have to be evaluated.

Different reactor configurations respond differently to the various operating problems (Table 3.7). To improve the performance of the biological air treatment systems, there is a continuous innovation in reactor configurations. For example, two sequential reactors have been used to treat sulfides and VOCs from a wastewater treatment plant. In the first reactor, inorganic sulfides are degraded in a biotrickling filter operating under acidic conditions, and a compost-based biofilter reduces the organic compounds that are sparsely reduced in the first reactor (Kraakman et al. 1996; Chitwood et al. 1999; Schroeder 2002). A combined system for the elimination of H_2S from waste gas has been described where, in the first step, H_2S is oxidized to elemental sulfur with ferric ion and, in the second step, the ferric ion is regenerated by *Thiobacillus ferroxidans* (Pagella et al. 1996). An advantage of this method is that the first reaction is very fast and that sulfur can be recovered from the medium.

The most important advantages of biological air treatment systems over physical and chemical technologies are that they are applicable for a wide range of pollutants, are effective at low concentrations, can be used under normal conditions (pressure, temperature and pH), and are very little energy- and material-intensive, simple to operate and economic. Although there are

Table 3.7. Operating problems associated with different bioreactor configurations

Type	Moisture control[a]	Nutrient addition and pH control	Biomass control, clogging	Transient response	Airflow channeling	Startup
Biofilter	++	++	+	+	++	+
Biotrickling filter	–	–	++	++	+	++
Rotating contactors	–	–	–	++	–	++
Bioscrubber	–	–	–	++	–	+
Suspended growth	–	–	+	+	–	+
Membrane reactors	–	–	++	+	–	++

[a] –, Not sensitive; +, sensitive; ++, highly sensitive

some disadvantages, research in this field has greatly expanded applicability and performance.

References

Acuña ME, Pérez-Guevara F, Auria R, Revah S (1999) Microbiological and kinetic aspects of a biofilter for the removal of toluene from waste gases. Biotechnol Bioeng 63:175–184

Aizpuru A, Malhautier L, Roux JC, Fanlo JL (2001) Biofiltration of a mixture of volatile organic emissions. J Air Waste Manage Assoc 51:1662–1670

Aizpuru A, Khammar N, Malhautier L, Fanlo JL (2003) Biofiltration for the treatment of complex mixtures of VOC – influence of the packing material. Acta Biotechnol 23:211–226

Alcántara S, Estrada I, Vásquez MS, Revah S (1999) Carbon disulfide oxidation by a microbial consortium from a trickling filter. Biotechnol Lett 21:815–819

Anguil GH (1998) Thermal oxidation. In: In: Rafson HJ (ed) Odor and VOC control handbook. McGraw-Hill, New York, pp 831–865

Auria R, Frere G, Morales M, Acuña ME, Revah S (2000) Influence of mixing and water addition on the removal rate of toluene vapors in a biofilter. Biotechnol Bioeng 68:448–451

Bell K (1988) Heat transmission: thermal design of heat transfer equipment. In: Perry R, Green D (eds) Perry's chemical engineering handbook, 6th edn. McGraw-Hill, New York, pp 1020–1023

Bielefeldt A (2001) Activated sludge and suspended growth bioreactors. In: Kennes C, Veiga MC (eds) Bioreactors for waste gas treatment. Kluwer, Dordrecht, pp 215–254

Bohn H (1992) Consider biofiltration for decontaminating gases. Chem Eng Prog 88(4):34–40

Bowker RP (1998) Biological systems: aeration tanks. In: Rafson H (ed) Odor and VOC control handbook. McGraw-Hill, New York, pp 8.192–8.202

Cadena FP, Peters RW (1988) Evaluation of chemical oxidizers for hydrogen sulphide control. JWPCF 60:1259–1263

Card TR (1998) Fundamentals: chemistry and characteristics of odors and VOCs. In: Rafson H (ed) Odor and VOC control handbook. McGraw-Hill, New York, pp 21–236

Cardenas-Gonzalez B, Ergas S, Switzenbaum M, Phillibert N (1999) Evaluation of full-scale biofilter media performance. Environ Prog 18:205–211

Chen KY, Morris JC (1972) Kinetics of oxidation of aqueous sulphide by oxygen. Environ Sci Technol 6:529–537

Cherry R, Thompson D (1997) Shift from growth to nutrient-limited maintenance kinetics during biofilter acclimation. Biotechnol Bioeng 56:330–339

Chitwood DE, Devinny JS, Reynolds FE (1999) Evaluation of a two-stage biofilter for treatment of POTW waste air. Environ Prog 18:212–221

Christen P, Domenech F, Michelena G, Auria R, Revah S (2002) Biofiltration of volatile ethanol using sugar cane bagasse inoculated with *Candida utilis*. J Hazard Mater 89:253–265

Corbitt RA (1990) Air quality control. In: Corbitt RA (ed) Standard handbook of environmental engineering. McGraw-Hill, New York, pp 4115

Cox H, Deshusses MA (1999) Biomass control in waste air biotrickling filters by protozoan predation. Biotechnol Bioeng 62:216–224

Cox H, Sexton T, Shareefdeen Z, Deshusses M (2001) Thermophilic biotrickling filtration of ethanol vapors. Environ Sci Technol 35:2612–2619

Davison B, Barton J, Klasson T, Francisco A (2000) Influence of high biomass concentrations on alkane solubilities. Biotechnol Bioeng 68:279–284

De Castro A, DG Allen, RR Fulthorpe (1997) Characterization of the microbial population during biofiltration and the influence of the inoculum source. In: Proc 90th Annu Meet Air and Waste Management Association, 11 June 1997, Toronto

Deshusses M, Johnson C (2000) Development and validation of a simple protocol to rapidly determine the performance of biofilters for VOC treatment. Environ Sci Technol 34:461–467

Devinny J, Hodge D (1995) Formation of acidic and toxic intermediates in overloaded ethanol biofilters. J Air Waste Manage Assoc 45:125–131

Devinny JS, Deshusses MA, Webster TS (1999) Biofiltration for air pollution control. CRC Press, Boca Raton

Dhamwichukorn S, Kleinheinz G, Bagley S (2001) Thermophilic biofiltration of methanol and α-pinene. J Ind Microbiol Biotechnol 26:127–133

Diks RMM, Ottengraf SPP (1994) Technology of trickling filters. VDI Berichte 1104. VDI Verlag, Düsseldorf, pp 19–35

Ensley B, Kurisko P (1994) A gas lift bioreactor for removal of contaminants from the vapor phase. Appl Environ Microbiol 60:285–290

Ergas S (2001) Membrane bioreactors. In: Kennes C, Veiga MC (eds) Bioreactors for waste gas treatment. Kluwer, Dordrecht, pp 163–177

Gabriel D, Deshusses M (2003) Retrofitting existing chemical scrubbers to biotrickling filters for H_2S emission control. Proc Natl Acad Sci USA 100:6308–6312

García-Peña I, Hernández S, Auria R, Favela E, Revah S (2001) Biofiltration of toluene by the fungus *Scedosporium apiospermum*. Biotechnol Bioeng 76:61–69

Heslinga D, van Groenestijn J (1997) Thermophilic biological waste gas cleaning. In: Prins WL, van Ham J (eds) Biological waste gas cleaning. VDI Verlag, Düsseldorf, pp 353–358

Hodge D, Devinny J (1995) Modeling removal of air contaminants by biofiltration. J Environ Eng 121:21–32

Kennes C, Veiga M (2002) Inert filter media for the biofiltration of waste gases – characteristics and biomass control. Rev Environ Sci Bio/Technol 1:201–214

Kennes C, Veiga C, Prado O (2001) Non biological treatment technologies. In: Kennes C, Veiga MC (eds) Bioreactors for waste gas treatment. Kluwer, Dordrecht, pp 17–46

Kirchner K, Wagner S, Rehm HJ (1996) Removal of organic air pollutants from exhaust gases in trickle bed reactors effect of oxygen. Appl Microbiol Biotechnol 45:415–419

Kraakman B, Oosting R, Koers B, Urlings L (1996) Biological odor removal with a biotrickling filter-biofilter combination. In: Proc 89th Annu Meet Exhibition Air and Waste Management Association, Nashville, Tennessee, Pap 96-RA87B04

Lee B, Apel W, Miller A (1999) Removal of low concentrations of carbon tetrachloride in compost-based biofilters operated under methanogenic conditions. Waste Manage Assoc 49:1068–1074

Leson G, Winer AM (1991) Biofiltration: an innovative air pollution control technology for VOC emissions. J Air Waste Mange Assoc 41:1045–1054

Lipski A, Altendorf K (2002) Identification of the degrading microorganisms in biofilters by in situ isotope labeling. In: Proc USC-TRG Conf Biofiltration, November 2002, Newport Beach, CA

Lu C, Chu W, Lin MR (2000) Removal of BTEX vapor from waste gases by a trickle bed biofilter. J Air Waste Manage Assoc 50:411–417

Mansfield LA, Melnyk PE, Richardson GC (1992) Selection and full scale use of a chelated iron absorbent for odor control. Water Environ Res 64:120–127

McGrath MS, Nieuwland JC, van Lith C (1999) Case study: biofiltration of styrene and butylacetate at a dashboard manufacturer. Environ Prog 18:197–204

Montalvo S, Guerrero L (2003) Tratamiento anaerobio de residuos: producción de biogas. Universidad Técnica Federico Santa Maria Press, Valparaiso

Morales M, Revah S, Auria R (1998) Start up and gaseous ammonia addition on a biofilter for elimination of toluene vapors. Biotechnol Bioeng 60:483–491

Morales M, Hernández S, Cornabé T, Revah S, Auria R (2003) Effect of drying on biofilter performance: modeling and experimental approach. Environ Sci Technol 37:985–992

Morgan-Sagastume JM, Ergas S, Revah S, Noyola A (2003) Changes in physical structure of a compost biofilter treating H_2S. J Air Waste Manage Assoc 53:1011–01021

Mukhopadhyay N, Moretti EC (1993) Current and potential future industrial practices for reducing and controlling volatile organic compounds. Center for Waste Reduction Technologies, American Institute of Chemical Engineers, New York

Ottengraf SPP (1986) Exhaust gas purification. In: Rehm HJ, Reed G (eds) Biotechnology, vol 8. VCH Verlag, Weinheim, pp 426–452

Ottengraf SPP, van den Oever AHC (1983) Kinetics of organic compound removal from waste gases with a biological filter. Biotechnol Bioeng 25:3089–3102

Pagella C, Silvestri P, de Faveri M (1996) H_2S gas treatment with *Thiobacillus ferroxidans*: overall process performance and chemical step. Trans Chem 74:123–132

Philip L, Deshusses MA (2003) Sulfur dioxide treatment from flue gases using a biotrickling filter-bioreactor system. Environ Sci Technol 37:1978–1982

Planker TW (1998) Masking and odor neutralization. In: Rafson HJ (ed) Odor and VOC control handbook. McGraw-Hill, New York, pp 818–824

Qasim RS (1999) Wastewater treatment plants: planning, design and operation. Technomic, Lancaster, PA

Särner E (1990) Removal of sulphate and sulphite in an anaerobic trickling (ANTRIC) filter. Water Sci Tech 22:395–404

Schroeder E (2002) Trends in applications of gas-phase bioreactors. Rev Environ Sci Bio/Technol 1:65–74

Shareefdeen Z, Baltzis BC (1994) Biofiltration of toluene vapor under steady-state and transient conditions: theory and experimental results. Chem Eng Sci 49:4347–4360

Shareefdeen Z, Baltzis B, Oh Y, Bartha R (1993) Biofiltration of methanol vapor. Biotechnol Bioeng 42:643–656

Shuler ML, Kargi F (1992) Bioprocess engineering - basic concepts. Prentice Hall, Englewood Cliffs

Smet E, van Langenhove H (1998) Abatement of volatile organic sulfur compounds in odorous emissions from the bio-industry. Biodegradation 9:273–284

Sologar VS, Lu Zijin, Allen DG (2003) Biofiltration of concentrated mixtures of hydrogen sulfide and methanol. Environ Prog 22:129–136

Song J, Kinney K (2000) Effect of vapor-phase bioreactor operation on biomass accumulation, distribution, and activity: linking biofilm properties to bioreactor performance. Biotechnol Bioeng 68:508–516

Spingo G, Pagella C, Daria F Molteni R, de Faveri M (2003) VOCs removal from waste gases: gas phase bioreactor for the abatement of hexane by *Aspergillus niger*. Chem Eng Sci 58:739–746

Thomson RB (1980) Catalytic removal of hydrogen sulfide from gases. US patent no 4189462

Van Groenestijn JW (2001a) Bioscrubbers. In: Kennes C, Veiga MC (eds) Bioreactors for waste gas treatment. Kluwer, Dordrecht, pp 133–162

Van Groenestijn JW (2001b) Combined advanced oxidation and biodegradation. In: Kennes C, Veiga MC (eds) Bioreactors for waste gas treatment. Kluwer, Dordrecht, pp 179–200

Van Groenestijn JW, Hesselink PG (1993) Biotechniques for air pollution control. Biodegradation 4:283–301

Van Groenestijn JW, Liu JX (2002) Removal of alpha-pinene from gases using biofilters containing fungi. Atmos Environ 36:5501–5508

Von Rohr R, Ruediger P (2001) Rotating biological contactors In: Kennes C, Veiga MC (eds) Bioreactors for waste gas treatment. Kluwer, Dordrecht, pp 201–214

Waldrop RE (1998) Condensation. In: Rafson HJ (ed) Odor and VOC control handbook. McGraw-Hill, New York, pp 825–830

West JR (1983) Sulfur recovery. In: Othmer K, Grayson M, Eckroth D (eds) Encyclopedia of chemical technology. Wiley, New York, pp 267–297

Woertz J, Kinney KA, McIntosh P, Szaniszlo PJ (2001a) Removal of toluene in a fungal vapor phase bioreactor containing the dimorphic black yeast *Exophiala lecanii-corni*. Biotechnol Bioeng 75:550–558

Woertz JR, Kinney KA, Szaniszlo PJ (2001b) A fungal vapor-phase bioreactor for the removal of nitric oxide from waste gas streams. J Air Waste Manage Assoc 51:895–902

4 Selection of Bioreactor Media for Odor Control

Rakesh Govind, Sandeep Narayan

4.1 Introduction

Support medium is one of the most critical elements of a biofiltration system. In this chapter, biofilters and biotrickling filters will be referred to as "biofilter", and this distinction will be discussed later in this chapter. The support medium's main function is to provide contact between the gas-phase contaminants and active microbial cultures either immobilized within and/or attached as a biofilm on the medium's surface. Other functions of the media are to distribute the gas flow evenly within the bed's cross-sectional area, with minimal gas-phase pressure drop, distribute any liquid nutrients sprayed on the bed's surface, and prevent accumulation of biomass to prevent clogging, which would eventually lead to channeling of the gas and liquid.

Biofilter media can be classified based on their mass transfer mechanism, and then subclassified according to other characteristics, as listed below:

1. Diffusive: contaminants diffuse toward and into the media perpendicular to the gas flow

1.1 Naturally bioactive

1.2 Synthetic

1.2.1 Randomly packed

1.2.1.1 Adsorbing

1.2.1.2 Non-adsorbing

1.2.2 Structured

1.2.2.1 Adsorbing

1.2.2.2 Non-adsorbing

1.3 Synthetic encapsulating

1.4 Membrane systems (flat sheet, hollow fibers)

2. Convective: contaminants and gas flow through the biofilm and media

Shareefdeen/Singh (Eds.)
Biotechnology for Odor and Air Pollution Control

4.2 Diffusive Versus Convective Media

Figure 4.1 shows the main differences between diffusive and convective biofilter support media. In diffusive media (Fig. 4.1a, b), the contaminants in the gas phase diffuse toward and/or into the media, containing immobilized active microbial cultures. The support medium itself is not porous and does not allow any gas flow through the medium. The diffusive flux of the contaminant is perpendicular to the direction of gas flow, and its value depends on the concentration difference between the bulk and the contaminant concentration at the surface of the gas–biofilm interface. This transport rate is dependant on the gas velocity, and in the case of porous media, there is a combination of film and pore diffusion. An effectiveness factor can be used to describe the effect of diffusion, and can be written as:

$$\eta = \frac{r_{\text{observed}}}{r_{\text{bulk conditions}}}, \qquad (4.1)$$

where r_{observed} is the actual measured reaction rate, and $r_{\text{bulk conditions}}$ is the reaction rate under bulk gas conditions.

In the case of non-adsorbing media (Fig. 4.1a), there is no adsorbed concentration of the contaminant on the medium's surface. However, if the support medium adsorbs the contaminant present in the gas phase, and since adsorption is much faster than biodegradation, there is a finite concentration of contaminant on the medium's surface, which results in back-diffusion of the contaminant (desorption) into the biofilm, as shown by the concentration profiles in Fig. 4.1b. While the thickness of biofilm in the case of non-adsorbing media is self-limiting (maximum effective biofilm thickness), there is no such thickness in the case of adsorbing media. The mathematics of these differences in concentration profile will be discussed later. However, it should be noted that, as the bulk contaminant concentration decreases, as in the case of odor biofiltration (contaminant concentration is usually low, combined with low

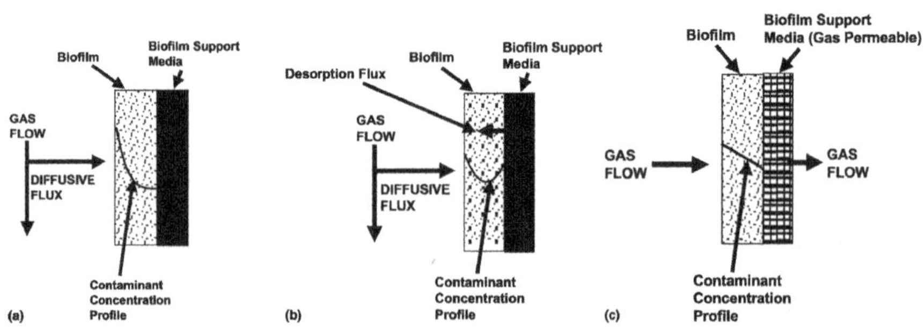

Fig. 4.1. Schematic of diffusive and convective support media in biofiltration: **a** non-adsorptive diffusive, **b** adsorptive diffusive, with back-diffusion of contaminant from the medium's surface into the biofilm, and **c** convective, with gas flow through the biofilm and porous support media

detection and recognition threshold concentrations), the diffusive flux of the contaminant also decreases, which results in low biofiltration treatment rates. Also, with increasing gas velocity, the diffusive fluxes increase, due to higher mass transfer coefficients.

In the case of convective media (Fig. 4.1c), the biofilm is supported by a porous support medium, and the contaminant is carried into the biofilm by convective transport, i.e., by actual gas flow, rather than by diffusion only, as in the case of diffusive biomedia. Since convective rates are significantly higher than diffusive fluxes, the rate of biofiltration treatment is consequently higher, and is not impacted by bulk contaminant concentrations. Of course, in convective systems, the gas has to flow from the feed side to the product side through the porous biofilm and support media. The effect of adsorbing versus non-adsorbing is the same, namely, that in the case of adsorbing, there is a finite back-diffusion into the biofilm, while there is no such back-diffusion in the case of non-adsorbing media. Clearly, convective biomedia, which use membranes or porous supports, offer higher biofiltration rates, especially at low bulk contaminant concentrations, as in the case of odor control. The need for convective biomedia is evident from close examination of a biofilm, as shown in Fig. 4.2. Diffusion of the contaminant occurs from the gas phase into the active biofilm, where simultaneous biodegradation and diffusion occur. Mathematically, this can be written as follows:

$$\frac{\partial^2 y}{\partial x^2} = \frac{\alpha y}{1+y} \tag{4.2}$$

$$\frac{\mathrm{d}y}{\mathrm{d}x} = 0\,, \quad \text{at} \quad x = 0 \tag{4.3}$$

$$y = y_1\,, \quad \text{at} \quad x = 1 \tag{4.4}$$

where the dimensionless variables and parameters are defined as follows:

$$y = \frac{C_\mathrm{b}}{K_\mathrm{C}} \tag{4.5}$$

$$x = \frac{z}{L_\mathrm{b}} \tag{4.6}$$

$$\alpha = L_\mathrm{b}^2 \frac{k}{K_\mathrm{C} D_\mathrm{biofilm}} \tag{4.7}$$

$$y_1 = \frac{C_\mathrm{bi}}{K_\mathrm{C}}\,. \tag{4.8}$$

In terms of these dimensionless variables, the effectiveness factor, defined in Eq. (4.1), becomes:

$$\eta = \frac{\left(\mathrm{d}y/\mathrm{d}x\right)_{x=1}}{\alpha y_1/\left(1+y_1\right)}\,. \tag{4.9}$$

A plot of this effectiveness factor, η, versus the dimensionless concentration at the air/biofilm interface, y_1, and the Thiele's modulus ($\sqrt{\alpha}$), is shown in

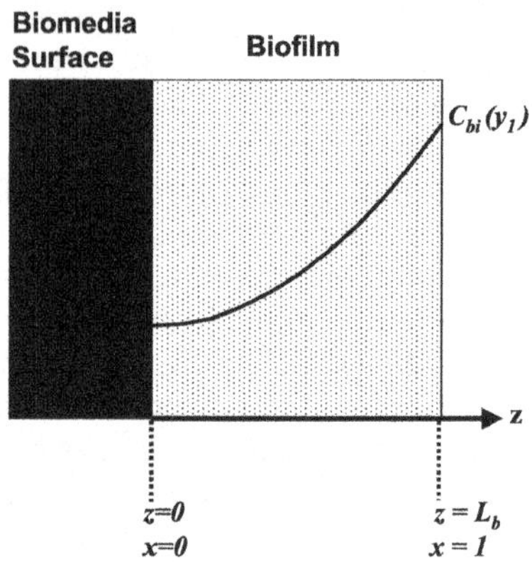

Fig. 4.2. Biomedia surface with attached biofilm

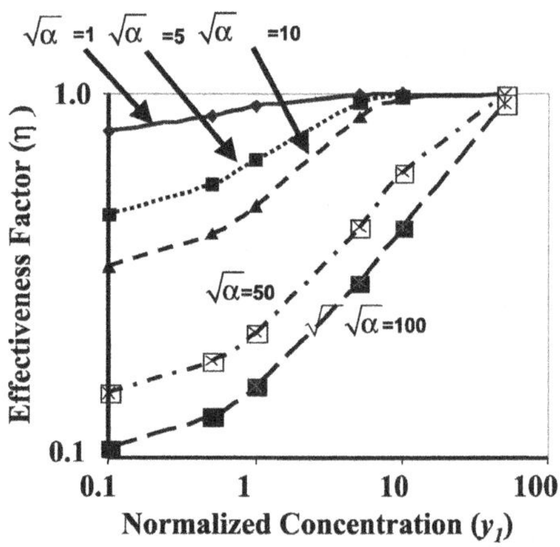

Fig. 4.3. Plot of biofilm effectiveness factor versus the normalized contaminant concentration, and as a function of Thiele's modulus ($\sqrt{\alpha}$)

Fig. 4.3. The effectiveness factor becomes significantly less than 1 when either the biofilm is thick, i.e., L_b is large, or the contaminant concentration at the air/biofilm interface is small, i.e., y_1 is small. Under these conditions, diffusive support media become ineffective, and the use of convective support media increases the biotreatment rates in the biofilter.

4.3 Naturally Bioactive Media

Naturally bioactive media consist of media, such as soil, compost, peat, cow manure, etc., which naturally contain bacteria dispersed within the structure of the medium itself, rather than as a distinct biofilm on the surface and within the pores. Most natural media contain clay and organic matter, and adsorption of contaminants on these materials occurs through various interactions, such as van der Waals' forces, H-bonding, dipole–dipole interaction, ion exchange, covalent bonding, ligand exchange, water bridging, and/or hydrophobic partitioning. The organic matter occurs primarily as amorphous, gel-like phases that allow for hydrophobic organic compound partitioning (Karickhoff et al. 1979; Chiou et al. 1983). The total sorption of the contaminant can be written as a combination of two mechanisms: (1) hydrophobic partitioning from the aqueous phase, in which the contaminant has solubilized from the gas phase; and (2) adsorption on the medium's surface, which tends to dominate at low contaminant concentrations:

$$q(C) = q_{\mathrm{p}}(C) + q_{\mathrm{ad}}(C) \,. \tag{4.10}$$

Such models have been referred to as "distributed reactivity" (Weber et al. 1999), "dual-mode" sorption (Xing and Pignatello 2001), or simply "adsorption partitioning" (Chiou et al. 2000; Accardi-Dey and Hschwend 2002). Hydrophobic partitioning of non-polar contaminants by amorphous organic matter is usually written as a partitioning constant:

$$K_{\mathrm{d}} = \frac{q_{\mathrm{p}}(C)}{C} \,. \tag{4.11}$$

K_{d} can also be normalized by the fraction of organic carbon in the media (f_{OC}):

$$K_{\mathrm{OC}} = \frac{K_{\mathrm{d}}}{f_{\mathrm{OC}}} \,. \tag{4.12}$$

Empirical regressions between contaminant properties (aqueous solubility (S), octanol–water partitioning coefficient (K_{OW})) provide an effective method for estimating K_{OC} values for non-polar chemicals in naturally bioactive media (Gao et al. 1996; Xia 1998; Seth et al. 1999).

Adsorption on the medium's surface can be described by standard adsorption isotherms, usually derived from the Polanyi adsorption potential for gas-phase adsorption (Polanyi 1916):

$$\varepsilon_{\mathrm{d}} = RT \ln\left(\frac{P^0}{P}\right) , \tag{4.13}$$

where ε_{d} is the differential work of adsorption, P/P^0 the partial pressure, T the temperature, and R the ideal gas constant. At low gas-phase concentrations, as in odor biofiltration, adsorption dominates the overall sorption process. As originally proposed by Dubinin and Ashtakhov (1971), the adsorbed volume of contaminant (q') can be written as follows:

$$\frac{q'}{q'_{\mathrm{max}}} = \exp\left[-\left(\frac{RT\ln(P^0/P)}{\beta_{\mathrm{i}}E_0}\right)^{d}\right] , \tag{4.14}$$

where q'_{max} is the maximum sorption capacity, the exponent d is a fitting parameter of the characteristic curve, and $\beta_{\mathrm{i}}E_0$ is the characteristic energy of adsorption.

Contaminants, once adsorbed into the organic fraction of the natural media, undergo a combination of abiotic transformations, such as hydrolysis, chemical binding (Ramani and Govind, unpubl. data), redox reactions, polymerization, often catalyzed by the clays and metal oxides present, and microbial breakdown due to aerobic and anaerobic biodegradation primarily by heterotrophic bacteria and actinomycetes, certain autotrophic bacteria, fungi and specific protozoa.

Mass transfer of the contaminants into the aggregates of the natural media, composed of smaller particles, impacts the performance of the media in biofiltration, and is dependant on the water content of the media, as shown in Fig. 4.4. Below a specific critical water content (W_i in Fig. 4.4), the bioactivity of the medium is negligible, and the medium loses its structure, developing cracks through which the gas can bypass the bed. Experiments have shown (Zhao and Govind 1997) that, in the case of compost, peat, and soil, this critical water content is an irreversible phenomenon, whereby once the medium goes below this water content, it does not recover its original bioactivity and structure, even if the water content is increased above this critical value. There is an "optimum" water content (maximum bioactivity; W_p in Fig. 4.4) above which the rate of biodegradation begins to decrease mainly due to mass transfer constraints arising from a distinct water film outside the media aggregates.

Naturally bioactive media, such as soils, compost, peat, etc., have a specific water retention capacity due to their inherent porosity, which depends on the media characteristics. While small pores in the support media cannot

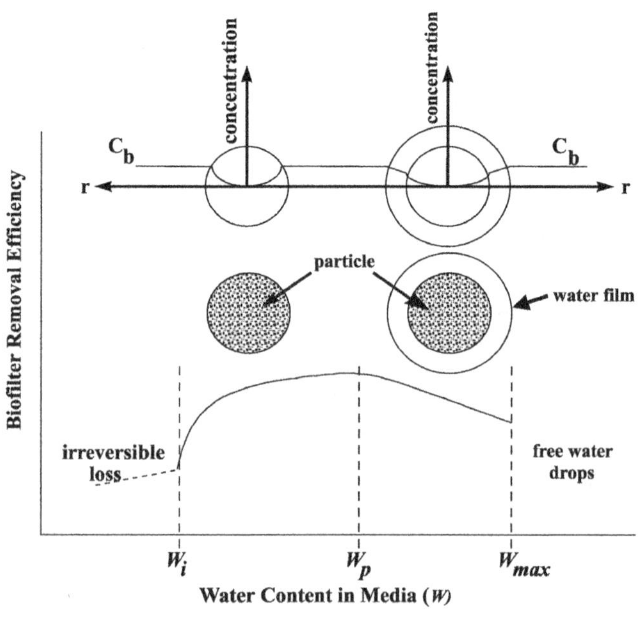

Fig. 4.4. Effect of water content on biofilter removal efficiency

accommodate biofilms, they can retain water that can be used by the biofilms attached to the media surface, thereby preventing drying of the active biofilms. Materials with significant water retention capacity can be used as biofilter media with humidified air, which reduces water evaporative losses. Usually, for biodegradative activity in naturally bioactive media, moisture content should be in the range 50–80% of the water holding capacity, which is equivalent to the "field capacity" (percentage of water remaining in the medium after it has been saturated and gravitational drainage has ceased). The water holding capacity is a function of the metric potential (measure of energy required to overcome capillary and adsorptive forces) of the medium water, and depends on the organic content of the naturally bioactive medium. A metric potential of −0.1 bar or greater is considered optimal for microbial activity in soils.

Table 4.1 summarizes the model equations that have been developed (Zhao and Govind 1997) to mathematically describe the performance of naturally bioactive media in biofiltration applications. The main advantages and disadvantages of using naturally bioactive media in biofiltration applications are summarized in Table 4.2.

4.4 Synthetic Media

Synthetic media consist of structures in which active bacteria have to be inoculated and are not inherently present in the medium's structure, as in the case of naturally bioactive media. Although synthetic media can be classified according to their structure, shape, size, etc., a more fundamental classification system would be based on their surface characteristics, rather than their geometry.

There are several types of synthetic support media that have been used in biofiltration systems. Synthetic media can be broadly classified as one of two types: (1) adsorbing, which exhibit significant physical adsorption of the contaminant(s); and (2) non-adsorbing, where there is negligible physical adsorption of the contaminants by the media. The major difference between adsorbing and non-adsorbing media is the occurrence of maximum effective biofilm thickness in non-adsorbing media but not in adsorbing media (Govind 2002). In non-adsorbing media, the concentration profile, assuming maximum biofilm thickness, is shown in Fig. 4.3, where the contaminant concentration at the biofilm–support media interface is zero. Since the contaminant cannot diffuse beyond this maximum biofilm thickness, any biofilm beyond this depth will begin to decay, eventually resulting in slough-off of the biofilm from the medium surface. However, if the contaminant adsorbs on the support medium's surface, back-diffusion of the contaminant from the medium's surface into the biofilm at the biofilm–support media interface, as shown in Fig. 4.3, prevents any such limitation of thickness from occurring, resulting in thicker biofilms. The impact of this back-diffusion is increased rates of biofiltration, mainly due to a combination of adsorption–desorption and biodegradation within the biofilm, resulting in increased amounts of active biomass within the biofilter.

Table 4.1. Model equations for a naturally bioactive media biofilter

Equation number	Model equation	Explanation
(a)	$\left(\frac{D_e}{r^2}\right)\frac{\partial}{\partial r}\left[r^2\frac{\partial C}{\partial r}\right] = R_B \quad 0 \le r \le R$	Diffusion and reaction within the porous compost particle of radius R
(b)	$R_B = \frac{KC}{K_C + C}\frac{(W - W_i)}{K_{CW} + (W - W_i)} \quad W_i < W < W_p$	Rate of biodegradation of contaminant in the porous compost particle, which accounts for water content
(c)	$r = 0 \quad \frac{dC}{dr} = 0$ $r = R; \quad D_e \left.\frac{\partial C}{\partial r}\right\vert_{r=R} = k_g(C_b - C_f)$	Boundary conditions. Symmetry condition at the center of the particle, and mass transfer balance at the outer boundary of the particle
(d)	$\frac{k_g P}{\varrho_g u_g} = 2.3\frac{(aR)^{-1.7}}{MW_s}\left(\frac{u_g \varrho_g R}{\mu_g}\right)^{-0.3}\left(\frac{\mu_g}{\varrho_g D_g}\right)^{-0.5}$	Gas-phase mass transfer coefficient for packed beds (Charpentier 1976)
(e)	$k_l = \frac{3D_l(1 + W_0)(1 - \varepsilon)}{\varrho_l R(W - W_p)} \quad W_p < W < W_{max}$	Mass transfer coefficient in the liquid-phase film outside the compost particle
(f)	$r = 0 \quad \frac{dC}{dr} = 0$ $r = R; \quad D_e \left.\frac{\partial C}{\partial r}\right\vert_{r=R} = K_g(C_b - C_f)$	Boundary conditions for the case of water film outside the compost particle. Symmetry condition at the center of the particle and mass transfer balance at the surface

Table 4.1. (continued)

Equation number	Model equation	Explanation
(g)	$\frac{1}{K_g} = \frac{1}{k_g} + \frac{1}{HNk_l}$	Overall mass transfer coefficient as a function of the gas and liquid mass transfer coefficients
(h)	$u_g\left(\frac{\partial C}{\partial Z}\right) + 3\left[(1-\varepsilon/\frac{D_e}{R}\right]\left\|\left(\frac{\partial C}{\partial r}\right)\right\|_{r=R} = 0$	Mass balance of contaminant along the biofilter height
(i)	$Z = 0 \quad C = C_{in}$	Boundary condition at biofilter inlet
(j)	$\left(\frac{D_{e0}}{r^2}\right)\frac{\partial}{\partial r}\left[r^2\frac{\partial C_0}{\partial r}\right] = R_{B0} \quad 0 \leq r \leq R$	Oxygen diffusion and utilization inside the compost particle due to aerobic biodegradation of contaminant
(k)	$R_{B0} = \beta R_B$	Stoichiometric balance between the rates of oxygen consumption and contaminant biodegradation

Table 4.2. Advantages and disadvantages of using naturally bioactive media in biofilters

Advantages	Disadvantages
1. Low cost	1. Settling of media due to biomass growth
2. Inherent bioactivity within media	2. Cannot separate biomass growth from the media
3. Physical adsorption of contaminant	3. Eventual consumption of bioavailable nitrogen and phosphorus, requiring medium replacement
4. Chemical complex of some organic contaminants	4. Water content has to be maintained to prevent drying or excess water in the media
5. Mimics natural ecosystem	5. Low biodegradation rates, requiring large volume of medium
	6. Limited bed height results in large cross-sectional area
	7. Large bed size increases cost of vessel
	8. Significant gas-phase pressure drop
	9. Inability to effectively neutralize acidic by-products
	10. Possible oxygen transfer limitations at high contaminant loadings
	11. Possible channeling of gas through the bed
	12. Significant bulk density of media

It should be noted that this increase in biofiltration rates is only observed when the contaminant can desorb effectively from the support media, and is not irreversibly adsorbed on the surface. Table 4.3 summarizes the advantages and disadvantages of various types of synthetic media that have been used in biofiltration.

Other important surface characteristics of the support media are surface roughness and charge. It was noted that the extent of microbial colonization increased with surface roughness (Characklis et al. 1990). This is mainly due to diminished shear forces and increased surface area of the rough surface. Most investigators have found that microorganisms attach more rapidly to hydrophobic, non-polar surfaces than to hydrophilic materials (Fletcher and Loeb 1979; Pringle and Fletcher 1983; Bendinger et al. 1993). Hydrophobic interaction between the cell's surface and the medium's surface enables the cell to overcome the repulsive forces active within a certain distance from

Table 4.3. Advantages and disadvantages of various types of synthetic biofilter media[a]

Packing configuration	Type of synthetic media	Advantages	Disadvantages
Randomly packed	Granular (gravel, lava rock, wood chips, etc.)	– Low cost and easily available – Rough surface for biofilm attachment – Has been used in old trickling filters (water treatment) – Has metal content that can react with hydrogen sulfide to form metal sulfides	– Low surface area – Low void fraction – can be easily plugged – High bulk density (heavy weight) – Gas channeling – High gas pressure drop, if small size
	Fibrous mesh pads	– High surface area – Low bulk density – Low gas-phase pressure drop – Biofilm surface area increases with biofilm thickness	– Small free passage diameter; hence can be easily clogged – Compacts due to weight of biomass growth, increasing pressure drop and causing clogging – Does not distribute trickling water evenly
	Open-cell polymeric foam	– Good retention of water – High surface area – Low bulk density – Retains biomass growth – High porosity	– Depending on number of pores per inch, can be easily clogged – Difficult to slough-off biomass growth – Compacts due to biomass weight – Does not distribute trickling water evenly

[a] "Synthetic" here means either manufactured or cut from natural materials without any inherent bioactivity

Table 4.3. (continued)

Packing configuration	Type of synthetic media	Advantages	Disadvantages
	Extruded plastic packing (Raschig rings, Berl saddles, etc.)	– Commercially available in many sizes and shapes – Good liquid distribution – Low gas pressure drop – Low bulk density – Some designs have high void fraction – Good gas–liquid contact	– Developed for gas absorption towers, where the liquid flow rates are much higher – Surface unsuitable for biofilm attachment and growth – Insufficient void fraction, especially with small sizes
Structured	Extruded plastic	– High surface area – Low gas-phase pressure drop – *High void fraction* – Low bulk density – Large free passage diameter – Easily supported	– Plastic surface unsuitable for biofilm attachment and growth – *Originally developed for trickling filter* (water treatment) – Have to be cut to fit vessel – High cost per unit volume
	Ceramic	– High surface area – Low gas-phase pressure drop – High void fraction – Low bulk density – Large free passage diameter – Easily supported – Can be used as convective media	– Have to be cut to fit vessel – High cost per unit volume – Not manufactured in large sizes – Few experimental studies in biofiltration

the medium's surface. Most bacteria are negatively charged, but still contain hydrophobic components on the surface (Rosenberg and Kjelleberg 1986).

However, any material surface exposed in an aqueous medium will inevitably and almost immediately become conditioned or coated by polymers present in the medium, and this resulting chemical modification affects the rate and extent of microbial attachment. Such conditioning films were first reported by Loeb and Neihof (1975). Other characteristics of the aqueous medium that are important for biofilm growth on the media are pH, nutrient levels, ionic strength, and temperature. Fletcher (1988a, 1988b) found that an increase in the concentration of several cations, such as sodium, calcium, and ferric ion, increased the attachment of cells to glass surfaces by reducing the repulsive forces between the negatively charged bacterial cells and the glass surface. Cowan et al. (1991) showed that an increase in nutrient concentration correlated with increases in the number of attached bacterial cells.

The hydrodynamics of liquid flow within the biofilter affect the slough-off of biomass from the support medium's surface. The relationship between the rate of shear loss of the biofilm ($R_{\text{slough-off}}$) and the biofilm thickness is given by Rittman (1989) as follows:

$$R_{\text{slough-off}} = 3.62 \times 10^{-6} X_f L_f \left[\frac{\tau_s}{1 + 0.0443(L_f - 30)} \right]^{0.58} \quad \text{when} \quad L_f > 30\,\mu\text{m} \tag{4.15}$$

$$= 3.62 \times 10^{-6} X_f L_f \tau_s^{0.58} \quad \text{when} \quad L_f < 30\,\mu\text{m}\,, \tag{4.16}$$

where $R_{\text{slough-off}}$ is the rate of biomass slough-off from the medium's surface ($\text{kg}\,\text{m}^{-2}\,\text{s}^{-1}$), τ_s the shear stress on the biofilm surface due to liquid flow ($\text{N}\,\text{m}^{-2}$), L_f the biofilm thickness (μm), and X_f the biofilm density ($\text{kg}\,\text{m}^{-3}$). Assuming flow occurs over a vertical plate, with only 40% of the available medium's surface area covered by the nutrient flow, a bed height of 3 ft, and a media surface area of 100 $\text{ft}^2\,\text{ft}^{-3}$, we get the following result for the biomass slough-off rate:

$$R_{\text{slough-off}} = 2.9 \times 10^{-7} X_f L_f \left[\frac{\sqrt{V(\text{gpm}\,\text{ft}^{-2})}}{1 + 0.0443(L_f - 30)} \right]^{0.58} \quad \text{when} \quad L_f > 30\,\mu\text{m} \tag{4.17}$$

$$= 2.90 \times 10^{-7} X_f L_f V^{0.29} \quad \text{when} \quad L_f < 30\,\mu\text{m}\,, \tag{4.18}$$

where gpm in Eq. (4.17) is gallon per minute.

Figure 4.5 shows a plot of the biomass slough-off rate versus the liquid flow rate per unit cross-sectional area of the biofilter. As liquid loading increases, the biomass slough-off rate also increases, but the rate of increase decreases at higher loadings. Also, included in this figure are inlet contaminant concentrations when the biomass generation rate is balanced by the biomass slough-off rate. Physical slough-off of the biomass is the only effective strategy to maintain biomass holdup within the biofilter and prevent accumulation of excess

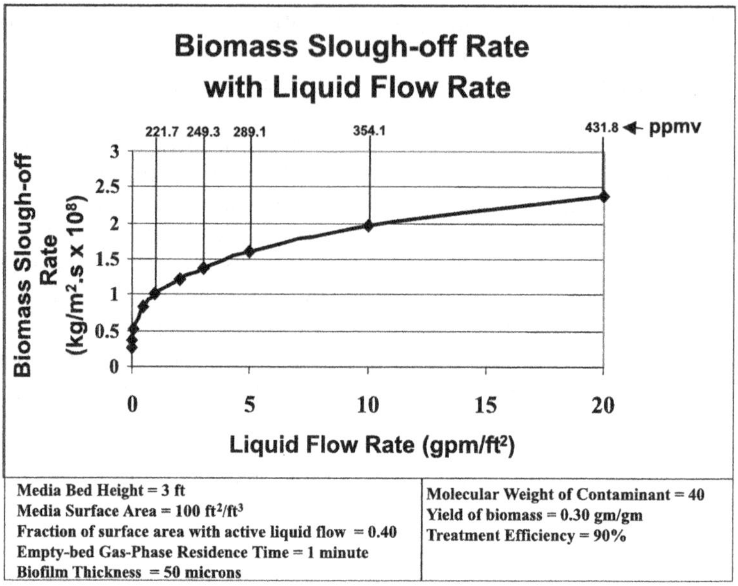

Fig. 4.5. Biomass slough-off rate versus the liquid flow rate in the biofilter

biomass, which eventually results in excessive gas-phase pressure drop through the biofilter bed. However, physical slough-off of the biomass from the media does not imply that the biomass has been removed. The design of the media has to allow this physical removal of this sloughed-off biomass from the bed, which otherwise would result in accumulation and clogging.

Table 4.4 gives the model equations for a synthetic media biofilter, which includes the effects of mass transfer and biokinetics. From Eq. (4.8), the fractional efficiency of a synthetic media biofilter can be approximated by the following equation:

$$E = \frac{A_S D_{\text{biofilm}} X_f \tau k}{u_g C_g^i \varepsilon} \,. \tag{4.19}$$

Using the following definitions for loading and elimination capacity:

$$\text{Loading } \left(\text{g m}^{-3}\,\text{h}^{-1}\right) = \frac{Q C_g^i}{V_{\text{bed}}} = \frac{C_g^i}{\tau} \tag{4.20}$$

$$\text{Elimination capacity } \left(\text{g m}^{-3}\,\text{h}^{-1}\right) = \frac{Q C_g^i (1-E)}{V_{\text{bed}}} = \frac{C_g^i (1-E)}{\tau} \tag{4.21}$$

and combining Eqs. (4.19), (4.20), and (4.21), we get

$$\text{Elimination capacity} = \text{Loading} - \frac{A_s D_{\text{biofilm}} X_f k}{u_g \varepsilon} \,. \tag{4.22}$$

Table 4.4. Model equations for a synthetic media biofilter

Equation number	Equation	Explanation
(l)	$J\|_{x=x_{AB}} = \frac{-D_{\text{biofilm}}}{(u_g + b)} R_B$	Flux of contaminant at the air (A) and biofilm (B) interface
(m)	$R_B = \frac{kX_f C_s}{K_C + C_s}$	Rate of biodegradation in biofilm
(n)	$J\|_{x=x_{AB}} = K_L \left(\frac{C_g}{HN} - C_I \right)$	Contaminant flux at the air–biofilm interface
(o)	$K_L \left(\frac{C_g}{HN} - C_I \right) = D_{\text{biofilm}} \left. \frac{dC_g}{dx} \right\|_{x=x_{AB}}$	Assuming that the gas velocity with no flow condition is small in comparison with u_g and equating the fluxes at the air–biofilm interface
(p)	$\frac{1}{K_L} = \frac{1}{k_l} + \frac{1}{HNk_g}$	Overall mass transfer coefficient in the liquid phase (K_L) written in terms of the individual gas- and liquid-phase mass transfer coefficients, determined using standard correlations (Treybal 1980)
(q)	$C_g = HNC_I - \frac{D_{\text{biofilm}} HNkX_f}{K_L u_g} \left(\frac{C_I}{K_C + C_I} \right)$	Interfacial concentration obtained by combining Eqs. (l), (n), and (o)

Table 4.4. (continued)

Equation number	Equation	Explanation
(r)	$u_g \frac{dC_g}{dh} = A_s D_{biofilm} \left. \frac{dC}{dx} \right\rvert_{x=x_{AB}}$	Mass balance along the height of the biofilter bed
(s)	$E = \frac{F_1 \tau k}{C_g^i \varepsilon} - \frac{F_2 k}{C_g^i} \ln \left(\frac{K_C HN + C_g^i}{K_C HN + C_g^0} \right) - \frac{(HNK_C - F_2 k)}{C_g^i} \ln \frac{C_g^i}{C_g^0}$ $F_1 = \frac{A_s D_{biofilm} X_f}{u_g}; \quad F_2 = \frac{X_f D_{biofilm} HN}{K_L u_g}$	Differentiating Eq. (q), substituting into Eq. (r), and integrating from the biofilter inlet to the outlet

If the biofilter operates at 100% removal efficiency, i.e., $E = 1$, then elimination capacity = loading, and hence:

$$E = 1.0 = \frac{A_s D_{\text{biofilm}} X_f \tau k}{u_g C_g^i \varepsilon} . \tag{4.23}$$

Using the definition of maximum elimination capacity when $E = 1$ and Eq. (4.23), we get:

$$EC_{\max} = \left(\frac{C_g^i}{\tau}\right)_{E=1} = \left(\frac{A_s}{\varepsilon}\right)\left(\frac{D_{\text{biofilm}} k X_f}{u_g}\right) . \tag{4.24}$$

The maximum elimination capacity ($EC_{\max}$) has been written as a product of two factors: (1) the biofilter elimination capacity media factor = $B_{EC} = (A_s/\varepsilon)$; and (2) the biokinetics factor = $(D_{\text{biofilm}} k X_f/u_g)$. For this maximum elimination capacity to be as high as possible, the biofilter synthetic medium should have the highest possible value of the media elimination capacity factor.

Another important consideration in biofiltration is gas-phase pressure drop. As mentioned earlier, gas-phase pressure drop depends on the ability of the media to allow sloughed-off biomass to exit the bed. It also depends on the thickness of the biofilms, bed void fraction, packing media characteristics, and gas flow rate. Since the gas-phase residence time in the biofilter bed, τ, can be written as H/u_g, Eq. (4.23) becomes:

$$\left(\frac{A_s}{\varepsilon}\right)\left(\frac{D_{\text{biofilm}} k X_f}{1}\right)\left(\frac{H}{C_g^i u_g^2}\right) = 1 . \tag{4.25}$$

The gas-phase pressure drop in the bed per unit height of biomedia can be written as follows (Treybal 1980):

$$\frac{\Delta P}{H} = C_D \varrho_g u_g^2 , \tag{4.26}$$

where ΔP is the gas-phase pressure drop through the biomedia bed of height H, and C_D is an empirical friction factor for the specific type of biomedia packed bed. Combining Eqs. (4.25) and (4.26), for 100% removal efficiency we get:

$$\Delta P = \left(\frac{A_s C_D}{\varepsilon}\right)\left(\frac{D_{\text{biofilm}} k X_f \varrho_g}{1}\right)\left(\frac{H^2}{C_g^i}\right) . \tag{4.27}$$

For minimum gas-phase pressure drop, the following biomedia pressure drop factor should be maximum:

$$\text{Biomedia pressure drop factor} = B_{\Delta P} = \left(\frac{\varepsilon}{A_s C_D}\right) . \tag{4.28}$$

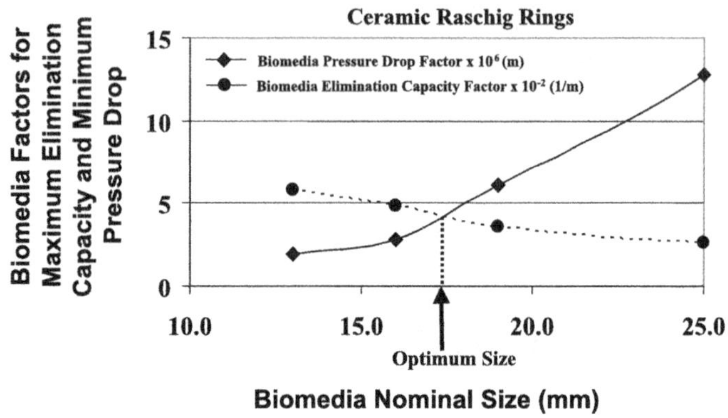

Fig. 4.6. Plot of maximum elimination capacity and minimum pressure drop biomedium factors versus the nominal size

Considering both the objectives of maximum elimination capacity and minimum gas-phase pressure drop, both the biomedia factors B_{EC} and $B_{\Delta P}$ must be maximized for a specific size of biomedia. Figure 4.6 shows a plot of these two biomedia factors for ceramic Raschig rings, using data on bed void fraction, specific area of packing per unit volume of packed bed, and empirical friction factor (Treybal 1980), and ignoring the effect of liquid holdup in the bed, presence of biofilm, etc. Clearly, there exists an "optimum" biomedia nominal size that will maximize the elimination capacity and pressure drop factors.

Ideal characteristics of synthetic media are as follows:

1. high specific surface area, which is the area of the biofilm exposed to the gas-phase contaminants; most synthetic media have specific areas of 30–300 ft.^{2} ft.$^{-3}$ (100–980 m^2 m^{-3});

2. high void fraction, which is the percentage of open space in the packing; void fractions generally have to be higher than 90% to be effective as a biofilter medium;

3. large free passage diameter, which is the diameter of the largest sphere that will pass through the packed media; large free passage diameter is necessary to allow excess biomass to not only slough-off the surface but also exit the bed, so that biomass accumulation does not result in media clogging and channeling;

4. rigid material that does not compact with biomass growth; fibrous mats and foam generally begin to compact due to biomass growth, resulting in lower void fractions and eventually high gas-phase pressure drop;

5. provide good biofilm attachment on the surface; generally, surface roughness, charge and hydrophobicity are important variables that need to be optimized for effective biofilm attachment;

6. low gas-phase pressure drop; the operating cost of the biofilter system depends mainly on the gas-phase pressure drop, which generally has to be less than 5 in. of water column to be cost-effective; and

7. low media cost per unit volume. Since biofilters are used to treat a wide range of gas flow rates, as high as 150,000 ft.^{3} min^{-1}, and empty-bed gas-phase residence times have to be less than 10 min for manageable vessel costs, the volume of media required can be very large and, hence, the cost of the media can be a critical factor in achieving acceptable biofiltration total costs per cubic foot of gas treated.

4.5 Randomly Packed Versus Structured Biomedia

Structured packing consists of regular elements that are combined together to form an insert block. Usually, these insert blocks can be made of any size and are manufactured from thin-walled materials, which reduces the bulk density of the media. Structured packing exhibits a higher surface area per unit volume than random packing, in which the packing elements are randomly packed in the bed. Since structured packing offers parallel flow arrangement, it exhibits lower gas-phase pressure drop and can handle higher biomass growth than random packings. Hence, structured packings are more suited for handling streams containing high biological oxygen demand (BOD), which result in increased biomass growth. Compared to random packings, structured packings are also easier to clean using the shear force of the liquid flow, which offers a tortuous path through the bed, and which usually results in holding the biomass within the bed, even after it is sloughed-off the medium surface. This makes structured packings more suitable when biomass clogging can occur due to increased biomass growth or the presence of solids in the gas or liquid stream.

Random packings consist of many shapes and sizes, ranging from simple Raschig rings to other shapes, such as Berl saddles, fibers, open-cell foams, porous ceramic cylinders, lava rock pieces, etc. Most of these media were developed for gas absorption towers, where it was desired to create a thin liquid film on the medium surface to maximize the gas absorption of the contaminant from the gas to liquid phase. However, in the case of biofilters, growth of biofilm on the surface of the medium is of paramount importance for efficient operation of the biofilter system. Since attached biofilms retain water within their structure, the emphasis in biofilter systems is to keep these biofilms moist, and prevent them from drying, rather than in creating a liquid film on the biofilm surface. Usually, if a liquid film is created on the biofilm, it would present additional resistance for mass transfer, especially if the contaminant(s) being treated has a low water solubility.

While surface area, bulk density, and cost are important parameters in the selection of randomly packed biomedia, the most important characteristics for short startup time in biofiltration was found to be water retention capacity.

Water retention capacity, defined as the volume of water held within the packing medium per unit volume of packed medium, consists of two terms: (1) static liquid holdup, which is a function of packing media design and packing factor; and (2) dynamic liquid holdup, which varies with gas velocity and pressure drop, especially in countercurrent flow. Since dynamic liquid holdup changes with direction of gas and liquid flows and gas flow rates, static liquid holdup was used as an important parameter in evaluating randomly packed media for biofiltration applications.

Static holdup occurs as a result of two factors:

1. design of the packing, leading to interstitial holdup at the points of contact between the various units of the packing medium and the contact points of the surfaces within the unit packing structure; and
2. adhesion of the liquid to the medium's surface, especially during startup when the biofilm has just begun to nucleate and grow.

Static holdup in packed beds is related to residual saturation, which is defined as the volume of liquid holdup after the packed bed is allowed to drain completely by gravity. Previous work has shown that a dimensionless quantity, the capillary number, effectively expresses the ratio of the water-draining driving force (gravity) and the liquid retention forces in the bed. The residual saturation, S_R, is defined as follows:

$$S_R = \text{Residual holdup of liquid in bed/bed void volume} . \tag{4.29}$$

The capillary number, defined by using the Carman-Kozeny equation, can be written as follows (Wakeman 1979):

$$Ca = \frac{\varepsilon^3 d_s^2 \varrho_f g}{(1-\varepsilon)^2 \sigma \cos\theta} , \tag{4.30}$$

where ε is bed voidage, d_s the equivalent diameter of a sphere with the same surface area as a single packing medium element, ϱ_f the fluid density, σ the surface tension of water, and θ the contact angle, which in the case of biofilters, would be 0. A correlation has been developed between the residual saturation and the capillary number, using data from packed beds and consolidated sands (Swindells 1982):

$$S_R = \frac{Ca^{-0.27}}{15.6} . \tag{4.31}$$

Figure 4.7 shows a plot of the residual saturation and the static holdup for two different types of media used in gas absorption towers (Treybal 1980). The residual saturation correlates strongly with the static holdup, and these two parameters represent the potential of the packing media to allow initial biomass growth, since it allows water to remain in the column for a longer period of time, and produces higher mass transfer coefficients due to higher tortuosity of the gas flow path through the packing media. As the size of

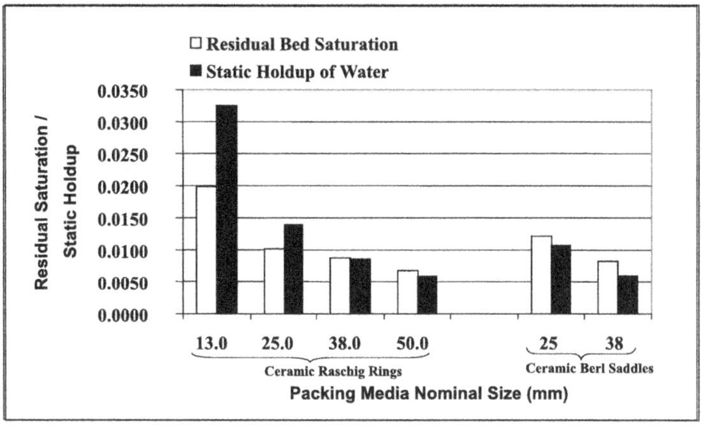

Fig. 4.7. Residual bed saturation and static water holdup in the biofilter bed for two different kinds of packings

the packing medium decreases, the residual saturation increases. However, as the medium size decreases, the potential for clogging of the medium due to biomass growth also increases. Hence, biofilter media have to be designed to offer high residual saturation, with large openings for excess biomass to be removed from the bed. Structured media have very low residual saturation, and hence are not suitable as biofilter media, even though they exhibit low gas-phase pressure drop and very high surface areas. The only exception is when the structured media are composed of porous medium walls, which exhibit water retention capacity, as in the case of ceramic monoliths. In addition, these monoliths have a very low gas-phase pressure drop, exhibit high specific surface areas per unit volume of bed, and excess biomass can be easily sloughed-off the walls and can exit the bed due to its straight, open passages. However, a major disadvantage of ceramic monoliths is their high cost.

4.6 Biofilter Versus Biotrickling Filter

In the literature, the term "biofilter" has been used to describe systems that traditionally have used naturally bioactive media, with water retention capacity, which has required the use of humidified air and periodic water addition, rather than a continuous flow of aqueous nutrients. Naturally bioactive media have bioavailable nutrients (nitrogen and phosphorus trace minerals, etc.) for biodegradation, and hence do not require continuous addition of nutrients. The term "biotrickling filter" is used for systems that have a continuous flow of aqueous nutrients flowing through a bed of synthetic media. A continuous flow of liquid is needed, since the media usually have low residual saturation and no or low water retention capability. In addition, a continuous flow of water allows excess biomass to be sloughed-off, provides the nitrogen and phosphorus nutrients needed for biodegradation, and buffers any pH changes that may occur due to the production of either acidic or basic by-products, such as sulfate, hydroxide, nitrate, etc. In the case of naturally bioactive media, continuous water flow will result in the segregation of smaller particles in the bottom section of the bed due to hydraulic transport, which would result in clogging of the media and an increase in gas-phase pressure drop.

The only difference between a biofilter and a biotrickling filter is that, in a biofilter, the medium has sufficient water holding capacity to maintain bioactivity using humidified air, while in a biotrickling filter, a continuous flow of water is necessary to maintain moisture content in the active biofilms, since the residual saturation or water holdup in the media is insufficient, especially since larger media size has to be used to minimize the potential for clogging, which reduces the water holdup in the media, as shown in Fig. 4.7.

4.7 Experimental Studies on Diffusive Biofilter Media

The objective of this study is to test the effect of support media on biofilter performance. Iso-pentane was chosen as the target in this study, since it is used as a blowing agent in the manufacture of foam products.

4.7.1 Experimental Setup

A schematic of the experimental system is shown in Fig. 4.8. A series of glass biofilter systems were assembled to test different support media. Each glass biofilter was made in two sections, with a total height of 90 cm and diameter of 5 cm. It had provisions for six side ports for withdrawing gaseous samples. Another small glass biofilter was built to carry out the temperature and water content effect tests. This small biofilter was 2.5 cm in diameter and had a total volume of 100 ml, with a water jacket outside to control temperature using a water bath. Compressed air was delivered to the bottom of the biofilter after passing through a countercurrent packed pre-humidifier column. The gas flow rate was controlled by a thermal mass flow controller (MKS Industries, type 1259, control channel type 247). The desired concentration of iso-pentane in the air stream was obtained by injecting the pure chemical into the air stream using a syringe pump. Nutrient solution (Govind and Bishop 1998) was pumped from the nutrient reservoir bottle to the top of the biofilter. The contaminated air stream flowed countercurrent to the trickling nutrient solution.

Biomass from a pilot-scale activated sludge plant treating hazardous waste was suspended in an aerated aqueous-phase bioreactor (column diameter 100 mm, height 700 mm). The bioreactor was fed daily with iso-pentane. Nutrients were also added periodically. The acclimated biomass from the bioreactor was later used to seed the synthetic support media.

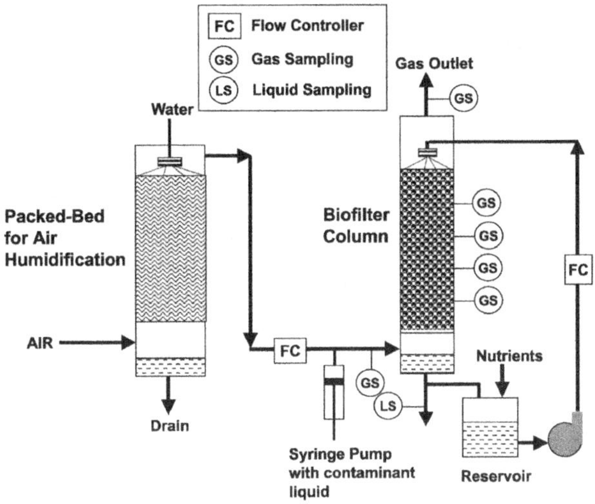

Fig. 4.8. Schematic of the experimental system for studying different kinds of biomedia

Five different support media were tested in this study. Peat and compost were chosen as the naturally bioactive media, with effective particle diameters of 3.5 and 2.5 mm, respectively. The water content of peat was 0.645 g water g^{-1} dry peat. The compost water content was 0.536 g water g^{-1} dry compost. Celite pellets (Manville Corp., Denver, CO), with a diameter of 6 mm, were packed into the biofilter. Cordierite monolith, obtained through Dow Corning, Corning, NY, with and without a coating of activated carbon, and with 50 square channels per square inch (channel dimension of 2×2 mm), was used as structured media. The biofilters were operated continuously, and inlet and outlet composition data were collected over a period of time. The concentration profiles of the compounds in the biofilter were obtained by withdrawing samples from the side ports and analyzing them using gas chromatography. Once the biofilter performance reached steady state, the concentration differences of carbon dioxide in the gas phase were measured to assess the mineralization of all compounds.

4.7.2 Analytical Procedure

The gas samples taken from the sampling ports were analyzed by gas chromatography in a Tracor 585 apparatus fitted with a DB-624 column (60 cm long, 0.53 mm in diameter), using photoionization and electrolytic conductivity detectors. Helium was used as the carrier gas. pH of the solutions was measured by combination pH electrode. Carbon dioxide concentration in gas streams was determined in a Fisher 1200 gas partitioner.

4.7.3
Results and Discussion

4.7.3.1
Adsorption Capacity

Fresh samples of support (400 ml) were taken for each medium. The contaminated gas was introduced at a flow rate of 100 ml min^{-1} with an inlet concentration of 350 ppmv. The breakthrough curves of all five support media are shown in Fig. 4.9. It was found that natural support media like peat and compost had limited adsorption capacity. The celite pellets and cordierite monolith systems had low adsorption capacity, while the activated carbon-coated monolith system exhibited higher adsorption capacity.

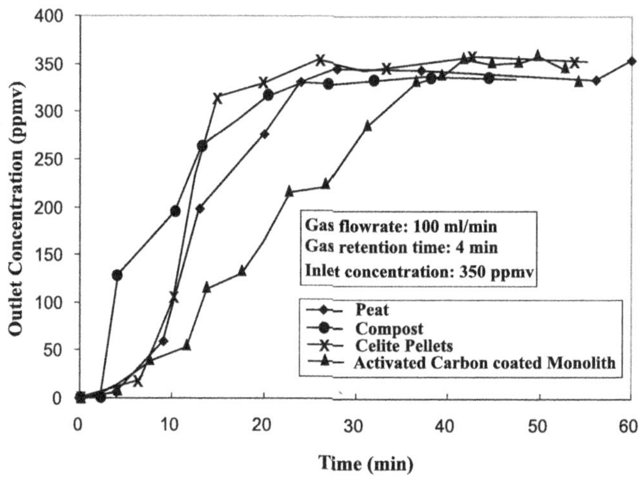

Fig. 4.9. Iso-pentane outlet concentration as a function of gas-phase residence time (empty bed) for different biomedia

4.7.3.2
Effect of Water Content

The removal efficiency of iso-pentane in peat and compost biofilters, using different media water contents, was tested in the small glass biofilter. The contaminated gas was introduced at 20 ml min^{-1} with an inlet concentration of 470 ppmv. Figure 4.10 shows the effect of water content on the operating efficiency of a peat and compost packed bed. The removal efficiency was maximized when the water content was in the range 0.62–0.67 g water g^{-1} dry media. When the water content was less than 0.57 g water g^{-1} dry peat and 0.58 g water g^{-1} dry compost, an irreversible loss of biofilter efficiency occurred, as shown by the dotted line. Above a water content of 0.67 g water g^{-1} dry media,

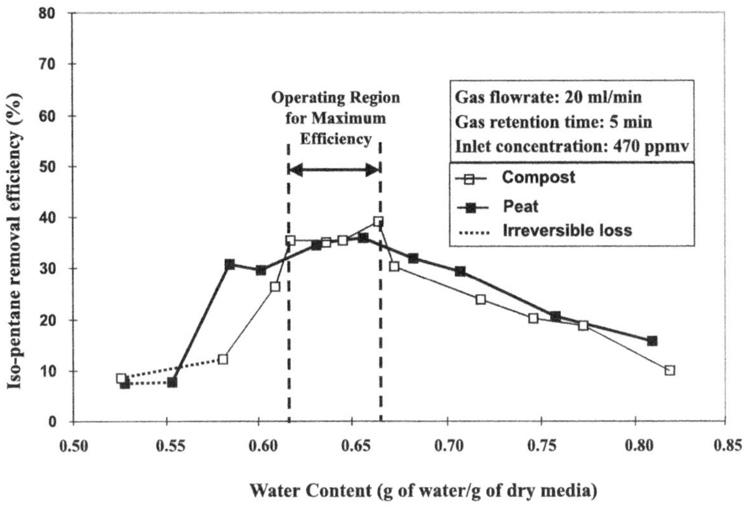

Fig. 4.10. Effect of water content on removal efficiency of iso-pentane in peat and compost biofilters

there was a significant decrease in operating efficiency, mainly due to the blanketing of the material macropores by free water, coupled with the low aqueous solubility of iso-pentane.

These results show the importance of maintaining water content within a narrow range in biofilters using these kinds of naturally bioactive media. There was an "optimum" water content, below and above which the biofilter removal efficiency declined. Typically shallow beds with large cross-sectional areas and inlet air humidification are used to effectively maintain water content in the media during operation.

4.7.3.3 Temperature Effect

As temperature increased in the range 25–40 °C, increased biodegradation rates were observed, which is typical for biochemical reactions. The data were used to obtain the activation energy for net growth of biomass in naturally bioactive systems, and this value was in the range 1.9–2.4 kJ mol^{-1}.

4.7.3.4 Steady-State Operation

Biofilters with different support media were operated under various operating conditions. Carbon dioxide measurements of the inlet and outlet gases were used to confirm the removal efficiencies. The removal efficiency increased at lower inlet iso-pentane concentrations and at higher gas retention times. Celite

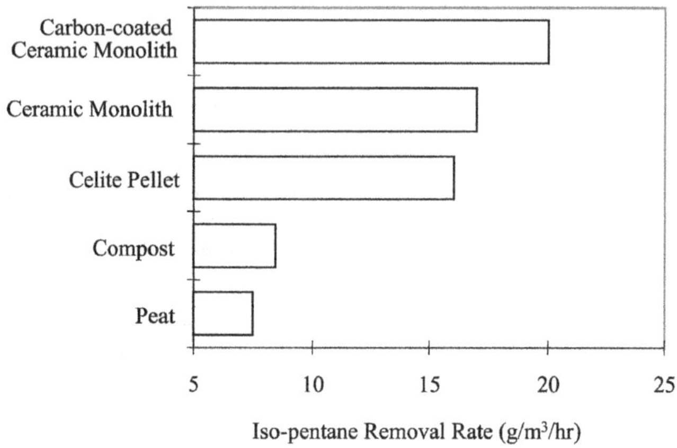

Fig. 4.11. Comparison of different biomedia in terms of the maximum elimination capacity for iso-pentane

pellets and cordierite monolith showed similar performance at steady state, since they had similar surface areas for biofilm growth. The carbon-coated monolith biofilter showed 10 to $\sim$20% higher removal efficiency, and this was attributed to iso-pentane adsorption and subsequent desorption into the biofilm, which resulted in thicker biofilms. Synthetic media biofilters exhibited higher removal efficiencies than the naturally bioactive media, peat and compost, and Fig. 4.11 shows a comparison of the different media, based on iso-pentane biotreatment rates.

4.8 Experimental Studies on Convective Biofilter Media

Two identical biofilters were constructed using ceramic monoliths, each with dimensions of 13-cm width by 15-cm length by 9-cm height. Each ceramic monolith was coated with a thin layer of activated carbon, and there were 10 square cells per square inch, with each cell being 0.45×0.45 cm in dimension. The porous walls between each cell had a thickness of 0.2 mm and an average pore size of 30 μm. In these monoliths, contaminated air would enter one end of the monolith and flow out through the other end, as shown in Fig. 4.12a. In this case, the contaminant in the gas phase would diffuse to the porous walls of the monolith, perpendicular to the direction of gas flow, and this would be a diffusive type of biofilter media. In the second monolith, alternate channels were plugged at the outlet end, as shown in Fig. 4.12b, which would force the contaminated gas to flow into the open channels, and then through the biofilm and porous walls of the cells to exit through the channels that were open at the exit end of the monolith. In this case, there would be a convective flow of contaminant through the biofilm, instead of diffusive, and this type of media would be classified as a convective type of biofilter media, as discussed earlier

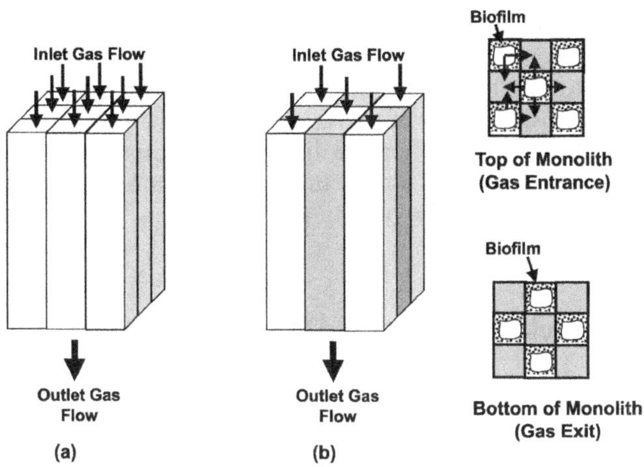

Fig. 4.12. Monolith support media used in biofilter: **a** diffusive, with gas entering and exiting through the straight passages; and **b** convective, with gas entering and then flowing through the biofilms and porous walls of the monolith, by alternate blocking of passages in the top and bottom surfaces of the media

in this chapter. Both monoliths were identical, except in the direction of flow of the contaminant into the biofilms, and both monoliths were operated under identical flow and contaminant concentration conditions.

The contaminants studied were ethanol and toluene. Both contaminants were analyzed using an HP 5710 gas chromatograph, equipped with a flame ionization detector (FID) and an Alltech 10% CBWX 20 M CWHP 80/100, 6′ × 1/8″ stainless-steel packed column (stock no. 12106PC, serial no. 1082-21753). The oven, injector and detector temperatures were set at 60, 150, and 300 °C, respectively, with nitrogen as the carrier gas. Analysis of carbon dioxide was performed using an HP 5890 Series II gas chromatograph equipped with

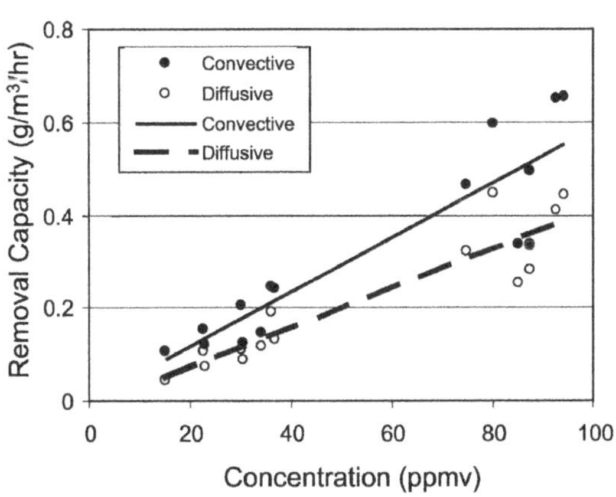

Fig. 4.13. Removal capacity of toluene in a diffusive and convective monolith support media biofilter as a function of inlet concentration. The *straight lines* are model fits

a thermal conductivity detector, using a 6-ft. Porapak 80/100 mesh packed column, with an oven temperature of 110 °C and a run time of 5 min.

Figure 4.13 shows the toluene elimination capacity versus loading for the convective and diffusive modes of operation. Clearly, the convective flow shows a higher elimination capacity than the diffusive, and this difference in performance has also been shown using a mathematical model (model lines are shown in Fig. 4.13; Fang and Govind, unpubl. data).

4.9 Studies on Encapsulated Biomass and Membrane Biofilters

Gel entrapment of active bacterial cultures was achieved using a modified procedure originally used by Fukushima et al. (1988). Colloidal silica SM grade (30% silica, 7-nm average particle size, 1.22 specific gravity) was pH adjusted to 7.0 using 5 N HCl, and the solution was maintained at room temperature with gentle stirring. Two grams of agarose (gelling temperature of 28 °C and melting point of 65 °C) was mixed with 50 ml water at room temperature, and then heated above melting point with gentle stirring. Low-viscosity alginate (1%) was separately dissolved in 50 ml water and mixed to obtain a clear solution. Centrifuged biomass was added to the alginate solution, and the resulting mixture was added to the agarose solution at 35 °C to completely disperse the biomass in the combined solution mixture. Just before gel pellets were synthesized, the combined alginate–biomass solution was added to the silica solution, and the resulting solution was dropped into a 5% calcium chloride solution to form beads with an average diameter of 3 mm. The bacterial cells are entrapped in the gel matrix inside the beads, and are active and capable of degrading contaminants.

The experimental system mainly consisted of a 2-in.-diameter (5 cm) glass column, packed with 10-cm height of 3-mm-diameter gel beads. The reactor had a liquid sump of about 100 ml and a headspace of 500 ml. Both the liquid and gas phases were recycled in a batch mode (liquid flow rate of

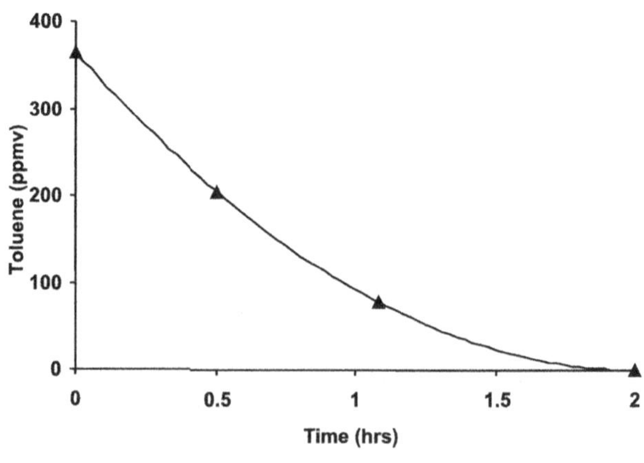

Fig. 4.14. Exit concentration of toluene as a function of time in a silica gel bead encapsulated biomass biofilter, operated in batch mode

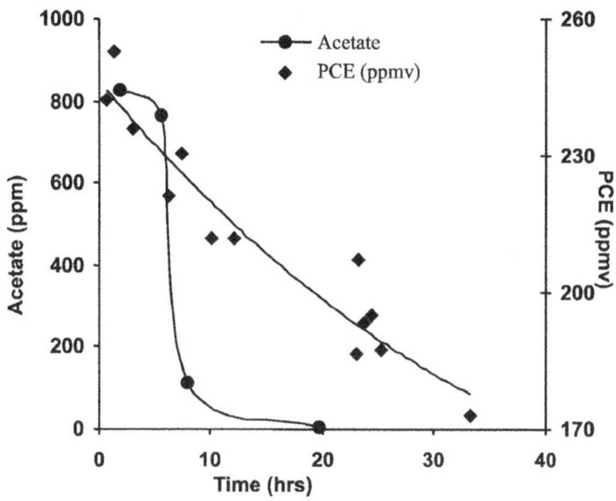

Fig. 4.15. Concentration of acetate (liquid phase) and perchloroethylene (PCE) in the gas phase of a silica gel, bead encapsulated biomass biofilter, operated in batch mode

10 ml min^{-1}, and gas flow rate of 30 ml min^{-1}), and the contaminant concentration in the gas phase was analyzed using gas chromatography. Figure 4.14 shows the degradation of toluene, which demonstrated the existence of active biomass encapsulated within the pellets. Figure 4.15 shows the degradation of perchloroethylene (PCE), with sodium acetate present in the liquid sump. While complete details of these results have not been provided here for the sake of brevity, corresponding measurements of gas-phase carbon dioxide and chloride ion in the liquid phase verified the biological breakdown of the perchloroethylene. The mechanism of PCE degradation is shown in Fig. 4.16, in which PCE diffuses into the anoxic zone within the gel bead and dechlorinates into less chlorinated by-products, which degrade into carbon dioxide and water in the outer region of the bead due to the diffusion of oxygen from the outer surface of the bead. The inner anoxic zone is surrounded by an aerobic zone, and these two zones – with widely different oxidation–reduction potentials – coexist within one and the same gel bead. The bead accomplishes the breakdown of PCE through a synchronous anaerobic–aerobic degradation

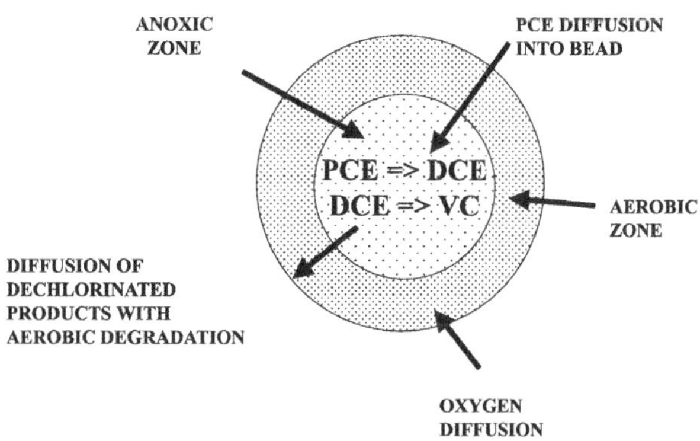

Fig. 4.16. Schematic showing the mechanism of PCE degradation within the anoxic and aerobic zones that simultaneously coexist within the gel bead

process, and detailed analysis of this mechanism has been presented elsewhere (Tabak et al. 2001).

Membranes offer high surface areas on which biofilms can be attached. In membrane biofilters, usually hollow fibers are used to develop biofilms outside the porous surfaces, by flooding the shell side with nutrients. Contaminated airflows on the inside of the porous hollow fibers, with the contaminant diffusing into the active biofilms through the pores. Membrane biofilters have been used to successfully biodegrade toluene and trichloroethylene (TCE; Parvatiyar et al. 1996a, b). Major problems encountered with membranes are pore plugging due to slime produced by aerobic organisms, and additional mass transfer resistance due to the membrane itself. However, they offer a compact design, with very high contact surface areas for biofiltration applications.

4.10 Conclusions

Media characteristics are of critical importance in the effective operation of any biofilter system. Broadly, biofiltration media can be classified as being either inherently bioactive (naturally bioactive media) or synthetic, which have to be initially inoculated to achieve biomass growth. Naturally bioactive media are suitable for odor biofiltration where the contaminant loading is low. At high contaminant loadings, as in biofiltration treatment of volatile organic emissions, synthetic media biotrickling filters are generally more effective. There is a wide variety of synthetic media, and in terms of their mass transfer characteristics, these can be classified as being either diffusive or convective. Very few studies on convective biofiltration have been conducted. Diffusive biofiltration becomes inefficient either when the biofilms are thick and/or when the bulk contaminant concentration is low, as usually is the case in odor biofiltration. Adsorption of the contaminants on the medium's surface offers an additional mechanism for biodegradation, namely, back-diffusion of the adsorbed contaminant into the biofilm, and eliminates the concept of maximum biofilm thickness, as in the case of non-adsorptive media.

There is an "optimum" synthetic media nominal size that maximizes the biotreatment efficiency and minimizes the gas-phase pressure drop. The residual saturation reflects the capacity of the media to hold water, and affects the startup time of the biofilter system. Several experimental studies have been conducted to obtain a better understanding of the role of biofilter media, and some of these results were presented in this chapter.

Encapsulated biomass biofilters offer a new type of biofilter medium, which can create synchronous aerobic–anoxic zones, especially for the breakdown of chlorinated compounds such as perchloroethylene (PCE), which are otherwise aerobically recalcitrant.

Membrane biofilters offer advantages of compact sizes, high surface areas, and effective contact between the contaminated gas and biofilms. However, they generally suffer from clogging of membrane pores due to slime, usually generated by aerobic organisms. More research work is needed on these systems to effectively use them for industrial and municipal biofilter applications.

Appendix

Nomenclature

A	specific area of the peat/compost particle ($m^2\,m^{-3}$)
A_s	surface area of biomedia per unit volume of packed bed (m^{-1})
b	contaminant velocity with no flow condition ($m\,s^{-1}$)
B_{EC}	biofilter elimination capacity media factor (m^{-1})
$B_{\Delta P}$	biomedium pressure drop factor (m)
C	concentration of contaminant ($g\,m^{-3}$)
Ca	capillary number (-)
C_b	bulk gas-phase concentration of contaminant ($kg\,m^{-3}$)
C_f	gas-phase contaminant concentration at air/biofilm interface ($kg\,m^{-3}$)
C_D	empirical friction factor for biomedia
C_g	gas-phase contaminant concentration ($kg\,m^{-3}$)
C_{in}	C_g^i inlet gas concentration of contaminant ($kg\,m^{-3}$)
C_g^0	concentration of contaminant in exit gas from biofilter ($kg\,m^{-3}$)
C_I	interfacial contaminant concentration in biofilm ($kg\,m^{-3}$)
C_0	oxygen concentration in particle ($kg\,m^{-3}$)
C_s	contaminant concentration in biofilm ($kg\,m^{-3}$)
d_s	equivalent diameter of a sphere with the same surface area as a single packing medium element (m)
$d_{biofilm}$	diffusivity of contaminant in biofilm ($m^2\,s^{-1}$)
D_e	effective diffusivity of contaminant in particle ($m^2\,s^{-1}$)
D_{eo}	effective diffusivity of oxygen in particle ($m^2\,s^{-1}$)
D_g	gas-phase diffusivity of contaminant ($m^2\,s^{-1}$)

D_l	liquid-phase diffusivity of contaminant ($m^2\,s^{-1}$)
E	fractional removal efficiency of biofilter (-)
E_0	characteristic energy of adsorption of a reference compound ($J\,mol^{-1}$)
EC_{max}	maximum elimination capacity of biofilter when $E = 1$ ($g\,m^{-3}\,s^{-1}$)
f_{OC}	fraction of organic carbon in biomedia (-)
F_1	factor in biofilter efficiency, Eq. (4.8)
F_2	factor in biofilter efficiency, Eq. (4.8)
g	gravitational constant ($m\,s^{-2}$)
h	direction along biofilter height (m)
H	height of biofilter bed (m)
HN	dimensionless Henry's law constant for contaminant (-)
J	contaminant flux ($g\,m^{-2}\,s^{-1}$)
k	Monod's maximum specific growth rate (s^{-1})
k_g	gas-phase mass transfer coefficient ($m\,s^{-1}$)
k_l	liquid-phase mass transfer coefficient ($m\,s^{-1}$)
K	maximum contaminant biodegradation rate $= kX_f/Y$
K_C	half-velocity constant ($kg\,m^{-3}$)
K_{CW}	half-velocity constant for water ($g\,water\,g^{-1}$ dry medium)
K_d	hydrophobic partitioning constant ($L\,kg^{-1}$)
K_g	overall gas-phase mass transfer coefficient ($m\,s^{-1}$)
K_L	overall liquid-phase mass transfer coefficient ($m\,s^{-1}$)
K_{OC}	hydrophobic partitioning constant normalized by organic carbon fraction ($L\,kg^{-1}$)
L_f	biofilm thickness (μm)
MW_s	molecular weight of contaminant ($g\,mol^{-1}$)
P	total gas pressure (Pa)
P/P_0	partial pressure (kPa)

$q(C)$	total sorption of contaminant by biomedia
q'	adsorbed volume (m^3)
q'_{max}	maximum sorption capacity ($cm^3\,g^{-1}$)
$q_P(C)$	hydrophobic partitioning of contaminant to biomedia from aqueous phase ($\mu g\,kg^{-1}$)
$q_{ad}(C)$	physical adsorption of contaminant on biomedium surface ($\mu g\,kg^{-1}$)
Q	gas flow rate through biofilter ($m^3\,s^{-1}$)
r	distance in radial direction
R	radius of particle
R_B	rate of biodegradation of contaminant in particle ($g\,m^{-3}\,s^{-1}$)
R_{BO}	rate of oxygen consumption in particle due to contaminant biodegradation ($g\,m^{-3}\,s^{-1}$)
$r_{observed}$	actual measured reaction rate ($kg\,m^{-3}\,s^{-1}$)
$r_{bulk\ conditions}$	reaction rate determined at bulk conditions ($kg\,m^{-3}\,s^{-1}$)
R	ideal gas constant ($J\,mol^{-1}\,K^{-1}$)
$R_{slough\text{-}off}$	rate of biomass slough-off from the media surface ($kg\,m^{-2}\,s^{-1}$)
S_R	residual saturation of biomedium packed bed (-)
T	absolute temperature (K)
u_g	empty-bed gas velocity in biofilter ($m\,s^{-1}$)
V	liquid flow rate per unit cross-sectional area of biofilter ($gpm\,ft.^{-2}$)
V_{bed}	volume of biomedium in biofilter (m^3)
W	water content of media (g water g^{-1} dry medium)
W_i	critical water content below which irreversible loss of biodegradative capacity occurs (g water g^{-1} dry medium)
W_0	original water content of media (g water g^{-1} dry medium)
W_P	water content above which a water film appears outside particle (g water g^{-1} dry medium)

W_{max} water content when free drops of water appear within the medium (g water g^{-1} dry medium)

X_f biofilm density (kg m^{-3})

x direction along biofilm thickness (m)

x_{AB} position at air/biofilm interface

Z dimension along height of biofilter bed (m)

Greek

β stoichiometric ratio of oxygen consumption and contaminant biodegradation (-)

β_i ratio of adsorption energies between the sorbate of interest and a reference compound

ε void fraction of biofilter packed bed

ε_d differential work of adsorption

μ_g gas viscosity (kg m^{-1} s^{-1})

ϱ_f fluid density (kg m^{-3})

ϱ_g gas density (kg m^{-3})

ϱ_l liquid density (kg m^{-3})

η biofilm effectiveness factor

σ liquid surface tension (N m^{-1})

τ empty-bed gas-phase residence time (s)

τ_S shear stress on the biofilm's surface due to liquid flow (N m^{-2})

θ contact angle (°)

ΔP gas-phase pressure drop across biofilter bed (Pa)

References

Accardi-Dey A, Gschwend PM (2002) Assessing the combined roles of natural organic matter and black carbon as sorbents in sediments. Environ Sci Technol 36:21–29

Bendinger B, Rijnaarts HHM, Altendorf K, Zehnder AJB (1993) Physicochemical cell surface and adhesive properties of coryneform bacteria related to the presence and chain length of mycolic acids. Appl Environ Microbiol 59:3973–3977

Characklis WG, McFeters GA, Marshall KC (1990) Physiological ecology in biofilm systems. In: Characlkis WG, Marshall KC (eds) Biofilms. Wiley, New York, pp 341–394

Charpentier JC (1976) Recent progress in two-phase gas-liquid mass transfer in packed beds. Chem Eng 11:161–167

Chiou CT, Porter PE, Schmedding DW (1983) Partition equilibria of nonionic organic compounds between soil organic matter and water. Environ Sci Technol 17:227–231

Chiou CT, Kile DE, Rutherford DW, Sheng G, Boyd SA (2000) Sorption of selected organic compounds from water to a peat soil and its humic-acid and humin fractions: potential sources of the sorption nonlinearity. Environ Sci Technol 34:1254–1258

Cowan MM, Warren TM, Fletcher M (1991) Mixed species colonization of solid surfaces in laboratory biofilms. Biofouling 3:23–34

Dubinin MM, Astakhov VA (1971) Development of concepts of volume filling of micropores in adsorption of gases and vapors by microporous adsorbents. Izv Akad Nauk SSSR Ser Khim 1:5–11

Fletcher M (1988a) The applications of interference reflection microscopy to the study of bacterial adhesion to solid surfaces. In: Houghton DR, Smith RN, Eggins HOW (eds) Biodeterioration 7. Elsevier, London, pp 31–35

Fletcher M (1988b) Attachment of *Pseudomonas fluorescens* to glass and influence of electrolytes on bacterium-substratum separation distance. J Bacteriol 170:2027–2030

Fletcher M, Loeb GI (1979) Influence of substratum characteristics on the attachment of a marine pseudomonad to solid surfaces. Appl Environ Microbiol 37:67–72

Fukushima Y, Okamura K, Imai K, Motai H (1988) A new immobilization technique of whole cells and enzymes with colloidal silica and alginate. Biotechnol Bioeng 32:584–594

Gao C, Govind R, Tabak H (1996) Predicting soil sorption coefficients of organic chemicals using neural network model. Environ Toxicol Chem 15:1089–1096

Govind R (2002) Biofiltration: new technology for the metal finishing industries. Finishers Manage 4:56–62

Govind R, Bishop DF (1998) Biofiltration for treatment of volatile organic compounds (VOCs) in air. In: Sikdar SS, Irvive RI (eds) Biodegradation technology developments, vol II. Technomic, Lancaster, PA, pp 403–460

Karickhoff SW, Brown DS, Scott TA (1979) Sorption of hydrophobic pollutants on natural sediments. Water Res 13:241–248

Loeb GI, Neihof RA (1975) Marine conditioning films. Adv Chem 145:319–335
Parvatiyar MG, Govind R, Bishop DF (1996a) Biodegradation of toluene in a membrane biofilter. J Membrane Sci 119:17–24
Parvatiyar MG, Govind R, Bishop DF (1996b) Treatment of trichloroethylene (TCE) in a membrane biofilter. Biotechnol Bioeng 50:57–64
Polanyi M (1916) Adsorption von Gasen (Dampfen) durch ein festes nichtfluchtiges Adsorbens. Ber Dtsch Physikal Ges 18:55–80
Pringle JH, Fletcher M (1983) Influence of substratum wettability on attachment of freshwater bacteria to solid surfaces. Appl Environ Microbiol 45:811–817
Rittmann BE (1989) Detachment from biofilms. In: Characklis WG, Wilderer PA (eds) Structure and function of biofilms. Bath Press, Bath pp 49–58
Rosenberg M, Kjelleberg S (1986) Hydrophobic interactions in bacterial adhesion. Adv Microb Ecol 9:353–393
Seth R, Mackay D, Muncke J (1999) Estimating the organic carbon partition coefficient and its variability for hydrophobic chemicals. Environ Sci Technol 33:2390–2394
Swindells RJ (1982) A mathematical model of a continuous sugar centrifuge. PhD Thesis, University of Queensland, Australia
Tabak HH, Govind R, Ramani M (2001) In-situ biorestoration of contaminated sediments using membranes and gel beads. In: Leeson A, Foote E, Banks K, Magar VS (eds) Phytoremediation, wetlands, and sediments. Battelle Press, Columbus, OH, pp 200–210
Treybal (1980) Mass transfer operations. McGraw-Hill, New York
Wakeman RJ (1979) The prediction and calculation of cake dewatering characteristics. Filtration Separation 11:655–660
Wakeman RJ, Rushton A, Brewis LN (1976) Residual saturation of dewatered filter cakes. Chem Eng 314:668–670
Weber WJ Jr, Huang W, LeBoeuf EJ (1999) Geosorbent organic matter and its relationship to the binding and sequestration of organic contaminants. Colloid Surf A Physicochem Eng Aspect 151:167–179
Xia G (1998) Sorption behavior of nonpolar organic chemicals on natural sorbents. PhD Thesis, Johns Hopkins University, Baltimore
Xing G, Pignatello JJ (2001) Detailed sorption isotherms of polar and apolar compounds in a high organic soil. Environ Sci Technol 35:84–94
Zhao W, Govind R (1997) Biofiltration of iso-pentane in peat and compost packed-beds. AIChE J 43:1348–1356

5 Microbiology of Bioreactors for Waste Gas Treatment

Ajay Singh, Owen Ward

5.1 Introduction

Over the past few decades, there have been a number of applications that utilized different types of biotreatment processes as effective and economical air pollution control strategies (Burgess et al. 2001). The three major types of bioreactors that currently dominate waste gas biotreatment are biofilters, biotrickling filters, and bioscrubbers. Alternative technologies such as photobioreactors, membrane bioreactors, and activated sludge bioreactors/diffusers have also been tested for contaminated-air treatment (Chung et al. 2001; Min et al. 2002; Morikawa and Erin 2003). Although microbial activity in bioreactors accounts for most of the degradation of organic contaminants, chemical and physical mechanisms can also provide significant transformation pathways for these compounds. Microbiological methods for soil, water and air remediation are generally considered a safe and less expensive method for the removal of hazardous contaminants, to nontoxic byproducts (van Hamme et al. 2003; Ward et al. 2003). A more complete understanding of these complex biodegradation processes and mechanisms is currently being gained through cellular and molecular research (Harayama 2001; Arp et al. 2002; Grommen and Verstraete 2002).

Biotechnological methods have been successfully implemented in soil bioremediation (Dua et al. 2002), and play an immense role in the development of contaminated-air treatment processes. Interest in microbial treatment of polluted air has increased in the last two decades, primarily because it was recognized that microbes were able to degrade toxic compounds that were earlier believed to be resistant to the natural biological processes. Biological treatments are cost and energy efficient, and particularly well suited to treating low concentration emissions when the pollutants are readily biodegradable (Burgess et al. 2001). This chapter will provide an overview of microbiological aspects of methods used in odor and air pollution control.

Shareefdeen/Singh (Eds.)
Biotechnology for Odor and Air Pollution Control

5.2 Microbial Communities Involved in Waste Gas Treatment

Biological air treatment systems are based on the efficient capacity of microorganisms to degrade air pollutants. The composition of the microbial community and the survival of microbes in a biofilter depend on physical and chemical conditions in the packing material. The diversity of the active microorganisms is to a large extent a function of the inlet gas stream composition and the biofilter medium, and will be dependent on the nature of the incoming pollutant and the prevailing environmental conditions. Use of microbial consortia, mixed cultures or wastewater sludge is common for the startup of bioreactors in large-scale applications for odor or VOC contaminants, instead of pure cultures (Cox et al. 1997). Biofilters using natural organic carriers are expected to contain a wide range of organisms, including bacteria, actinomycetes, fungi, yeasts, algae and protozoa, as compared to synthetic or inert carriers that may have a less diverse population. There is one report on utilizing a co-culture of fungi and mites for biofiltration of hydrophobic pollutants (van Groenestijn 2001). However, biofilter inoculation may speed up the startup period (Veiga and Kennes 2001). Various microbial cultures reported for treatment of air pollutants are shown in Table 5.1.

Various species of bacteria present in biofilters and belonging to the genera *Pseudomonas*, *Alcaligenes*, *Bacillus*, *Corynebacterium*, *Sphingomonas*, *Xanthomonas*, *Nocardia*, *Mycobacterium*, *Rhodococcus*, *Xanthobater*, *Clostridium* and *Enterobacter* have been identified (Arnold et al. 1997; Juneson et al. 2001; Kennes and Veiga 2001). Use of white-rot fungi such as *Phanerochaete chrysosporium*, *Tramtes versicolor*, *Pleurotus ostreatus* and *Bjerkandera adusta*, and other fungal species like *Exophiala jeanselmei* has been reported in biofiltration studies for the elimination of volatile organic compounds (Braun-Lüllemann et al. 1997; Cox et al. 1997). Fungi are generally resistant to extreme environmental conditions, and are suitable for the removal of hydrophobic compounds, such as alkylbenzenes and styrene.

The selection of the microbial population is determined by the characteristics of the pollutant to be treated (Deshusses and Johnson 2000). In the case of VOCs, heterotrophic microorganisms derive energy from the oxidation of the organic molecule, and part of the carbon is incorporated into biomass. When the pollutant contains sulfur, nitrogen or halogens, some of these elements accumulate in the system. Reduced inorganic pollutants such as H_2S and NH_3 are treated with autotrophic microorganisms that derive their energy from the oxidation of the molecule and use CO_2 as a carbon source. In the case of highly hydrophobic compounds, fungal cultures offer some interesting advantages over flat bacterial biofilms. Fungal mycelia occupy the internal void space in the reactor, increasing substantially the transfer surfaces (García-Peña et al. 2001; Woertz et al. 2001; van Groenestijn and Liu 2002; Spingo et al. 2003), which are higher than those reported for bacterial films.

According to one estimate, microbiological populations in biofilters are of the order of 1 billion microorganisms per gram of organic material (Bohn

1992). Although microbes can be isolated and identified by classical microbiological methods, only a limited number of microbes present in a bioreac-

Table 5.1. Microbial cultures reported for the treatment of waste gas contaminants

Contaminants	Microorganisms	References
Benzene	*Pseudomonas putida*, *Phanerochaete chrysosporium*	Braun-Lüllemann et al. (1997)
Toluene	*Pseudomonas putida*, *Phanerochaete chrysosporium*, *Scedosporium apiospermum*	Ergas et al. (1995); Oh et al. (1998); García-Peña et al. (2001)
Styrene	*Exophiala jeanselmei*, *Phanerochaete chrysosporium*, *Sphingomonas capsulate*, *Xanthomonas* sp., mixed culture	Arnold et al. (1997); Braun-Lüllemann et al. (1997); Cox et al. (1997); Juneson et al. (2001)
Acetone	*Rhodococcus* sp., *Corynebacterium* sp.	Kirchner et al. (1987)
Butanol	*Pseudomonas fluorescens*, *Corynebacterium rubrum*, *Micrococcus luteus*, mixed culture	Kirchner et al. (1987); Wübker and Friedrich (1996)
Isopropanol	*Pseudomonas fluorescens*, *Rhodococcus*	Kirchner et al. (1987)
Methanol	*Pseudomonas fluorescens*, mixed culture consortium	Kirchner et al. (1987); Shareefdeen et al. (1993)
Ethanol	*Candida guilliermondii*, *Saccharomyces cerevisiae*, mixed microbial consortium	Cox et al. (2001); Granström et al. (2002)
Dichloromethane	*Hyphomicrobium* sp., *Pseudomonas putida*	Ergas et al. (1994); Veir et al. (1996)
Methane	*Methylomonas fodinarum*	Sly et al. (1993)
Trichloroethane	*Methylosinus trichsporium*	Speitel and McLay (1993)
Dimethyl sulfide	*Hyphomicrobium* sp.	Smet et al. (1997)
Hydrogen sulfide	*Chlorobium limicola*, *Thiobacillus* spp., *Xanthomonas* sp., activated sludge	Cho et al. (1992); Jensen and Webb (1995); Chung et al. (1997)
Ammonia	*Alcaligenes*, *Pseudomonas*, *Arthrobacter*, *Nitrosomonas*, *Nitrobacter*, *Vibrio*	Stepanov and Korpela (1997); Kim et al. (2000)

tor is cultivable. Advanced molecular tools have recently been developed for strain identification and to obtain the useful information on microbial ecology (van Elsas et al. 1998; Theron and Cloete 2000). Phospholipid fatty acid analysis (PLFA), fluorescent in situ hybridization (FISH) and 16S rRNA characterization are some of the techniques used in such studies. However, such investigations have not been popular in characterizing microbial populations associated with bioreactors treating air contaminants (Sakano and Kerkhof 1998).

It is important to analyze the microbial community structure and its changes during biodegradation processes. In addition, it is important to characterize the differences in the community from inlet to outlet through the depth of the biofilter, since the concentration of oxygen and contaminant will decline in the matrix at increasing distances from the biofilter gas entry point (Mallakin and Ward 1996). The challenge of characterizing the roles of a range of organisms in degrading contaminants present in natural organic media is clearly substantial. However, such studies can also provide major insights into important biochemical and physiological aspects of biodegradation. Culture-dependent and culture-independent methods are being applied to microbial community characterization. The temporal and spatial changes in bacterial populations, and the diversity of the microbial community during biodegradation processes can be determined using sophisticated molecular methods (Stapleton et al. 1998; Widada et al. 2002). During treatment, odorants or VOC contaminants introduced into a bioreactor shift the distribution of the existing microbial populations toward a specific culture that can metabolize the target odor or VOC pollutants.

Since the majority of the biological waste gas treatment processes rely on the activities of complex microbial communities, we have much to learn about the interactive and interdependent roles played by individual species in these communities. Recent advances in molecular techniques, combined with genomic information, are greatly assisting microbiologists in unraveling the diverse roles of microbes in these communities, as well as determining the importance of lateral gene transfer in complex microbial communities. Catabolic genes have the ability to spread through a microbial community in high frequencies (Top and Springael 2003; van der Meer and Senchilo 2003).

5.3 The Nature of Microbial Biofilms

Much of the microbiological degradation activity in air biofilters occurs in microbial biofilms, complex associations of microorganisms and microbial products that attach to the biofilter matrix surfaces (Lappin-Scott and Costerton 1995). Frequently, the cells are embedded in extracellular microbial polymeric matrices (Costerton et al. 1995). In the non-sterile confines of a typical biofilter, the microbial population constitutes a mixed consortium of cells located at different depths in the biofilm, which will be exposed to different environments with respect to concentration of oxygen, carbon dioxide, air contaminants, and metabolites from other organisms as well as cell components from dying and lysing organisms. Clearly, understanding the biological states and functions of different species in such a complex system is daunting. Nevertheless, substantial scientific information is being gleaned by the use of methods including scanning electron microscopy and confocal scanning laser microscopy, and other tools characterizing biofilm architecture and surface topography to facilitate three-dimensional computerized reconstruction of investigated biofilms. The modified Robbins device provides opportunities to investigate the abilities of microbes to colonize surfaces having different degrees of roughness (Elvers and Lappin-Scott 2000).

Engesser and Plaggemeier (2000) characterized the establishment of a biofilm as follows: macromolecules are first adsorbed to the surface, and microbes are then transported to the surface and bind through initial adhesion. Growth of these microbes results in generation of adhesive polysaccharide structures, leading to biofilm formation. Under conditions such as exist in biofilters, the biofilm has been characterized as having a rugged filamentous shape, which accommodates bacteria in matrix-enclosed microcolonies, whereas other locations may have regions of lower densities, even containing channels, through which water, substrates (contaminants), and oxygen may more easily access biofilm organisms (Costerton et al. 1994; deBeer et al. 1994; Zhang et al. 1995). Typical mixed-species biofilms contain complex associations of bacteria, algae and fungi, which are all excellent colonizers of surfaces. Fungal hyphal slimes and bacterial exopolysaccharides (EPS) contribute to the anchorage and desiccation–retardation functions of biofilms, and serve as nutrients and support for other organisms.

Hence, in general, it is challenging to place fine control on microbial processes occurring in air biofilters, and to a large extent microbiological techniques are applied to monitoring the microbes that become established under the particular circumstances. Provided the general environmental conditions are present to support microbiological activity, this strategy can be expected to work for easily degraded air odors and contaminants. For contaminants that are more difficult to degrade, optimizing conditions for appropriate microbial selection, biomass growth, and contaminant metabolism by transforming strains will require a greater understanding of the contaminant-degrading microbial species, including both their nutritional and physicochemical requirements. Such information will direct the biofilter researcher to attempt to aspire to achieving conditions/parameters in the biofilter that support growth and metabolism by the required strains. In the case of air biofilters, gaining such a scientific understanding is not trivial, as we are dealing with several levels of poorly understood complexities, the mixed culture community and the biofilm environment. Nevertheless, with contaminants that are difficult to degrade, there is an opportunity to achieve greater biodegradation effectiveness by defining and controlling the required microbial parameters.

5.4 Biodegradation of Air Pollutants

5.4.1 Biokinetics

Microorganisms involved in biological air treatment systems exhibit responses similar to those observed in other bioreactor systems. In mixed microbial populations, as found in biological waste gas treatment bioreactors, the different species grow or even die at different rates, and the profile of the consortium evolves with time. New species possibly arrive through the air and establish in the reactors, depending on their adaptability to the prevailing conditions. The growth rate of each microbial species depends on the concentration of the carbon and energy source, and environmental conditions such as temperature, pH, water activity, and available nutrients. Figure 5.1 shows the biodegradation process in a contaminated-air treatment bioreactor. The relation between growth and the substrate concentration can be described by the Monod model (Monod 1942):

$$\mu = \frac{\mu_{\max} C_L}{K_s + C_L} \tag{5.1}$$

where C_L is the substrate concentration ($\mathrm{g\,l^{-1}}$), $\mu_{\max}$ maximum cell growth ($\mathrm{h^{-1}}$), and K_s saturation constant.

According to the Monod model, at high pollutant concentration ($C_L \gg K_s$) growth attains $\mu_{\max}$ but at low concentration ($C_L \ll K_s$), growth is a first-order reaction. In biological air treatment systems, the interest is to sustain high elimination capacity (EC), so the pollutant degradation rate is more important than the growth rate. Pollutant uptake is directly correlated with biomass concentration, provided that all the biomass is active. Therefore, increasing biomass

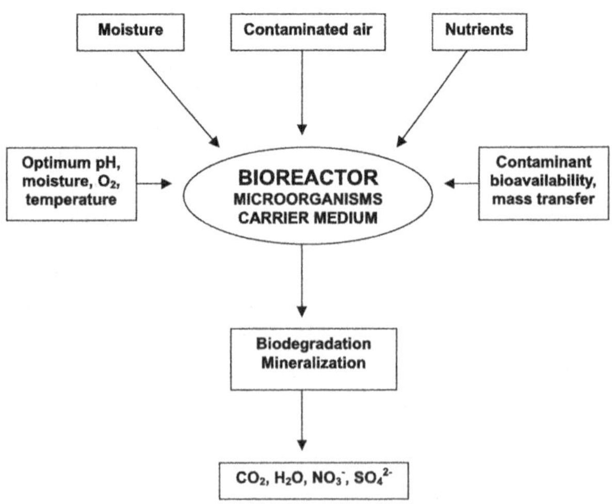

Fig. 5.1. Biodegradation process in waste gas treatment bioreactors

concentration in waste gas bioreactor systems allows improved performance to a point where other factors become limiting.

5.4.2 Biodegradation of Organic Compounds

Both bacteria and fungi have been widely studied for their ability to degrade a range of environmental pollutants, and biodegradation involves the breakdown of organic compounds either through biotransformation into less complex metabolites or through mineralization into inorganic minerals, H_2O, CO_2 (aerobic) or CH_4 (anaerobic). The biochemical pathways/enzymes required for the initial transformation stages are often specific for particular, target environmental contaminants, converting them to metabolites that can be assimilated into more ubiquitous, central bacterial pathways. Naturally produced compounds, such as those found in odorous air (water treatment plants, composting facilities, food and feed factories, etc.), are in general highly biodegradable in low concentrations. They can be treated using the natural population found in organic media, or can be inoculated in other systems with a diverse mixed population such as that found in activated sludge.

Complete oxidation of pollutants results in mineralization. These are the preferred reactions, as the end products are not toxic. Mineralization requires supply of stoichiometric amounts of oxygen that may be deficient when the pollutant is in high concentration in the liquid. In some cases, partial oxidation may result in the production of volatile intermediates that exit in the exhaust gas. In the absence of adequate oxygen, the microbes may derive some energy from the partial oxidation of certain pollutants, thus producing intermediates that may accumulate in the liquid or get carried away with the exiting air if they are volatile (Christen et al. 2002).

Organic compounds that can be degraded by all the major contaminated-air treatment techniques include alkanes, alkenes, aldehydes, alcohols, ketones, carboxylic acids, esters, ethers, and aromatic, cyclic and heterocyclic compounds as well as compounds containing $-NH_2$ (amines), $-SH$ (mercaptans) and $-Cl$ (van Groenestijn and Hesselink 1993; Kennes and Thalasso 1998). Various bacterial and fungal cultures have been identified for the degradation of organic pollutants, for example, in the case of acetone (*Rhodococcus, Corynebacterium*), butanol (*Pseudomonas, Rhodococcus*), dichloromethane (*Hyphomicrobium*), 1,2-dichloroethane (*Xanthobacter*), ethanol (*Candida, Saccharomyces*), isopropanol (*Pseudomonas*), methanol (*Pseudomonas*), methane (*Methylomonas*), MEK (*Rhodococccus*), phenol (*Pseudomonas*), and styrene (*Nocardia, Exophiala jeanselmei, Phanerochaete*).

The extent and rate of biodegradation depend on many factors including the pH, temperature, oxygen, microbial population, degree of acclimation, accessibility of nutrients, chemical structure of the compound, cellular transport properties, and chemical partitioning in the growth medium. Some recalcitrant chemicals contain novel structural elements that seldom occur in nature and that may be incompletely transformed, as microbes lack the degradative pathway for complete degradation of these xenobiotics. While microbes may not have the metabolic pathways for mineralization of certain newly introduced synthetic chemicals, there is evidence that microorganisms have the capacity to evolve such catabolic systems over time.

The inability of microbes to mineralize particular contaminants but rather to partially transform them typically means that these organisms require other substrates to support their growth. In such situations, the contaminants are transformed by 'co-metabolic' processes. Co-metabolism is metabolic transformation of a substrate where a second substrate serves as primary energy or carbon source.

There is no unanimously accepted mechanism for the biodegradation of organic pollutants. In a mixture of organic pollutants, the biological reactions can be competitive or noncompetitive, depending of the chemical structure of the pollutants (Devinny et al. 1999; Aizpuru et al. 2001). In some cases, one pollutant may be preferentially degraded due to its inherently greater biodegradability and/or bioavailability, while elimination of the second pollutant may be retarded. For example, the ester butyl acetate can inhibit styrene degradation in a biofilter treating butyl acetate and styrene (McGrath et al. 1999). In some cases, overall elimination is diminished by the presence of the mixture, as in BTEX (Lu et al. 2000). In some cases, proper conditions can be established to allow the simultaneous elimination of very different compounds, such as methanol and H_2S emissions found in the pulp and paper industry (Sologar et al. 2003).

5.4.3 Biodegradation of Inorganic Compounds

Microbiological methods are becoming more popular than physical and chemical methods, due to their efficient and economical applications, and the capacity of microbes to operate at around ambient temperature and at atmospheric pressure.

Removal of sulfur compounds from waste gases is required for reasons of health, safety, and corrosion during emission. Species belonging to the genera *Thiobacillus*, *Desulfobacter*, *Hyphomicrobium*, *Xanthomonas*, *Chlorobium* and others are known for their capability of transforming H_2S and SO_2-contaminated waste gases (Jensen and Webb 1995). *Xanthomonas* sp. can remove H_2S as a single gas or in the presence of sulfur-containing compounds such as dimethyl sulfide and methanethiol (Cho et al. 1992). The green sulfur bacterium *Chlorobium limicola* is capable of transforming hydrogen sulfide (H_2S) into elemental sulfur, using the following reaction under anaerobic conditions (Kim et al. 1990):

$$2nH_2S + nCO_2 \xrightarrow{h\nu} 2nS + n(CH_2O) + nH_2O \,.$$

High quantities of H_2S and SO_X, produced in various petrochemical waste gas streams, require treatment, and bacterial processes are now being commercialized that can simultaneously purify these streams and convert these by-products to elemental sulfur. The Thiopaq process ($H_2S + 1/2O_2 \rightarrow S^0 + H_2O$)

describes a desulfurization process for the production of elemental sulfur from H_2S-containing gas streams by sulfur-oxidizing bacteria (Janssen and Buisman 2001; Sipma et al. 2003). In this process, gas streams are first scrubbed with an aqueous washing liquid, where sulfur components are dissolved in an aqueous phase. Sulfide-oxidizing bacterial species such as *Thioalcalobacteria*, *Thiobacilli* or *Thiocalovibrio* can then convert sulfides to elemental sulfur in the presence of an electron acceptor at neutral pH. The reaction uses the following steps:

$$H_2S + OH \rightarrow HS^- + H_2O \quad \text{(scrubber)}$$
$$HS^- + 1/2O_2 \rightarrow S^0 + OH^- \quad \text{(sulfur-oxidizing bacteria)}$$

A two-step process (BIO-SR) involving an absorber and a bioreactor containing *Thiobacillus ferroxidans* to carry out biological oxidation has been described using the following reaction (Satoh et al. 1988):

$$H_2S + Fe_2(SO_4)_3 \rightarrow S\downarrow +2FeSO_4 + H_2SO_4 \quad \text{(absorber)}$$
$$2FeSO_4 + H_2SO_4 + 1/2O_2 \rightarrow SO_4^{2-} + S^0 + 2H_2O \quad (\textit{T.ferroxidans})$$

The biodegradation of ammonia takes place through the process of nitrification, which is a sequential transformation of NH_3 to nitrite (NO_2^-) and then nitrate (NO_3^-) in a two-step process. First, ammonia-oxidizing autotrophic bacteria (e.g., *Nitrosomonas* sp.) derive reductant required for energy and biosynthesis by oxidizing NH_3 to nitrite (NO_2^-), and carbon for growth from CO_2. The NO_2^- produced by ammonia-oxidizing bacteria is subsequently oxidized to NO_3^- by NO_2^--oxidizing bacteria (e.g., *Nitrobacter* sp.). Trace amounts of NO, N_2O and nitrogen may also be produced under suitable operating conditions. However, nitrification rates are reduced under adverse conditions such as low oxygen, high ammonia concentration, and high temperature (Chung et al. 1997; Lee et al. 2001). Oxidation of ammonia to nitrate by autotrophic bacteria can be described as follows:

$$NH_4^+ + 0.5O_2 \rightarrow NO_2^- + 2H^+ + H_2O + \text{energy} \quad (\textit{Nitrosomonas})$$
$$NO_2^- + 0.5O_2 \rightarrow NO_3^- + \text{energy} \quad (\textit{Nitrobacter})$$

Heterotrophic nitrifiers can also degrade ammonia to nitrate. However, the process is not coupled to energy generation, and the nitrification rate of heterotrophs is significantly lower than that of autotrophs. Heterotrophic bacteria use the following pathway:

$$NH_4^+ \rightarrow NH_2OH \rightarrow NO_2^- \rightarrow NO_3^-$$
$$(\textit{Alcaligenes, Pseudomonas, Arthrobacter})$$

Sometimes the process does not go to completion, which may result in the final product being nitrite that is available to the autotrophic nitrifiers, offering the possibility of simultaneous autotrophic and heterotrophic biodegradation (Stepanov and Korpela 1997). Nitrate produced during the nitrification process can be reduced to N_2 in a process called denitrification. This process occurs in the areas of low oxygen potential or anoxic zones usually formed in the inner regions of biofilms.

Denitrifying organisms are capable of utilizing NO as an electron acceptor, oxidizing different types of organic or inorganic substrates. NO is an intermediate in the sequential reduction pathway of nitrate to N_2. The enzyme NO reductase, present in typical denitrifiers, is relatively insensitive to O_2, and NO consumption by denitrification can take place in both anaerobic and aerobic conditions. The availability of reductant, mostly organic carbon compounds, is important during the denitrification process. The rate of NO uptake under aerobic condition is generally lower than that in the absence of oxygen.

$$2NO + 2H^+ + 2e^- \Leftrightarrow N_2O + H_2O$$

Microalgae have the capability of simultaneously removing NO_x and CO_2 from waste gases. Several strains of marine microalgae have been found to effectively remove NO from flue gas in bubble column photobioreactors (Yoshihara et al. 1996; Nagase et al. 1997). NO in the gas phase is first dissolved in the aqueous phase, and then oxidized and assimilated by the algal cells. NO permeates directly into the algal cells by diffusion and, for cell growth, NO taken up in the algal cells is preferentially utilized as a nitrogen source as compared to nitrate.

5.5 Factors Affecting Microbial Degradation of Air Contaminants

In bioreactors, the pollutants diffuse into the liquid phase where indigenous or inoculated microorganisms degrade them into products such as CO_2, H_2O, inorganic metabolites, or elemental nitrogen or sulfur. Natural packing material contains a sufficient number of different microorganisms to initiate biodegradation of contaminants, and process efficiency is generally enhanced following the growth of active organisms during the adaptation or acclimation period, which varies from less than 10 days to more than 1 month. As with any bioprocess, optimizing the environmental conditions in bioreactor processes is a central goal in order that the microbial, physiological, and biochemical activities are directed toward biodegradation of the target contaminants. Factors that influence microbial growth and biodegradation potential include bioavailability, structure and toxicity of the compound, nutrients, and environmental conditions.

5.5.1 Bioavailability

For efficient biodegradation of air contaminants to occur, the contaminants must be bioavailable to the degrading cells. The biodegradation rate of a contaminant depends on the rate of contaminant uptake and mass transfer by the microorganisms. The bioavailability of a contaminant is influenced by a number of factors such as desorption, diffusion, and dissolution. The decrease in bioavailability due to long-term contamination of soil, often referred to as aging or weathering, is a result of chemical oxidation reactions, and slow chemical diffusion of the contaminant into small pores incorporating contaminants into the organic matter. The use of chemical surfactants or biosurfactants during the biodegradation process may help overcome bioavailability problems (van Hamme et al. 2003). Chemical surfactants have the ability to emulsify or pseudosolubilize poorly water-soluble compounds, thus potentially improving their accessibility to microorganisms. Properties of chemical surfactants that influence their efficacy include charge (nonionic, anionic or cationic), hydrophilic–lipophilic balance (HLB, a measure of surfactant lipophilicity), and critical micellar concentration (CMC, the concentration at which surface tension reaches a minimum and surfactant monomers aggregate into micelles). Micellarization of the contaminant (at or above the surfactant CMC) may sometimes prevent access to the contaminant by the microorganism. Nonylphenol ethoxylated surfactants with HLB 12 and 13 can substantially enhance biodegradation of hydrocarbons at surfactant concentrations greater than CMC value (van Hamme and Ward 1999). The low molecular weight biosurfactants (glycolipids, lipopeptides) are more effective than the high molecular weight biosurfactants (amphipathic polysaccharides, proteins, lipopolysaccharides, lipoproteins) in lowering the interfacial and surface tensions (Rosenberg and Ron 1999).

The hydrophobicity and molecular structure of some compounds may also affect pollutant uptake by the microorganisms, and understanding the mechanisms of their uptake by the degrading microbes and developing strategies to promote or accelerate their accession represent important aspects of effective remediation processes. We need to develop strategies for improving the bioavailability of the many hydrophobic contaminants that have extremely low water solubility and tend to adsorb to the media particles and persist there.

5.5.2 Nutritional

Microorganisms use the pollutants to draw chemical energy for growth and the maintenance of biological activity. Microbial activity can be realized under aerobic or anaerobic conditions but as the pollutants are in air, oxygen plays a pivotal role in the biotransformation (Wackett and Hershberger 2001). Several types of reactions take place in biological air treatment systems. Microorganisms require a range of nutrients for growth and activity to carry out biodegradation of pollutants. Nutrients may be naturally present in organic filter beds, but are added to the synthetic or inert beds. Carbon and energy sources required for microbes may be derived from the contaminant, while other nutrients such as nitrogen, phosphorus, minerals and trace elements should be supplied to microorganisms in the case of artificial packing material. Minerals and trace metals needed for growth include N, S, P, Ca, K, Na,

Table 5.2. Important nutrients required for the growth and activity of microorganisms employed in waste gas treatment

Nutrients	Nutrient sources	References
Nitrogen	Ammonium sulfate, ammonium nitrate, ferrous ammonium citrate	Wübker and Friedrich (1996); Cox et al. (1997); Juneson et al. (2001); Granström et al. (2002)
Phosphate	Dipotassium hydrogen phosphate, potassium dihydrogen phosphate	
Minerals	Magnesium sulfate, calcium chloride, ferrous sulfate	
Trace metals	Zinc sulfate, cobalt chloride, manganese chloride, copper sulfate, ferric chloride, sodium molybdate, borate, nickel chloride	
Vitamins	Nicotinic acid, Ca-pentothenate, cyanocobalamine, inositol, *p*-aminobenzoate, thiamine-HCl, pyridoxine-HCl, biotin, riboflavin, folic acid, thioctic acid	

Mg, Mn, Fe, Co, Zn, and Mo (Table 5.2). Biomass yield often depends on the nature of nutrients.

Microbial growth is a consequence of uptake and metabolism of the pollutants. These nutrients can be found in the traditional supports in biofilters, and are available in the incoming water in certain suspended growth reactors but need to be added in the other systems to sustain growth. When nutrients are limited, new biomass is formed at low rates from the dead cells. Excessive growth and undesirable plugging in biofilters or biotrickling filters occur when nutrients are continually supplied. Clogging due to extensive biomass formation is more common with inert material where a nutrient solution is fed at regular interval. Limiting the supply of some nutrients, such as phosphate and potassium ion or nitrogen, to the bioreactor (Wübker and Friedrich 1996) or control using protozoan predation (Cox and Deshusses 1999) are possible means to reduce biomass growth.

5.5.3 Environmental

In general, microorganisms have a specific optimum pH range for their activities: pH 5–9 for bacteria, and pH 2–7 for fungi. Microbial activity in mixed populations generally slows down below pH 4 or above pH 8. Compost beds generally have a pH between 7 and 8, a range preferred by most microorganisms. Carbon dioxide or sulfate evolved in metabolic activities of aerobic microorganisms have a tendency to reduce the pH of the medium, and therefore additional pH control is accomplished by addition of an alkali solution. pH fluctuations of more that 2–3 pH units are generally harmful to microbial activity.

Temperature control is important in biofiltration to avoid thermal shocks. There are three general temperature classes of aerobic microorganisms: psychrophilic (grow best < 20 °C), mesophilic (achieve highest growth rates at 20–40 °C), and thermophilic (grow best > 45 °C). Biological activity roughly doubles for each 10 °C rise in temperature, up to an optimum of about 37 °C for mesophilic bacteria. Although temperatures of contaminated air from industrial plants are often higher than the mesophilic range, most air contaminant treatment studies have been done under mesophilic (15–40 °C) conditions. The recommended temperature range for biofiltration is between 15 and 40 °C. Therefore, if the pollutant gas temperature is above 40 °C, the gas should be cooled before it enters the biofilter. Similarly, for cold air below 10 °C, heating of the gas stream to a desirable temperature is needed. The cost of controlling temperature to within the mesophilic range is often not economical and for practical applications, bioreactors operating efficiently under psychrophilic and thermophilic conditions need to be developed for treating waste gases at a wide range (10–65 °C) of temperature (Cox et al. 2001; Kong et al. 2001; Datta et al. 2004).

The moisture content of the bioreactor media is a critical factor for microorganisms, which require water to carry out their normal metabolic activities. Moisture levels in a biofilter are often maintained through humidification of the inlet airstream. Too little moisture content causes drying of the bed and channeling, which deprives microbes of water, causing a significant reduction in the contaminant degradation rate. Optimal water levels vary with different filter media, depending on media surface area, porosity and other factors, and moisture content for optimal operation of a biofilter should be within 40–60%. Excessive water supply inhibits oxygen transfer and the availability of hydrophobic pollutants to the biofilms. It also increases backpressure due to the reduced void volume, and creates channeling within the bed.

The predominant microorganisms in waste air treatment are aerobic and require oxygen for metabolism, and therefore oxygen is another parameter vital to the biofilter operation. At least 5–15% oxygen in the inlet gas stream is required for aerobic heterotrophic bacteria present in filter beds. Biofiltration can be oxygen-limited in high contaminant-loading or high moisture-containing systems. Generally, air is not limited in bioreactors for air pollution treatment systems. However, in overloaded filters it may be a limitation, resulting in the formation of undesirable intermediates.

Microbial activity is also influenced by the accumulation of inorganic products and intermediates of incompletely oxidized organic compounds. A long lag phase, diauxie or catabolite repression may also result in poor biodegradation of the contaminants (van Groenestijn and Hesselink 1993). Intermittent feeding creates the problems with lag phases, whereas diauxie may affect degradation of a poorly degradable compound if an easily degradable substrate is present.

5.6 Genetic Approaches for Improved Microorganisms

In order to design improved contaminant-degrading microbes, we need to understand current metabolic processes, and perhaps create new metabolic routes that are characterized by exhibiting improved substrate fluxes through the pathways, so as to avoid accumulation of inhibitory intermediates (Pieper and Reineke 2000). Substrate specificities of the catabolic enzymes need to be broadened, and abilities of microbes to better access more hydrophobic contaminants need to be improved. Newly created strains need to be genetically stable with respect to their biodegradative abilities. Such a rational optimization of known catabolic pathways, and creation of new pathways will rely on the rapidly growing knowledge based on biochemical/pathway genetics.

The available information on the structure–function relationship of important catabolic enzymes offers further possibilities for improvements in their activities. Novel strategies are being investigated with a view to designing new biocatalysts for environmental remediation (Parales et al. 2002). Approaches may be directed to increasing expression or modifying the specificity of a target enzyme, or could involve optimization of an entire pathway through directed evolution. This knowledge can provide us with the requisite information to test, design and engineer biocatalysts with improved substrate specificities, reaction rates or other desired catabolic properties, and ultimately to engineer improved catabolic pathways for waste gas remediation. Limiting factors in biodegradation can be overcome by improving the activities of indigenous organisms, isolating or engineering better-degrading strains, and constructing microbial consortia of suitable organisms in addition to overcoming contaminant bioavailability.

The disposal of recombinant organisms in nature is a big concern, given the potential for recombinant strain proliferation, modification or recombinant gene transfer processes in the environment (Watanabe 2001). Since the potential to use recombinant organisms in any remediation process has been constrained, a better understanding of the nature of bioreactor filter beds as host media for the retention, survival and propagation of recombinant microorganisms is required.

Understanding the biochemical and physiological aspects of the processes will provide us with the requisite knowledge and tools to optimize these processes, to control key parameters, and to make the processes more reliable. Combined physiological and genetic approaches provide us with a deeper understanding of many aspects of contaminant biodegradation. New information on the metabolic diversity of biodegradative organisms has lead to the discovery of new metabolic routes, and approaches to optimizing metabolic activities through manipulation of rate-limiting steps. Advances in molecular techniques and availability of sophisticated methods for physiological manipulation of microorganisms will provide opportunities for further strain and process improvement.

5.7 Monitoring of Microbial Processes

Microbiological processes in waste gas treatment bioreactors can be monitored by characterization of the morphological, physiological, biochemical or molecular characteristics of microorganisms. Traditional methods include isolation and cultivation of microbes on solid media, most probable number (MPN) assays, and the relatively new BIOLOG method that determines substrate utilization patterns. Respirometry tests and pollutant mineralization studies involving measurements of oxygen consumption or CO_2 production can provide useful information on the biodegradability potential of microorganisms. Pure culture experiments in the laboratory are essential for detailed analysis of the physiology, metabolism and genetics of microorganisms. Bioassays can be used to measure the toxicity of a contaminant, or inhibition of enzymes such as dehydrogenases, lipases, ureases, acid and alkaline phosphatases, and catalases. The rates of contaminant degradation and formation of products can also be followed to monitor the process.

Physiological conditions in a laboratory culture are generally different from those in the natural environment. Molecular approaches such as denaturing gradient gel electrophoresis (DGGE), terminal restriction fragment length polymorphism (T-RFLP), fluorescence in situ hybridization (FISH), and polymerase chain reaction (PCR)-amplified 16sRNA can be used for the screening of dominant phylogenetic signatures from the contaminated source (Watanabe and Hamamura 2003). Fragments of genes coding catabolic enzymes in the degradation pathways can be amplified from environmental DNA samples. Use of catabolic genes is considered to be an attractive method for analyzing the diversity of potential catabolic microbes and monitoring in situ pollutant degradation. Reverse sample genome probing (RSGP) has been used to track specific microorganisms in a quantitative manner (Hubert et al. 1999).

Sakano and Kerkhof (1998) used molecular techniques to identify bacteria in a laboratory-scale ammonia biofilter. Both 16S rRNA and ammonia monooxygenase (*amoA*) genes were used to characterize the heterotrophic and ammonia-oxidizing bacteria collected from the biofilter during a 102-day experiment. The results indicated that the overall diversity of the heterotrophic microbial population decreased by 38% at the end of the experiment. The community structure of the heterotrophic population shifted from predominantly members of two subdivisions of the *Proteobacteria* (the beta and gamma subdivisions) to members of one subdivision (the gamma subdivision).

Domain-, class-, and subclass-specific rRNA-targeted probes have been applied to investigate the microbial communities of three industrial- and three laboratory-scale biofilters (Friedrich et al. 1999). *Actinobacteria* (gram-positive) and α-*Proteobacteria* were the most dominant groups, and in situ hybridization proved to be a practical tool for microbiological studies of biofiltration systems.

Microorganisms utilize various cooperative interactions with other species for their growth in natural ecosystems. One population or a few different populations that best adapt to the environmental conditions become predominant, while others either exist as minorities or eventually die off. In situ biodegradation of contaminants can also be assessed by the detection of contaminant-specific transformation products, metabolic biomarkers (Smets and Pritchard 2003). Metabolic biomarkers have been detected during transformation of alkanes and aromatic hydrocarbons (Gieg and Suflita 2002; Reusser et al. 2002). The detection of specific catabolic genes using real-time PCR also provides a possible biomarker, which can be correlated with pollutant transformation rates (Johri et al. 1999).

5.8 Conclusions

Significant opportunities for application of microbial techniques for odor and air pollution control will continue to evolve with the changing demands of environmental regulations and social acceptance of biotechnological methods. We need to continue to unravel the complex metabolic pathways that microbes have evolved to degrade a range of contaminants, and to understand the nature of rate-limiting steps, bottlenecks, and underlying genetic and biochemical regulatory mechanisms. We must recognize that some chemical species are inherently intractable to enzyme transformation, and we should be open to combining chemical or physical strategies with biological systems to achieve overall effective treatment.

Better methods for monitoring and assessing the progress and effectiveness of microbial biodegradation processes are needed to be devised on a continuous basis at both the laboratory- and full-scale levels. However, the implementation of contaminated-air remediation technologies will require a combination of expertise from microbiologists, chemists and environmental engineers. Current research on molecular and biochemical aspects of microbiological processes will provide the foundation for the development and implementation of new, improved and more dependable biotechnological processes for waste gas treatment.

References

Aizpuru A, Malhautier L, Roux JC, Fanlo JL (2001) Biofiltration of a mixture of volatile organic emissions. J Air Waste Manage Assoc 51:1662–1670

Arnold M, Reittu A, von Wright A, Martikainen PJ, Sihko M-L (1997) Bacterial degradation of styrene in waste gases using a peat filter. Appl Microbiol Biotechnol 48:738–744

Arp DJ, Sayavedra-Soto LA, Hommes NG (2002) Molecular biology and biochemistry of ammonia oxidation by *Nitrosomonas europaea*. Arch Microbiol 178:250–255

Bohn HL (1992) Considering biofiltration for decontaminating gases. Chem Eng Prog 88:34–40

Braun-Lüllemann A, Aajcherczyk A, Hüttermann A (1997) Degradation of styrene by white-rot fungi. Appl Microbiol Biotechnol 47:150–155

Burgess JE, Parsons SA, Stuetz RM (2001) Development in odor control and waste gas treatment biotechnology: a review. Biotechnol Adv 19:35–63

Cho K-S, Hirai M, Shoda M (1992) Degradation of hydrogen sulfide by *Xanthomonas*sp. strain DY44 isolated from peat. Appl Environ Microbiol 58: 1183–1189

Christen P, Domenech F, Michelena G, Auria R, Revah S (2002) Biofiltration of volatile ethanol using sugar cane bagasse inoculated with *Candida utilis*. J Hazard Mater 89:253–265

Chung YC, Huang C, Tseng CP (1997) Biotreatment of ammonia from air by an immobilized *Arthrobacter oxydans* CH8 biofilter. Biotechnol Prog 13:794–798

Chung YC, Huang C, Liu CH, Bai H (2001) Biotreatment of hydrogen sulfide and ammonia containing waste gases by fluidized bed bioreactor. J Air Waste Manage Assoc 51:163–172

Costerton JW, Lewandowski Z, deBeer D, Caldwell D, Korber D, James G (1994) Biofilms, the customized microniche. J Bacteriol 176:2137–2142

Costerton JW, Lewandowski Z, Caldwell DE, Korber D, Lappin-Scott HM (1995) Microbial biofilms. Annu Rev Microbiol 49:711–745

Cox H, Deshusses MA (1999) Biomass control in waste air biotrickling filters by protozoan predation. Biotechnol Bioeng 62:216–224

Cox H, Moerman RE, van Baalen S, van Heiningen WNM, Doddema HJ, Harder W (1997) Performance of a styrene-degrading biofilter containing the yeast *Exophiala jeanselmei*. Biotechnol Bioeng 53:259–266

Cox H, Sexton T, Shareefdeen Z, Deshusses M (2001) Thermophilic biotrickling filtration of ethanol vapors. Environ Sci Technol 35: 2612–2619

Datta I, Fulthorpe R, Allen DG (2004) High temperature biofiltration of H_2S. In: Proc Air and Water Management Association Conf, 22–25 June 2004, Indianapolis, IN

DeBeer D, Stoodley P, Roe F, Lewandowski Z (1994) Effects of biofilm structure on oxygen distribution and mass transport. Biotechnol Bioeng 43:1131–1138

Deshusses M, Johnson C (2000) Development and validation of a simple protocol to rapidly determine the performance of biofilters for VOC treatment. Environ Sci Technol 34:461–467

Devinny JS, Deshusses MA, Webster TS (1999) Biofiltration for air pollution control. Lewis, Boca Raton

Dua M, Singh A, Sethunathan N, Johri AK (2002) Biotechnology and bioremediation: successes and limitations. Appl Microbiol Biotechnol 59:143–152

Elvers KT, Lappin-Scott HM (2000) Biofilms and biofouling. Encyclopedia of microbiology, vol 1, 2nd edn. Academic Press, San Diego, pp 471–485

Engesser KH, Plaggemeir T (2000) Microbiological aspects of biological waste gas purification. In: Rehm H-J, Reed G (eds) Biotechnology, vol 11C, 2nd edn. Wiley/VCH, Weinheim, pp 275–302

Ergas SJ, Kinney K, Fuller ME, Scow KM (1994) Characterization of a compost biofiltration system degrading dichloromethane. Biotechnol Bioeng 44:1048–1054

Ergas SJ, Schroeder ED, Chang DPY, Morton RL (1995) Control of volatile organic compound emission using a compost biofilter. Water Environ Res 67:816–821

Friedrich U, Naismith MM, Altendorf K, Lipski A (1999) Community analysis of biofilters using fluorescence in situ hybridization including a new probe for the *Xanthomonas* branch of the class *Proteobacteria*. Appl Environ Microbiol 65:3547–3554

García-Peña I, Hernández S, Auria R, Favela E, Revah S (2001) Biofiltration of toluene by the fungus *Scedosporium apiospermum*. Biotechnol Bioeng 76:61–69

Gieg LM, Suflita JM (2002) Detection of anaerobic metabolites of saturated and aromatic hydrocarbons in petroleum-contaminated aquifers. Environ Sci Technol 36:3755–3762

Granström T, Lindberg P, Nummela J, Jokela J, Leisola M (2002) Biodegradation of VOCs from printing press air by an on-site pilot plant bioscrubber and laboratory scale continuous yeast cultures. Biodegradation 13:155–162

Grommen R, Verstraete W (2002) Environmental biotechnology: the ongoing quest. J Biotechnol 98:113–123

Harayama S (2001) Environmental biotechnology. Curr Opin Biotechnol 12: 229–230

Hubert C, Shen Y, Voordouw G (1999) Composition of toluene-degrading microbial communities from soil at different concentrations of toluene. Appl Environ Microbiol 65:3064–3070

Janssen AJH, Buisman CJN (2001) Process for biological removal of sulphide. US patent no 6,221,652

Jensen AB, Webb C (1995) Treatment of H_2S-containing gases: a review of microbiological alternatives. Enzyme Microb Technol 17:2–10

Johri AK, Dua M, Singh A, Sethunathan N, Legge RL (1999) Characterization and regulation of catabolic genes. Crit Rev Microbiol 25:245–273

Juneson C, Ward OP, Singh A (2001) Microbial treatment of styrene contaminated air stream in a biofilter with high elimination capacities. J Ind Microbiol Biotechnol 26:196–202

Kennes C, Thalasso F (1998) Waste gas biotreatment technology. J Chem Technol Biotechnol 72:303–319

Kennes C, Veiga C (2001) Conventional biofilters. In: Kennes C, Veiga C (eds) Bioreactors for waste gas treatment. Kluwer, Dordrecht, pp 47–98

Kim BW, Kim IK, Chang HN (1990) Bioconversion of hydrogen sulfide by free and immobilized cells of *Chlorobium thiosulfatophilum*. Biotechnol Lett 12:381–386

Kim NJ, Sugano Y, Hirai M, Shoda M (2000) Removal characteristics of high load ammonia gas by a biofilter seeded with a marine bacterium, *Vibrio alginolyticus*. Biotechnol Lett 22:1295–1299

Kirchner K, Hauk G, Rehm H-J (1987) Exhaust gas purification using immobilized monocultures (biocatalysts). Appl Microbiol Biotechnol 26:579–587

Kong Z, Farhana L, Fulthorpe R, Allen DG (2001) Treatment of volatile organic compounds in a biotrickling filter under thermophilic conditions. Environ Sci Technol 35:4347–4352

Lappin-Scott HM, Costerton JW (1995) Microbial biofilms. Cambridge University Press, Cambridge

Lee DH, Lau AK, Pinder KL (2001) Development and performance of an alternative biofilter system. J Air Waste Manage Assoc 51:78–85

Lu C, Chu W, Lin MR (2000) Removal of BTEX vapor from waste gases by a trickle bed biofilter. J Air Waste Manage Assoc 50:411–417

Mallakin A, Ward OP (1996) Degradation of BTEX compounds in liquid media and in peat biofilters inoculated with cultures isolated from gasoline contaminated peat. J Ind Microbiol 16:309–318

McGrath MS, Nieuwland J-C, van Lith C (1999) Case study: biofiltration of styrene and butylacetate at a dashboard manufacturer. Environ Prog 18:197–204

Min K-N, Ergas SJ, Harrison JM (2002) Hollow-fiber membrane bioreactor for nitric oxide removal. Environ Eng Sci 19:575–583

Monod J (1942) Recherches sur la croissance des cultures bactériennes. Hermann et Cie, Paris

Morikawa H, Erkin OC (2003) Basic processes in phytoremediation and some applications to air pollution control. Chemosphere 52:1553–1558

Nagase H, Yoshihara K-I, Eguchi K, Yokota Y, Matsui R, Hirata K, Miyamato K (1997) Characteristics of biological NO_X removal from flue gas in a *Dunaliella tertiolecta* system. J Ferment Bioeng 83:461–465

Oh YS, Choi SC, Kim YK (1998) Degradation of gaseous BTX by filtration with *Phaneroxhaete chrysosporium*. J Microbiol 36:34–38

Parales RE, Bruce NC, Schmid A, Wackett LP (2002) Biodegradation, biotransformation and biocatalysis. Appl Environ Microbiol 68:4699–4709

Pieper DH, Reineke W (2000) Engineering bacteria for bioremediation. Curr Opin Biotechnol 11:262–270

Reusser DE, Istok JD, Beller HR, Field JA (2002) In situ transformation of deutered toluene and xylene to benzylsuccinic acid analogues in BTEX-contaminated aquifers. Environ Sci Technol 36:4127–4134

Rosenberg E, Ron EZ (1999) High- and low-molecular-mass microbial surfactants. Appl Microbiol Biotechnol 52:154–162

Sakano Y, Kerkhof L (1998) Assessment of changes in microbial community structure during operation of an ammonia biofilter with molecular tools. Appl Environ Microbiol 64:4877–4882

Satoh H, Yoshizawa J, Kamentani S (1988) Bacteria help desulfurize gas. Hydrocarb Proc Int Edn 76:76D–76F
Shareefdeen Z, Baltzis BC, Oh YS, Bartha R (1993) Biofiltration of methanol vapor. Biotechnol Bioeng 41:512–524
Sipma J, Janssen AJH, Hulshoff Pol LW, Lettinga G (2003) Development of a novel process for the biological conversion of H_2S and methanethiol to elemental sulfur. Biotechnol Bioeng 44:1–11
Sly LI, Bryant LJ, Cox JM, Anderson JM (1993) Development of a biofilter for the removal of methane from coal mine ventilation atmosphere. Appl Microbiol Biotechnol 39:400–404
Smet E, Van Langenhove H, Verstraete W (1997) Isobutaraldehyde as a competitor of the dimethyl sulfide degrading activity in biofilters. Biodegradation 8:53–59
Smets BF, Pritchard PH (2003) Elucidating the microbial component of natural attenuation. Curr Opin Biotechnol 14:283–288
Sologar VS, Lu Zijin, Allen DG (2003) Biofiltration of concentrated mixtures of hydrogen sulfide and methanol. Environ Prog 22:129–36
Speitel GE, McLay McLay DS (1993) Biofilm reactors for treatment of gas stream containing chlorinated solvents. J Environ Eng 119:658–678
Spingo G, Pagella C, Daria F, Molteni R, de Faveri M (2003) VOCs removal from waste gases: gas phase bioreactor for the abatement of hexane by *Aspergillus niger*. Chem Eng Sci 58:739–746
Stapleton RD, Ripp S, Jimenez L, Cheol-Koh S, Fleming JT, Gregory IR, Sayler GS (1998) Nucleic acid analytical approaches in bioremediation: site assessment and characterization. J Microbiol Methods 32:165–178
Stepanov AL, Korpela TK (1997) Microbial basis for the biotechnological removal of nitrogen from flue gases. Biotechnol Appl Biochem 25:97–104
Theron J, Cloete TE (2000) Molecular techniques for determining microbial diversity and community structure in natural environments. Crit Rev Microbiol 26:37–57
Top EM, Springael D (2003) The role of mobile genetic elements in bacterial adaptation to xenobiotic organic compounds. Curr Opin Biotechnol 14:262–269
Van der Meer JR, Senchilo V (2003) Genomic islands and the evolution of catabolic pathways in bacteria. Curr Opin Biotechnol 14:248–254
Van Elsas JD, Duarte GF, Rosado AS, Smalla K (1998) Microbiological and molecular biological methods for monitoring microbial inoculants and their effects in the soil environment. J Microbiol Methods 32:133–154
Van Groenestijn JW (2001) Bioscrubbers. In: Kennes C, Veiga MC (eds) Bioreactors for waste gas treatment. Kluwer, Dordrecht, pp 133–162
Van Groenestijn JW, Hesselink PGM (1993) Biotechniques for air pollution control. Biodegradation 4:283–301
Van Groenestijn JW, Liu JX (2002) Removal of alpha-pinene from gases using biofilters containing fungi. Atmos Environ 36:5501–5508
Van Hamme J, Ward OP (1999) Influence of chemical surfactants on the biodegradation of crude oil by a mixed bacterial culture. Can J Microbiol 45:130–137

Van Hamme JD, Singh A, Ward OP (2003) Recent advances in petroleum microbiology. Microbiol Mol Biol Rev 7:503–549

Veiga MC, Kennes C (2001) Parameters affecting performance and modeling of biofilters treating alkylbenzene-polluted air. Appl Microbiol Biotechnol 55:254–258

Veir JK, Schroeder ED, Chang DPY, Scow KM (1996) Interaction between toluene and dichloromethane degrading populations in a compost biofilter. In: Proc 89th Annu Meet Exhibition Air and Waste Management Association, Tennessee, Pap 96-WP87A.07

Wackett LP, Hershberger CD (2001) Biocatalysis and biodegradation: microbial transformation of organic compounds. ASM Press, Washington, DC, pp 39–69

Ward OP, Singh A, van Hamme J (2003) Accelerated biodegradation of petroleum hydrocarbon waste. J Ind Microbiol Biotechnol 30:260–270

Watanabe K (2001) Microorganisms relevant to bioremediation. Curr Opin Biotechnol 12:237–241

Watanabe K, Hamamura N (2003) Molecular and physiological approaches to understanding the ecology of pollutant degradation. Curr Opin Biotechnol 14:289–295

Widada J, Nojiri H, Omori T (2002) Recent developments in molecular techniques for identification and monitoring of xenobiotic-degrading bacteria and their catabolic genes in bioremediation. Appl Microbiol Biotechnol 60:45–59

Woertz JR, Kinney KA, Szaniszlo PJ (2001) A fungal vapor-phase bioreactor for the removal of nitric oxide from waste gas streams. J Air Waste Manage Assoc 51:895–902

Wübker S-M, Friedrich CG (1996) Reduction of biomass in a bioscrubber for waste gas treatment by limited supply of phosphate and potassium ions. Appl Microbiol Biotechnol 46:475–480

Yoshihara K-I, Nagase H, Eguchi K, Hirata K, Miyamato K (1996) Biological elimination of nitric oxide and carbon dioxide from flue gas by marine microalgae NOA-113 cultivated in a long tubular photobioreactor. J Ferment Bioeng 82:351–354

Zhang TC, Fu Y-C, Bishop PL (1995) Competition for substrates and space in biofilms. Water Environ Res 67:992–1003

Part II

Biological Reactor Technologies

6 Biofilter Technology

INDRANI DATTA, D. GRANT ALLEN

6.1 Introduction

Over the past few decades, the impact of air pollution on human health and the environment has received an increasing amount of attention from the public, the government, and industry. As a result of increasingly stringent regulations governing releases of volatile organic compounds (VOCs), hazardous air pollutants (HAPs), and reduced sulfur compounds (RSCs), research into developing new air pollution control technologies and improving existing technologies has grown dramatically. Currently, biofiltration of air emissions is gaining attention as a low-cost, energy-efficient, and effective treatment for emissions containing low concentrations of biodegradable compounds. It has been used extensively for over the last 40 years for odor control. During the past few years, it has been used increasingly in North America for treating high-volume airstreams containing low concentrations of a range of pollutants.

The term biofilter is generally applied to the reactor for conversion of gas-phase chemical compounds to common biological degradation products of carbon dioxide, water, and mineral salts. In the reactor, contaminated air is passed through a bed of porous and moist medium, and the contaminants are sorbed to the medium surface where they are degraded by microorganisms in the medium. Microbial reactions have been used throughout the 20th century to treat wastewater and solid waste, but it is only since 1950 that this technique has been employed to treat waste gases (Pomeroy 1957).

This chapter provides an overview of the fundamentals and applications involved in biofiltration technology. Some of the basic terminology and methods to characterize system performance are described, followed by a discussion of the key operating parameters and what is known about the microbiology of these systems. The chapter ends with a discussion of some of the applications of biofilters, and where some future work is required.

6.2 Overall Process Description

In the past, biofilters were commonly constructed as open, single-bed systems. A common construction method is to excavate a 1-m-deep, appropriately sized

Shareefdeen/Singh (Eds.)
Biotechnology for Odor and Air Pollution Control

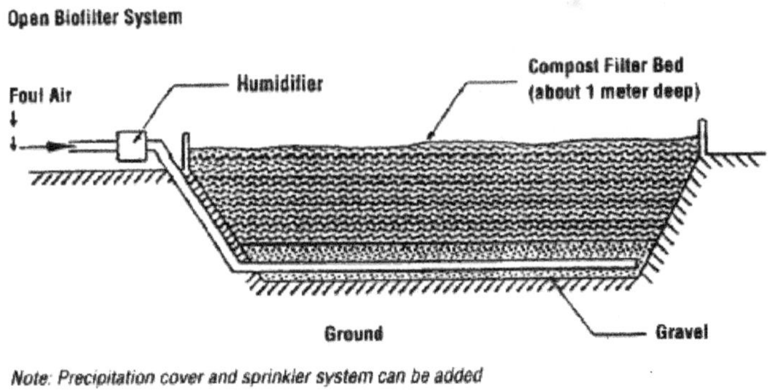

Fig. 6.1. Open biofilter system. (Williams and Miller 1992)

hole, place a gravel-covered distribution network at the bottom, and fill with the excavated soil, or alternatively with sand or compost (Fig. 6.1). Recently, fully enclosed biofilters have become more popular, in part, because of the need to comply with emission monitoring requirements. Although enclosed systems are more costly, they tend to be more engineered systems that allow for a greater contaminant loading over a given footprint area, and also provide more precise control of biofilter operation and moisture. An enclosed system consists of a humidifier and a packed filter bed populated with microorganisms through which a waste airstream is passed. Influent air is pre-humidified to maintain adequate moisture in the filter bed. Alternatively or in addition, water may be trickled over the bed. This water may contain nutrients required for the growth of microorganisms. Liquid trickling from the bottom of the filter bed may be bled off or recycled (Fig. 6.2).

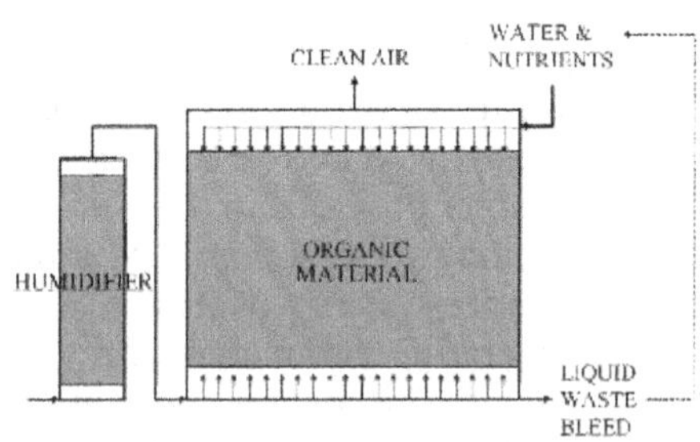

Fig. 6.2. Closed biofilter system. (McNevin and Baford 2000)

6.3 Biofiltration Terminology

The following section describes some of the common terms that are used to describe biofilters and their performance.

6.3.1
Empty Bed Residence Time

The empty bed residence time (EBRT, in seconds), defined below as the empty bed filter volume divided by the airflow rate, is a very common term used to describe biofilters, because it readily allows one to estimate the filter size for a given airflow:

$$\text{EBRT} = V_f/Q\,,$$

where V_f is the filter bed volume (m^3), and Q the airflow rate (m^3/s).

The true residence time, which is the actual time air remains in the filter bed, is defined as the total filter bed volume multiplied by the bed porosity of the filter medium, divided by the airflow rate:

$$\tau = \left(V_f \times \theta\right)/Q\,,$$

where τ is the true residence time (s), and θ = porosity = volume of void space/ volume of filter material.

6.3.2
Surface (or Volumetric) and Mass Loading Rate

The surface loading rate is defined as the volume of the gas per unit area of filter material per unit time. Volumetric loading rate is defined as the volume of gas per unit volume of filter material per unit time.

$$\text{Surface loading} = Q/A\,,$$

where A is the filter cross-sectional area (m^2).

$$\text{Volumetric loading} = Q/V_f = 1/\text{EBRT}\,.$$

The mass loading rate (surface or volumetric) is the mass of the contaminant entering the biofilter per unit area or volume of filter material per unit time. If flow remains constant through a filter, the mass loading along the length of the bed will decline as contaminant is removed. An overall mass loading rate for a system is defined based on the inlet as:

$$\text{Mass loading (surface)} = \left(Q \times C_{Gi}\right)/A\,,$$

where C_{Gi} is the inlet concentration (g/m^3).

$$\text{Mass loading (volumetric)} = \left(Q \times C_{Gi}\right)/V_f$$

6.4 Mechanism of Operation

The underlying process for biofiltration is comprised of a series of steps beginning with the transfer of contaminants from the air to the water phase, adsorption to the medium or absorption into a water film, and finally biodegradation of contaminants within a biofilm. Understanding and enhancing the steps that are rate-limiting provide opportunities to predict and improve the performance of biofilters.

6.4.1 Transfer and Partitioning of Contaminants to the Biofilm

The transfer of contaminant from the air to the water phase is generally not a rate-limiting step, and so one frequently assumes that the gas and liquid are at equilibrium. At equilibrium, the partition between air and water is generally described by Henry's law, which is

$$p = HC_{\mathrm{W}} ,$$

where p is the partial pressure of contaminant in the gas phase (Pa), C_{W} is the equilibrium concentration of contaminant in the water phase ($\mathrm{mol/L_{water}}$ or $\mathrm{g/l_{water}}$), and H is Henry's law constant (Pa L/mol or $\mathrm{Pa\,L_{water}/g}$).

Given that the total pressure of the gas phase (generally air) is approximately constant throughout the biofilter, the partial pressure of the contaminant is directly proportional to the gas-phase concentration, and so one can instead express Henry's law as a dimensionless air–water partition coefficient, K_{AW}, as follows:

$$K_{\mathrm{AW}} = C_{\mathrm{A}}/C_{\mathrm{W}} = H/(RT) ,$$

where C_{A} is the concentration of contaminant in the air phase ($\mathrm{mol/l_{air}}$ or $\mathrm{g/l_{air}}$) $= p/(RT)$ (via the ideal gas law), R is the gas constant (8.314 J/mol K), and T is the temperature (K).

There are several sources of data on the Henry's law constant, including some available on the internet (e.g., http://webbook.nist.gov/chemistry) as well as in several books (e.g., Mackay et al. 1991–1997; Yaws 2001) and on CD-ROM (Mackay et al. 2000).

In general, the elimination capacity of a biofilter declines with increasing Henry's law constant or, equivalently, K_{AW}, since this indicates a tendency to partition away from the liquid/biofilm phase where degradation is taking place. For example, Deshusses and Johnson (2000) reported on several pollutants, and showed a decline in maximum elimination capacity and critical load with an increasing Henry's Law constant. In spite of the general tendency for compounds with high K_{AW} to have lower removal rates, biofiltration can still be effective for compounds that are very hydrophobic (i.e., have relatively high K_{AW} values), such as α-pinene (e.g., Mohseni and Allen 2000) and hexane

(Spigno et al. 2003). Some researchers have suggested that such high rates of removal can be explained and modeled by replacing the air/water partition coefficient with an air/biofilm partition coefficient; for hydrophobic pollutants, this air/biofilm partition coefficient is much lower than the air/water partition coefficient, due to the presence of organic material in the biofilm that can sorb the pollutant (Davison et al. 2000; Mohseni and Allen 2000; Spigno et al. 2003). However, recent studies (Miller 2004; Miller and Allen 2004) suggest that higher partitioning to a biofilm does not lead to higher transport rates through the film, since the organic phase is not mobile and so it is only the partitioning to the aqueous (mobile) phase that matters. Further elucidation of the mechanism by which hydrophobic pollutants are transported and degraded in biofilms suggests that enhanced solubilization of a compound via rapid reaction (e.g., via exoenzymes) or other mechanisms need to be explored to explain the transport and degradation of hydrophobic pollutants in biofilms (Miller 2004).

6.4.2 Biodegradation

The biofilm is the key element of the biofilter involved in destroying contaminants. This is the mass of organisms growing on the surface of the solid medium, and carrying out the metabolic activities that transform the contaminant to harmless products (Devinny et al. 1999).

The thickness of biofilm is influenced by several factors (Cohen 2001). These include the rate of flow through the biofilter, the bedding material construction, and different treatment system designs. Biofilm thickness usually varies from tens of micrometers to more than 1 cm, although an average of 1 mm or less is usually observed (Wanner and Gujer 1984). The activity increases with the thickness of the biofilm, up to a level termed the 'active thickness'. Above this level, the diffusion of nutrients becomes a limiting factor, thus differentiating an 'active' biofilm from an 'inactive' one (Lazarova and Manem 1995).

6.5 Characterizing Biofilter Performance

6.5.1 Removal Efficiency

The removal efficiency (*RE*) is the fraction of the contaminant removed by the biofilter, expressed as a percentage:

$$\text{Removal efficiency} = \left(C_{\mathrm{Gi}} - C_{\mathrm{Go}}\right)/C_{\mathrm{Gi}} \times 100\,,$$

where C_{Gi} is the inlet concentration, and C_{Go} outlet concentration.

6.5.2 Elimination Capacity

The elimination capacity (*EC*) is the mass of contaminant degraded per unit volume of filter material per unit time. An overall elimination capacity is generally defined as:

$$\begin{aligned}\text{Elimination capacity} &= \left(C_{\text{Gi}} - C_{\text{Go}}\right) \times Q/V_{\text{f}} \\ &= \text{Volumetric mass loading} \times RE\,,\end{aligned}$$

where Q is the airflow rate (m^3/h) and V_f the filter bed volume (m^3).

6.5.3 Maximum Elimination Capacity

In most biofiltration processes, at low loadings, the elimination capacity increases with increased loading but at high loadings the rate of degradation becomes independent of loading (Fig. 6.3). The plateau observed in the plot is often referred to as the maximum elimination capacity. Pushing the biofilter above this value leads to a decrease in removal efficiency.

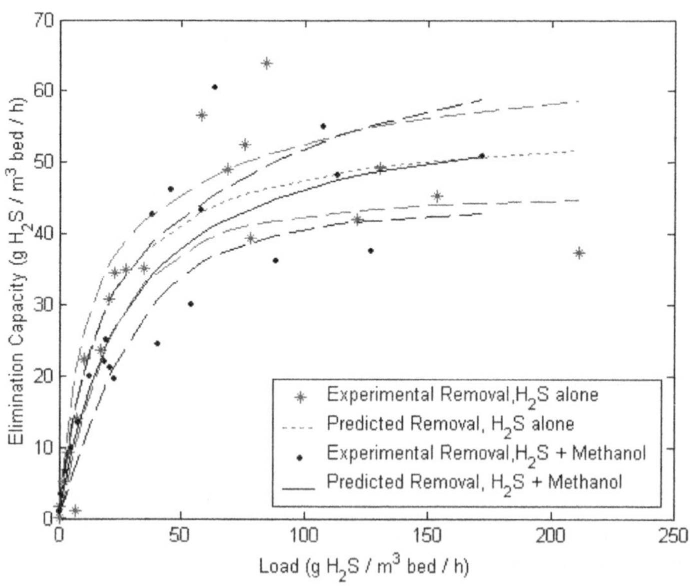

Fig. 6.3. Elimination capacity vs. loading for a biofilter treating hydrogen sulfide. The plateau at high loadings is indicative of the maximum elimination capacity for this biofilter. (Sologar et al. 2003)

6.6 Factors Affecting Biofilter Performance

6.6.1 Packing Media

To maximize the biodegradation of airborne contaminants, several factors can be optimized, many of which focus on the packing media. Many biofiltration companies have proprietary media that are designed to provide optimal performance through optimizing: high surface area for biofilm growth, long-term physical stability, low pressure drop, good moisture retention, pH buffering capacity, and nutrients. Traditional natural medium components that are frequently used for simple biofilters include compost, peat, wood chips, fertilizer, and soil. Some biofilters are also packed with adsorbents such as activated carbon. Although these adsorbents can be helpful in that they can reduce the quantity of contaminant that escapes during the microbiological acclimation period (Bishop and Govind 1995), and have potential to dampen peak loads if the adsorbent is not coated with a deep biofilm, they do not improve performance during steady-state operation (Mohseni et al. 1998).

6.6.2 Moisture Content

The moisture content of the filter bed is a critical factor for biofilter effectiveness, because microorganisms require water to carry out their normal metabolic activities (Ottengraf 1986; Shimko et al. 1988; Marsh 1994). Too little moisture content causes drying of the bed, along with the development of fissures that cause channeling and short circuiting. This also deprives microorganisms of water, causing a significant reduction in the biodegradation rate. Too much water inhibits transfer of oxygen and hydrophobic pollutants to the biofilm, thereby promoting the development of anaerobic zones within the bed and limiting the reaction rate. Too much water can also result in foul smelling emissions due to the lack of oxygen, increasing backpressure due to the reduced void volume, and channeling of the gas within the bed.

Optimal water levels vary with different filter media, depending on medium surface area, porosity, and other factors (Hodge et al. 1991). Filter moisture content for optimal operation of the biological filter should be within 30–60% by weight, depending on the medium used (Ottengraf 1986, 1987; van Lith et al. 1990). Moisture levels in a biofilter are often maintained through prehumidification of the inlet gas stream. Also, it is often necessary to provide direct application of water to the bed through a sprinkler system at the top of the bed. More advanced controls include the use of load cells that sense the weight of filter bed (van Lith et al. 1990; Rozich 1995) and are connected to sprinkler controls. Supplemental moisture adjustments may be required because bio-oxidation is an exothermic reaction, and so drying can occur within the bed. Drying of the packing material can lead to localized dry spots, and can result in non-uniform gas distribution and reduction in the activity of microorganisms.

From the perspective of adding moisture, it is often advantageous to have the flow of waste gas downward (van Lith et al. 1990). Since most of the drying occurs at the entrance to the filter bed, drying at the top is easily handled through direct water addition and flow from the top to the bottom. Downward flow also helps when too much water is added (either directly or due to humid gas cooling), since the water will flow by gravity co-currently with the gas out the bottom of the filter.

6.6.3 Temperature

Temperature control is also very important in biofiltration to avoid thermal shocks. There are three general temperature classes of aerobic microorganisms: psychrophilic microorganisms, which grow best below a temperature of 20 °C; mesophilic microorganisms, which achieve highest growth rates at 20–40 °C; and thermophilic organisms, which grow best above 45 °C. Biological activity roughly doubles for each 10 °C rise in temperature, up to an optimum of about 37 °C for mesophilic bacteria. The recommended temperature range for biofiltration is between 15 and 40 °C (Leson and Winter 1991; Bohn 1992). This means that if the pollutant gas temperature is above 40 °C, then the gas should be cooled before it enters the biofilter. Similarly, for cold air below 10 °C, the heating of the gas stream to a desirable temperature is needed because microorganisms are relatively inactive at low temperatures.

The cost of controlling temperature to within the mesophilic range often means that it is not economical to treat emissions that are relatively cold or hot. However, some recent studies suggest that we may be able to expand the temperature range of biofilters, and treat emissions at the temperature at which they are emitted. For example, Giggey et al. (1994) reported on the biofiltration of odor below 0 °C with snowfall. Also, a recent study by Kong et al. (2001) has shown that treatment of methanol and α-pinene can be conducted at temperatures up to 70 °C, and Datta et al. (2004) have shown that hydrogen sulfide can also be effectively treated at 70 °C. Future work at the laboratory, pilot and full scale in expanding the temperature range can open the economic application of biofilters to a range of emission sources.

6.6.4 Oxygen Content

Oxygen is vital to the operation of biofilters because the predominant microorganisms used in biofiltration are aerobic, and require oxygen for metabolism. Aerobic heterotrophic bacteria present in filter beds require at least 5–15% oxygen at the inlet gas stream to survive (Dharmavaram 1991). Yang et al. (2002) showed that biofiltration can be oxygen-limited in highly loaded systems, by showing that biofiltration with 63% oxygen in the inlet stream increased the maximum removal rate of methanol from 120 to 145 $g\,m^{-3}\,h^{-1}$ over regular air (21% oxygen); a further increase in oxygen content up to 80% did not lead to a further improvement in biofilter performance. Generally, for most air pollution control systems, oxygen supply is not an issue because it is abundant in the incoming airstream and the biofilm is relatively thin. In overloaded filters, however, it may be a limitation resulting in the formation of acidic and other intermediaries.

6.6.5 pH

Microorganisms have a specific, optimum pH range for their activities. The pH within the biofilters can be maintained by the addition of solid buffer agents to the packing material at the beginning of the operation, and once this buffering capacity is exhausted, the filter bed is removed and replaced with fresh material.

Compost beds generally have a pH between 7 and 8, a range preferred by most microorganisms. Carbon dioxide or SO_4^{2-} evolved in the metabolic activities of aerobic microorganisms tend to depress the system pH. So, if the waste gas or its intermediate byproducts do not provide sufficient buffering capacity, additional pH control has to be accomplished by addition of a base such as sodium or magnesium hydroxides.

Although hydrogen sulfide gas can be biofiltered effectively at low pH (Yang and Allen 1994), other odorous gases like methyl sulfide may not be removed effectively (Pomery 1982; Tanji et al. 1989).

6.6.6 Nutrients

Carbon and energy required for microorganisms may be derived from the contaminant gas, while other nutrients such as nitrogen, phosphorus, minerals, and trace elements should be supplied to microorganisms in the biofilter for good performance (Auria et al. 1996). Natural packing materials (e.g., peat, compost) contain nutrients to support biomass growth but, in the case of artificial packing material, nutrients should be provided for better performance (Weckhuysen et al. 1993; Morgenroth et al. 1996). Yang et al. (2002) studied nitrogen requirements for biofiltration of methanol and found that, at low nitrogen levels, removal rate increased with increasing N : C ratio for both NH_3 and NO_3. At high concentrations, however, NH_3 had an inhibitory effect on the removal rate while the removal rate simply reached a plateau at high NO_3 concentrations.

6.6.7 Pressure Drop

In a biofilter, the synthesis of biomass leads to accumulated growth of microbial mass over time, which has been related to an increase in airflow resistance in the bed (Kinney et al. 1996; Mohseni et al. 1998). Biomass accumulation is greater at the inlet sections of the biofilters (Corsi and Seed 1995; Swanson and Loehr 1997), and leads to a change in bed characteristics e.g., reduction in interparticle void space, and the compaction of natural packing material like wood chips; these changes cause channeling and increased pressure drop.

In general, there is an approximately linear increase in pressure drop with increasing gas flow rate (Yang and Allen 1994), which begins to become exponential at higher flow rates (Morgan-Sagastume et al. 2001). In addition, at a given gas flow rate, the pressure drop increases exponentially with increased biomass (Morgan-Sagastume et al. 2001) and with decreasing particle size, especially for particles less than 1 mm. The exponential increase with biomass means that a lower overall pressure drop will be obtained if biomass growth is distributed along the entire filter, as opposed to being localized in specific regions (Morgan-Sagastume et al. 2001). Compaction of the filter bed over extended periods of usage and due to overwatering will also give rise to high pressure drops (Pinnette et al. 1994).

Several researchers have developed predictive equations to describe pressure drops across biofilters for various particles, but there is no universal correlation that we are aware of that can accurately predict pressure drop for packed beds of varying particle sizes as well as the impact of biomass. Higgins et al. (1982) have developed equations for predicting pressure losses through compost piles. However, Williams (1988) found that the equation did not accurately predict headlosses through a compost biofilter that utilized screened sludge compost. Morgan-Sagastume et al. (2001) developed an equation to take into account biomass growth that fits their experimental data. While these predictive tools can provide guides for media development, the variability in the physical characteristics of the media (particle size distribution, available pore space, moisture content, bulk density, etc.), biomass growth, and the characteristics of the gas (loading, particulates) are so variable from one medium to another that onsite measurements are required over a long term to thoroughly assess the pressure drop associated with a particular application.

6.6.8
Medium Depth

Biofilter medium depth has ranged from less than 0.5 to 2.5 m. A depth of approximately 1 m appears to be most common, to allow sufficient residence time while minimizing filter land area requirements. Some manufacturers recommend the use of multiple layers of biofilter media, since these will need less land area for high loading rates (Leson and Winter 1991). Greater filter depth could be used, too, but the system headloss will increase in that case, and there is also the potential for compaction of the bed at the bottom, with subsequent increased pressure drop and channeling.

6.6.9
Waste Gas Pretreatment

The microbial communities in biofilters can be poisoned by the presence of toxic contaminants, excessive concentration of the contaminant, or excursion in environmental conditions like pH, temperature, and moisture content. In order to meet the basic requirements for optimal operation of the biofilter, waste gas conditioning is often required. A sufficient supply of oxygen and humidity, and an acceptable range of temperature and pH levels in the filter bed, are indispensable for the survival of the microbial community present in the bed (Werner et al. 1986; Beerli and Rotman 1989; van Lith et al. 1990). High particulate loads in the waste gas can adversely affect the operation of a biofilter by clogging the air distribution system and the filter material itself (William and Miller 1992; Bohn 1993). Pretreatment options can include humidification for temperature and humidity control and/or the use of devices for particulate control, such as a wet scrubber or a wet electrostatic precipitator.

6.6.10
Maintenance

The timing and frequency of routine or periodic maintenance of a biofilter depend upon a number of factors including waste gas temperature and relative humidity, filter bed moisture content, medium stability, temperature, pH, and backpressure (Leson and Winter 1991; Yang and Allen 1994). Fully engineered, enclosed systems with optimized packing generally reduce maintenance requirements. However, no matter how carefully a biofilter is designed, aging due to the bio-oxidation of organic substrates of the medium and buildup of minerals occur in most systems, which often require medium replacement.

Biofilters can fail to achieve their designed removal efficiencies for various reasons, such as inadequate assessment of the waste gas stream for its contaminants, particulates and the concentration levels, variations in temperature, pH, moisture, and oxygen content within the filter bed (Goldstein 1996; Standerfer and Willingham 1996; van Lith et al. 1996). Channeling in the filter bed, bed drying, generation of acid metabolites, and system upsets due to improper gas conditioning are the probable problems encountered during operation (Ottengraf 1986, 1987; Leson and Winter 1991; Leson et al. 1995; Allen and van Til 1996).

6.7
Microbiology of Biofilters

Bohn (1992) estimates microbiological populations in biofilters are of the order of 1 billion microorganisms per gram of organic material. Several groups of microorganisms are known to be involved in the degradation of air pollutants in biofilters, including bacteria, actinomycetes and fungi (Ottengraf 1987). There is one report on utilizing a co-culture of fungi and mites for the biofiltration of hydrophobic pollutants (Van Groenestijin et al. 2001). The composition of the microbial community and their survival in a biofilter depend on physical and chemical conditions in the packing material. The diversity of the active microorganisms is a function of the inlet gas stream composition and media.

Natural packing material like compost contains a sufficient number of different microorganisms to initiate biodegradation of contaminants. Initially, it takes time for the microorganisms to adapt, this time interval being known as the acclimatization period. The efficiency of the process is generally enhanced following the growth of active organisms during the adaptation phase. For easily biodegradable organic compounds, acclimatization can typically take less than 10 days, and for less biodegradable compounds and those contaminants for which the microorganisms are less likely to be initially present in the biofilter material, the period can be longer (Ottengraf 1986; Leson and Winter 1991). If the filter bed is inoculated with a specific culture that is known to degrade that particular pollutant, the adaptation time can be reduced to only a couple of days. Microorganisms can survive for fairly long periods when the biofilter is not loaded (Ottengraf and van der Oever 1983) – up to 2 months, if sufficient nutrients are available from the filter material (Leson and Winter 1991).

6.8 Advantages and Disadvantages

Biofiltration has its advantage over other technologies, as it has a simple, flexible design, with low capital costs. It is an excellent choice for high volumes of low-concentration pollutants, due to its low energy costs in comparison with competing technologies like incineration that involve fuel addition (Tables 6.1 and 6.2). There is no secondary waste stream generated in this case, like in scrubbing or adsorption and, since fuel is not added for combustion, there are relatively lower greenhouse gas (i.e., carbon dioxide) emissions. The packing material is very inexpensive, and sometimes it is available on-site or close by, for example, in the use of wood chips and/or bark in the forest products industry.

There also some disadvantages to use this technology. Its design criteria are still being developed, and packing media may need regular replacement (i.e.,

Table 6.1. Cost comparison of VOC control technologies (Bohn 1992)

Control technology	Total cost (US$/$10^6$ ft.^{3} air)
Incineration	130
Chlorine	60
Ozone	60
Activated carbon	20
Biofiltration	8

Table 6.2. Cost comparison of odor control technologies. (Vaith et al. 1996)

Parameter	Unit	Control technology Packed tower wet scrubber	Mist scrubber	Biofilter
Capacity	(scfm)[a]	155,000	3,000	400
Construction cost	(US$)	454,000	140,000	10,000
	(US$/scfm)	29.29	46.67	25.00
Operating cost	(US$/year)	74,000	22,000	280
	(US$/scfm per year)	4.77	7.33	0.70
H_2S removal efficiency	(%)	> 99.66	> 99.0	> 99.5

[a] Standard cubic foot per minute

every few years). Moreover, a large land area is needed for its setup, and so this is often a restriction where land is costly or not available. A typical footprint for a single, 1-m-deep, compost-based biofilter to treat 85,000 m^3/h stream would be approximately 25 × 30 m. The weight of biofilters often precludes their placement on a building roof, as is sometimes done with thermal oxidizers. Due to channeling, compaction of the compost-based media, and the relatively fragile nature and variability of the biological systems, this technology may sometimes seem unattractive to industries, particularly those without prior experience with the technology.

6.9 Applications of Biofilters

There are a wide range of pollutants that are biodegradable and that can therefore be treated in biofilters. The most successful removal of pollutants in biofilters occurs for low molecular weight and highly soluble organic compounds with simple bond structures. Originally applied to the treatment of odorous gases from sewage treatment plants, rendering and composting facilities, biofiltration is adding to its list of applications rapidly, as a result of continuous efforts to expand the useful domain of this cost-effective and environmentally friendly gas cleaning technology. The ability of biofilters to act as a dual control technology for both VOC and odor makes it appealing for its application at publicly owned treatment works (POTWs), where large volumes are combined with low concentrations. In these cases, biofiltration often represents the most economical way to satisfy statutory regulations.

Biofilters have also been used to treat a wide variety of organic and inorganic pollutants in industrial and municipal exhaust streams. Among those are odorous gas (ammonia, hydrogen sulfide, mercaptan, disulfides, etc.) from food-processing wastes, wastewater treatment facilities, composting operations and others, and VOC (propane, butane, styrene, phenols, methylene chloride, methanol, etc.) from industries. Biofilters have been shown to be effective (Table 6.3) for treating aromatics such as benzene, toluene, styrene, phenols, etc. (Cox et al. 1993; Shareefdeen et al. 1993; Zilli et al. 1993; Tonga and Skladany 1994; Ergas et al. 1995), aliphatics such as dichloromethane, propane, isobutane, etc. (Ebinger et al. 1987; Kampbell et al. 1987; Ergas et al. 1994, 1995), more easily biodegradable organics such as alcohols (Shareefdeen et al. 1993), ketones and esters (Windsperger 1991; Devinny et al. 1994), hydrophobic terpenes like α-pinene (Apel et al. 1995; Lee et al. 1996; Mohseni and Allen 1996), and odorous reduced sulfur gases such as carbon disulfide, hydrogen sulfide, mercaptans and methyl sulfides (Takase et al. 1976; Wada et al. 1986; Smith and Kelly 1988; Kangawa and Mikami 1989; Cho et al. 1991; Sinitsyn et al. 1991; Zhang et al. 1991; Cho et al. 1992; Park et al. 1993; Ostojik and O'Brien 1994; Bora and Allen 1996). Besides these contaminants, biofiltration has also achieved $> 90\%$ removal efficiency for odorous nitrogenous pollutants (Togashi et al. 1986; Barnes et al. 1995; Hartikainen et al. 1996; Martin et al. 1996). In response to the 1990 US Clean Air Act Amendments, bench/pilot-scale research has shown that 60 out of 189 hazardous air pollutants (HAPs) can be successfully treated with biofiltration (Fouhy 1992; Hussey 1995).

In Europe, more than 600 chemical processing industries are using biofilters for deodorization and treatment of VOCs from waste gas (Fouhy 1992). Bohn (1992) has presented a list of successful full-scale biofilter installations (Table 6.4) treating air toxics and VOCs at chemical processing industries in North America. The new areas of application for this technology are pulp and

Table 6.3. Degradability of pollutant gases. (Bohn 1992)

Rapidly degradable VOCs	Rapidly reactive VICs	Slowly degradable VOCs	Very slowly degradable VOCs
Alcohols	H_2S	Hydrocarbons[a]	Halogenated hydrocarbons[b]
Aldehydes	NO_x	Phenols	Polyaromatic hydrocarbons
Ketones	(but not N_2O)	Methylene chloride	CS_2
Ethers	SO_2		
Esters	HCl		
Organic acids	NH_3		
Amines	PH_3		
Thiols	SiH_4		
Other molecules containing O, N, S functional groups	HF		

[a] Aliphatics degrade faster than aromatics such as xylene, toluene, benzene, and styrene
[b] Such as trichloroethane, carbon tetrachloride and pentachlorophenol

Table 6.4. Biofilter applications. (Bohn 1992)

Company	Location	Application
S.C. Johnson & Son Inc.	Racine, WI	Propane and butane removal from room air; 90% RE, about 30,000 cfm[a]
Monsanto Chemical Co.	Springfield, MA	Ethanol and butyraldehyde removal from dryer air; 99% RE, 28,000 cfm; styrene removal from production gas
Dow Chemical Co.	Midland, MI	Chemical process gases
Hoechst Celanese Corp.	Coventry, RI	Process gases
Sandoz	Basel, Switzerland	Chemical process gases
Esso of Canada	Sarnia, Ontario	Hydrocarbon vapors from fuel storage tanks (proposed)
Mobil Chemical Co.	Canadaigua, NY	Pentane from polystyrene-foam molding (proposed)
Upjohn Co.	Kalamazoo, MI	Pharmaceutical production odors; 60,000 cfm (proposed)

[a] Cubic foot per minute

paper mills, petrochemicals, petroleum processing and transportation, wood processing, and site remediation.

In North America, more than 21 million biofilters are operating continuously – but are unrecognized (Bohn 1992). Biofilters are that part of household and industrial wastewater leach fields (septic systems) that completely reliably remove the VOCs emitted by the wastewater.

6.10 Conclusions

Biofiltration has clearly been shown to be a cost- and energy-effective technology for a range of emissions, and its future for growth is bright. Although it is well established for odor control for many emissions and for VOC control in others, there are many emission sources, particularly in the industrial sector, that have yet to be developed. There is scope for expanding the temperature range into thermophilic (high temperature) and psycrophilic (low temperature) conditions, and expanding the range of pollutants tested through both laboratory and field studies. There is also a need to work on innovative strategies (e.g., pretreatments to remove particulates and/or enhance biodegradability, separately optimized biofilters) to treat more challenging airstreams. In addition, there is a need to develop a more scientific foundation to help design and optimize biofilters, and improve their reliability. Improvements in our understanding of the links between biofilm properties (structure, depth, composition, and microbial community), media composition (size, shape, composition), and performance are needed. It is also necessary to get regulators to recognize the inherent net environmental advantages (e.g., low energy consumption) of this technology, and to adopt regulations that encourage its application, rather than expecting the same stringent removal efficiency delivered by other, incineration-based technologies. Overall, expanded testing and application is important so that this will lead to increased confidence and reliability for each given industry.

References

Allen PJ, van Til TS (1996) Installation of a full-scale biofilter for odor reduction at a hardboard mill. In: Proc Conf Biofiltration: an Air Pollution Control Technology, 1996. USC, LA, pp 31–38

Apel WA, Lee BD, Walton MR, Cook LL, Dinerstein KB (1995) Removal of α-pinene from off gases by biofiltration. In: Proc 88th Annu Meet Air and Waste Management Association, San Antonio, Pap no 95-TP9C.07

Auria R, Morales M, Acuna ME, Perez F, Revah S (1996) Biofiltration of toluene vapors: start up and gaseous ammonia addition. In: Proc Conf Biofiltration: an Air Pollution Control Technology, 1996. USC, LA, pp 134–141

Barnes JM, Apel WA, Barret KB (1995) Removal of nitrogen oxides from gas stream using biofiltration. J Hazard Mater 41:315–326

Beerli M, Rotman A (1989) Biofilter – a unique method to reduce and/or eliminate VOCs. In: Proc Environcon 89, 1st Int Conf Environmental Issues for Converters, Jacksonville, Florida, pp 1–32

Bishop DF, Govind R (1995) Development of novel biofilters for treatment of volatile organic compounds. In: Hinchee RE, Skeen RS, Sayles GD (eds) Biological unit processes for hazardous waste treatment. Proc 3rd Int In Situ and On-Site Bioreclamation. Battelle Press, Columbus, OH, 219 p

Bohn HL (1992) Considering biofiltration for decontaminating gases. Chem Eng Prog 88:34–40

Bohn HL (1993) Biofiltration: design and pitfalls. In: Proc 86th Annu Meet Air and Water Management Association, Denver, Pap no 93-TP-52A.01

Bora A, Allen ER (1996) Control of organo sulfur compound emissions using biofiltration: dimethyl disulfide. In: Proc 89th Annu Meet Air and Waste Management Association, Nashville, Pap no 96-RP87C.02

Cho KS, Hirai M, Shoda M (1991) Degradation characteristics of hydrogen sulfide, methanethiol, dimethyl sulfide and dimethyl disulfide by *Thiobacillus thioparus* DW44 isolated from peat biofilter. J Ferment Bioeng 71:384–389

Cho KS, Hirai M, Shoda M (1992) Enhanced removal efficiency of malodorous gases in a pilot-scale peat biofilter inoculated with *Thiobacillus thioparus* DW44. J Ferment Bioeng 73:46–50

Cohen Y (2001) Biofiltration – the treatment of fluids by microorganisms immobilized into the filter bedding material: a review. Bioresour Technol 77:257–274

Corsi RL, Seed L (1995) Biofiltration of BTEX: media, substrate and loading effects. Environ Prog 14:151–158
Cox HHJ, Houtman JHM, Doddema HJ, Harder W (1993) Enrichment of fungi and degradation of styrene in biofilters. Biotechnol Bioeng 49:391–398
Datta I, Fulthorpe R, Allen DG (2004) High temperature biofiltration of H_2S. In: Proc Air and Water Management Association Conf, 22–25 June 2004, Indianapolis, IN, Pap no 772
Davison BH, Barton JW, Klasson KT, Francisco AB (2000) Influence of high biomass concentrations on alkane solubilities. Biotechnol Bioeng 68:279–284
Deshusses MA, Johnson CT (2000) Development and validation of a simple protocol to rapidly determine the performance of biofilters for VOC Treatment. Environ Sci Technol 34:461–467
Devinny JS, Medina VF, Hodge DS (1994) Biofiltration for treatment of gasoline vapors In: Hinchee RE, Alleman BC, Hoeppel RE, Miller RN (eds) Hydrocarbon bioremediation. Lewis, Ann Arbor, MI, pp 12–19
Devinny JS, Deshusses MA, Webster TS (1999) Biofiltration for air pollution control. Lewis, Boca Raton
Dharmavaram S (1991) Biofiltration: a lean emission abatement technology. In: Proc 84th Annu Meet Exhibition Air and Waste Management Association, Pittsburgh, PA, Pap 91/103.2
Ebinger MH, Bohn HL, Puls RW (1987) Propane removal from propane-air mixtures by soil beds. J Air Waste Manage Assoc 37:1486–1489
Ergas SJ, Kinney K, Fuller ME, Scow KM (1994) Characterization of a compost biofiltration system degrading dichloromethane. Biotechnol Bioeng 44:1048–1054
Ergas SJ, Schroeder ED, Chang DPY, Morton RL (1995) Control of volatile organic compound emissions using a compost based biofilter. Water Environ Res 67:816–821
Fouhy K (1992) Cleaning waste gas naturally. Chem Eng 99:41–46
Giggey MD, Dwinal CA, Pinnette JR, O'Brien MA (1994) Performance testing of biofilters in a cold climate. In: Proc Water Environment Federation Speciality Conf Odor and Volatile Organic Compound Emission Control for Municipal and Industrial Wastewater Facilities, Jacksonville, FL, pp 29–39
Goldstein N (1996) Odor control experiences: lessons from the biofilter. Biocycle 37:70–75
Hartikainen T, Ruuskanen J, Vanhatalo M, Martikainen PJ (1996) Removal of ammonia from air by a peat biofilter. Environ Technol 17:45–53
Higgins A, Chen S, Singley M (1982) Airflow resistance in sewage sludge composting aeration systems. Trans ASAE 25:1010–1018
Hodge DS, Medina VF, Islander RL, Devinny JS (1991) Treatment of hydrocarbon fuel vapors in biofilters. Environ Technol 12:655–662
Hussey F (1995) Emission reduction options for painting operations. Ind Paint Powder 71:12–16
Kampbell DH, Wilson JT, Read HW, Stocksdale TT (1987) Removal of volatile aliphatic hydrocarbon in a soil bioreactor. J Air Waste Manage Assoc 37:1236–1240

Kangawa T, Mikami E (1989) Removal of methanethiol, dimethyl sulfide, dimethyl disulfide and hydrogen sulfide from contaminated air by *Thiobacillus thioparus* TK-m. Appl Environ Microbiol 55:555–558

Kinney KA, Chang DPY, Schroeder Ed, Scow KM (1996) Performance of a directionally-switching biofilter treating toluene contaminated air. In: Proc 89th Annu Meet Exhibition Air and Waste Management Association, Nashville

Kong Z, Farhana L, Fulthorpe R, Allen DG (2001) Treatment of volatile organic compounds in a biotrickling filter under thermophilic conditions. Environ Sci Technol 35:4347–4352

Lazarova V, Manem J (1995) Biofilm characterization and activity analysis in water and wastewater treatment. Water Res 29:2227–2245

Lee BD, Appel WA, Cook LL, Nichols KM (1996) Effect of bed moisture on α-pinene removal by biofilters. In: Proc Conf Biofiltration: an Air Pollution Control Technology, 1996. USC, LA, pp 214–222

Leson G, Winter AM (1991) Biofiltration: an innovative air pollution control technology for VOC emissions. J Air Waste Manage Assoc 41(8):1045–1054

Leson G, Chavira R, Winter A, Hodge D (1995) Experience with a full scale biofilter for control of ethanol emission. In: Proc 88th Annu Meet Air and Waste Management Association, San Antonio, Pap no 95-MP9A.04

Mackay D, Shiu WY, Ma KC (1991–1997) Illustrated handbook of physical-chemical properties and environmental fate for organic chemicals. (1991) Monoaromatic hydrocarbons, chlorobenzenes and PCBs, vol I, pp 1–697. (1992) Polynuclear aromatic hydrocarbons, polychlorinated dioxins and dibenzofurans, vol II, pp 1–597. (1993) Volatile organic chemicals, vol III, pp 1–916. (1995) Oxygen, nitrogen and sulfur containing compounds, vol IV, pp 1–962. (1997) Pesticides, vol V, pp 1–812. Lewis/CRC Press, Boca Raton

Mackay D, Shiu WY, Ma KC (2000) Physical-chemical properties and environmental fate and degradation handbook. CRCnetBASE 2000. Chapman and Hall, CRCnetBASE, CRC Press, Boca Raton (CD-ROM)

Marsh A (1994) Biofiltration for emission abatement. Eur Coat J 7/8:528–536

Martin G, Lemasle M, Taha S (1996) The control of gaseous nitrogen pollutants in a fixed peat bed reactor. J Biotechnol 46:15–21

McNevin D, Baford J (2000) Biofiltration as an odour abatement strategy. Biochem Eng J 5:231–242

Miller MJ (2004) Transport and degradation of hydrophobic pollutants in biofilms in biofilters. PhD Thesis, University of Toronto

Miller MJ, Allen DG (2002) Transport of hydrophobic pollutants through biofilms in biofilters. In: Proc 95th Annu Conf Exhibition Air and Waste Management Association, 23–27 June 2002, Baltimore, MD, pp 1271–1286

Miller MJ, Allen DG (2004) Transport of hydrophobic pollutants through biofilms in biofilters. Chem Eng Sci 59(17):3515–3525

Mohseni M, Allen DG (1996) Biofiltration of α-pinene using wood waste and activated carbon media. In: Proc Conf Biofiltration: an Air Pollution Control Technology, 1996. USC, LA, pp 45–54

Mohseni M, Allen DG (2000) Biofiltration of mixtures of hydrophilic and hydrophobic volatile organic compounds. Chem Eng Sci 55:1545–1558

Mohseni M, Allen DG, Nichols KM (1998) Biofiltration of α-pinene and its application to the treatment of pulp and paper air emissions. TAPPI J 81:205–211

Morgan-Sagastume F, Sleep BE, Allen DG (2001) Effects of biomass growth on gas pressure drop in biofilters. J Environ Eng 127:388–396

Morgenroth E, Schroeder ED, Chang DPY, Scow KM (1996) Nutrient limitation in a compost biofilter degrading hexane. J Air Waste Manage Assoc 46:300–308

Ostojik N, O'Brien M (1994) Biology VS chemistry: choice between biofilter and wet scrubbing for control of odors. J N Engl Water Environ Assoc 28:146–153

Ottengraf SPP (1986) Exhaust gas purification. In: Rehm HJ, Reed G (eds) Biotechnology, vol 8. VCH, Weinheim, pp 427–452

Ottengraf SPP (1987) Biological system for waste gas elimination. TIBTECH 5:132–136

Ottengraf SPP, van der Oever AHC (1983) Kinetics of organic compound removal from waste gases with a biological filter. Biotechnol Bioeng 25:3089–3102

Park SJ, Hirai M, Shoda M (1993) Treatment of exhaust gases from a night soil treatment plant by a combined deodorization system of activated carbon filter reactor and peat biofilter with *Thiobacillus thioparus* DW44. J Ferment Bioeng 76:423–426

Pinnette JR, Giggey MD, Marcy GJ, O'Brien MA (1994) Performance of biofilters at two agitated bin composting facilities. In: Proc 87th Annu Meet Air and Waste Management Association, Pap 94-TP52.02

Pomeroy RD (1957) De-odorizing of gas streams by the use of microbial growth. US patent 2,793,096

Pomeroy RD (1982) Biological treatment of odorous air. J Water Pollut Contr Fed 54:1541–1545

Rozich A (1995) Tackle airborne organic vapours with biofiltration. Environ Eng World 1:32–34

Shareefdeen Z, Baltzis BC, Oh Y-S, Bartha R (1993) Biofiltration of methanol vapors. Biotechnol Bioeng 41: 512–524

Shimko IG, Spasov VA, Chinennaya SK, Zakirova RI, Tananina IN, Perchugor GY, Pavlova OI (1988) Biochemical methods of freeing gas-air mixture from sulphur containing compounds. Fibre Chem 19:373–378

Sinitsyn VY, Donstov AG, Gubinov NA (1991) Purification of sulfur containing gaseous emissions of a kraft mill using solid phase biofilter. Bum Promst 6/7:17

Smith NA, Kelly DP (1988) Mechanism of oxidation of dimethyl disulfide by *Thiobacillus thioparus* strain E6. J Gen Microbiol 134:3031–3039

Sologar VS, Lu Z, Allen DG (2003) Biofiltration of concentrated mixtures of hydrogen sulfide and methanol. Environ Prog 22:129–136

Spigno G, Pagella C, Fumi MD, Molteni R, de Faveri DM (2003) VOCs removal from waste gases: gas-phase bioreactor for the abatement of hexane by *Aspergillus niger*. Chem Eng Sci 58:739–746

Standerfer S, Willingham R (1996) Experience with pilot and full scale biofilter operations. In: Proc Conf Biofiltration: an Air Pollution Control Technology, 1996. UCS, LA, pp 102–107

Swanson WJ, Loehr RC (1997) Biofiltration: fundamentals, design and operations principles and applications of biological APC technology. J Environ Eng ASCE 123:538–546

Takase K, Kenmoku T, Hashimoto M (1976) Soil beds for reduction of malodorous gases in kraft pulp mills. Jpn Tappi J 30:252–256

Tanji Y, Kangawa T, Mikami E (1989) Removal of dimethyl sulphide, methyl mercaptan and hydrogen sulphide by immobilized *Thiobacillus thioparus* TK-m. J Ferment Bioeng 67:280–285

Togashi IM, Suzuki M, Hirai M, Shoda M, Kubota H (1986) Removal of NH_3 by a peat biofilter without and with nitrifier. J Ferment Technol 64:425–432

Tonga AP, Skladany GJ (1994) Field pilot-scale vapor-phase treatment of styrene using biofiltration. In: Flathman PE, Jerger DE, Exner JH (eds) Bioremediation: field experience. Lewis, Ann Arbor, MI, pp 507–521

Vaith K, Cannon M, Milligan D, Heydron J (1996) Comparing scrubbing technologies: packed tower, mist scrubbers and biofilters. Water Environ Tech 8:35–38

Van Groenestijin JW, van Heiningen WNM, Kraakman NJR (2001) Biofilters based on the action of fungi. Water Sci Technol 44(9):227–232

Van Lith C, David SL, Marsh R (1990) Design criteria for biofilters. In: Van Lith C, David SL, Marsh R (eds) Effluent treatment and waste disposal. ClairTech, Utrecht, Netherlands Institution of Chemical Engineers Symp Ser 116, pp 127–132

Van Lith C, Leson G, Michelsen R (1996) Evaluating design options for biofilters. In: Proc Conf Biofiltration: an Air Pollution Control Technology, 1996. USC, LA, pp 77–84

Wada A, Shoda M, Kubota H, Kobayashi T, Fujimura YK, Kurasihi H (1986) Characteristics of hydrogen sulfide oxidizing bacteria inhibiting a peat biofilter. J Ferment Technol 64:161–167

Wani AH, Branion RMR, Lau AK (1997) Biofiltration: a promising and cost-effective control technology for odors, VOCs and air toxics. J Environ Sci Health A32:2027–2055

Wanner O, Gujer W (1984) Competition in biofilms. Water Sci Technol 17:27–44

Weckhuysen B, Vriens L, Verachtert H (1993) The effect of nutrient supplementation on the biofiltration removal of butanol in contaminated air. Appl Microbiol Biotechnol 39:395–399

Werner W, Liebe HG, Striefler B (1986) Emission control with biofilters: application, design and operational experience. In: Proc 7th World Clean Air Congr, Sydney, Australia, pp 537–544

Williams TO (1988) Use of a compost biofilter for odour control. Report E and A Environmental Consultants Inc, Canton, Massachusetts

Williams TO, Miller FC (1992) Biofilters and facility operations. Biocycle 33:75–79

Windsperger A (1991) Use of biofilters for the purification of gases containing solvents. Radex Rundsch 3/4:457–464

Yang Y, Allen ER (1994) Biofiltration control of hydrogen sulphide. 1. Design and operational parameters. J Air Waste Manage Assoc 44:863–868

Yang H, Minuth B, Allen DG (2002) Effects of nitrogen and oxygen on biofilter performance. J Air Waste Manage Assoc 52:279–286

Yaws CL (2001) Matheson gas data book. McGraw-Hill, New York

Zhang L, Kuniyoshi I, Hirai M, Shoda M (1991) Oxidation of dimethyl sulfide by *Pseudomonas acidovorans*DMR-11 isolated from peat biofilter. Biotechnol Lett 13:223–228

Zilli M, Converti A, Lodi A, Del-Borghi M, Ferraiolo G (1993) Phenol removed from waste gases with a biological filter by *Pseudomonas putida*. Biotechnol Bioeng 41:693–699

7 Biotrickling Filter Technology

MARC A. DESHUSSES, DAVID GABRIEL

7.1 Introduction

Biotrickling filtration is one of the many promising biological techniques for odor and VOC control. Contrary to biofilters that are usually packed with some sort of organic medium and are operated with a minimum of free water (Devinny et al. 1999), biotrickling filters are almost exclusively packed with inorganic or manufactured media, over which a distinct liquid phase is trickled. The trickling liquid in biotrickling filters provides a convenient means to control pH, salt or metabolite concentration, and to supplement nutrients to the process culture. Biotrickling filters have several advantages over other biotreatment technologies for air pollution control (see Table 7.1; Oh and Bartha 1997; Mpanias and Baltzis 1998; Cox et al. 2000; Gabriel and Deshusses 2003a). Some book chapters and reviews published previously contain comprehensive discussions of biotrickling filtration technology (Cox and Deshusses 1998, 2001, 2002a; Deshusses 2004). In the present chapter, selected aspects of biotrickling filtration for air pollution control are presented and discussed. Background information on biotrickling filter is presented in Sect. 7.1, while Sect. 7.2 deals with biotrickling filter design and operation. In Sect. 7.3, the focus is placed on high-performance biotrickling filters for H_2S and odor control – in particular, the conversion of chemical scrubbers to biological trickling filters is presented and discussed. The latter issue has received considerable attention, as it has been shown that biotrickling filters can achieve H_2S elimination rates equivalent to those observed in chemical scrubbers.

The basic principle of a biotrickling filter is illustrated in Fig. 7.1. In biotrickling filters, foul or contaminated air is forced through a packed bed, either downflow or upflow. The packed bed is almost always made of an inert material such as a random dump plastic packing, lava rocks, structured plastic packing, or open-pore synthetic foam. Other materials that have been used include glass or rock wool, shredded tires, glass beads, or ceramics. The packing provides the necessary surface for biofilm attachment and for gas–liquid contact. During treatment, an aqueous phase is recycled over the packing. It provides moisture, mineral nutrients to the process culture, and a means to control basic operating parameters. In most cases, the trickling

Shareefdeen/Singh (Eds.)
Biotechnology for Odor and Air Pollution Control

Table 7.1. Advantages of biotrickling filter technology

Parameter	Characteristics
Bed height	Little bed height limitation, since packings are resistant to crushing. Bed height is only limited by occurring in liquid channeling (usually beds taller than 2–3 m)
Footprint	Since taller beds can be constructed and high volumetric performance can be obtained, biotrickling filters can be built such that they use a relatively small footprint
Packing longevity	Packing life is usually over 10 years. No need for replacement schedule
Pressure drop	Biotrickling filter beds often have a porosity greater than 70%, and therefore exhibit low pressure drops, even at very high air velocities
Controls	The existence of a free liquid phase allows for easy control of moisture, temperature, pH, salt, and possible metabolite accumulation
Treatment of air with dust or grease	Treatment of air with dust or grease is often possible, since blow-down of the trickling phase allows for purging of dust or grease
Range of pollutants that can be controlled	Since metabolites (e.g., sulfate in H_2S treatment) can easily be purged from the system, biotrickling filters can be used to treat a wider range of pollutants than biofilters

liquid is continuously supplemented with essential mineral nutrients such as nitrogen, phosphorus, potassium, and trace elements. Nutrient requirements depend on the pollutant being treated, the pollutant loading and concentration, and general bioreactor operation strategy. In general, most of the pollutant is biodegraded in the biofilm, but part may also be removed by microorganisms suspended in the recycle liquid (Cox et al. 2000). Possible biodegradation metabolites will leave the system via the liquid purge, along with small amounts of biomass and dissolved pollutant, if any. Usually, the amount of pollutant leaving the biotrickling filter via the purge is negligible compared to the amount degraded in the system (Cox et al. 2000).

Biotrickling filters work because of the action of the pollutant-degrading microorganisms. In the case of the removal of hydrocarbon vapors, the primary degraders are aerobic heterotrophic organisms that use the pollutant as a source of carbon and energy (see Chap. 5). For H_2S or ammonia removal, the primary degraders are autotrophs, and they will use the pollutant as a source of energy, and carbon dioxide as a source of carbon for growth. The removal

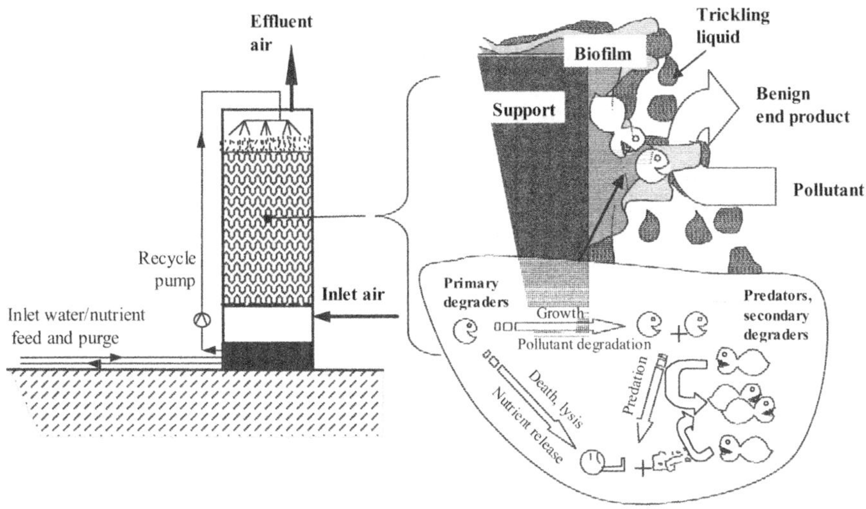

Fig. 7.1. Basic setup and principle of a biotrickling filter

of compounds such as dimethyl sulfide or dimethyl disulfide will require both autotrophs and heterotrophs to be present. In any case, the biotrickling filter will host a wide variety of microorganisms, similar to those encountered in wastewater treatment operations. The microorganisms responsible for pollutant removal in biotrickling filters are usually aerobic, because biotrickling filters are well-aerated systems. However, it has been proposed that the deeper parts of the biofilm (see Fig. 7.1), where anaerobic conditions probably prevail, can be utilized to perform anaerobic biodegradation (e.g., reductive dechlorination, or NO_X reduction) for the treatment of pollutants that are recalcitrant under aerobic conditions (Devinny et al. 1995). Anaerobic treatment in aerobic biotrickling filters remains a challenge.

As illustrated in Fig. 7.1, if the biofilm grows to a certain thickness, the fraction of the biofilm against the substratum will become inactive due to mass transfer limitations. Thus, in those cases, the active primary degraders constitute only a minor fraction of the total population in the biofilm. Secondary degraders feeding on metabolites, biopolymers, or predators feeding on the primary degraders include bacteria, fungi, and higher organisms such as protozoa, rotifers, mites, even mosquito or fly larvae, worms, or small snails. The importance of the secondary degraders and higher organisms should not be underestimated. They have been shown to play an important role in reducing the rate of biomass accumulation, and in recycling essential inorganic nutrients (Cox and Deshusses 1999; Woertz et al. 2002; Won et al. 2002). In fact, comparison of fermentation broth design with biotrickling filter recycle liquid composition reveals that most biotrickling filters used for VOC control are operated under some degrees of inorganic nutrient limitation, suggesting that nutrient recycle through the action of secondary degraders within the reactor is a common process.

7.2 Biotrickling Filter Design and Operation

Typical design and operating parameters of biotrickling filters are listed in Table 7.2. Note that these are average values, and that some biotrickling filters are designed or operated outside these ranges. Many factors influence the performance of biotrickling filters. A brief discussion of the effects of selected parameters is presented below.

The question of nutrient supply, in particular in the case of VOC control in biotrickling filters, has been investigated for some years. Even so, there is no agreement or universal rule to determine the optimum nutrient supply. It is generally admitted that excessive nutrient supply (usually combined with high concentrations of the pollutant undergoing treatment) will result in excess biomass formation and plugging of the bed, leading to failure of the system (Weber and Hartmans 1996; Wübker and Friedrich 1996; Holubar et al. 1999). Note that the issue of biomass overgrowth is typical of VOC control. Biotrickling filter for odor and H_2S control are usually not subject to plugging problems, most probably because of the relative inefficiency of autotrophic organisms,

Table 7.2. Typical design and operating parameters of biotrickling filters

Parameter	Value
Bed height	1–4 m
Bed cross-sectional area	Depends on application. A priori, only limited by space available
Packing nominal size	Highly variable: e.g., 5–30 mm diameter lava rocks, 10–100 mm nominal size random dump packing, open-pore foam with 4–10 pores cm^{-1}
Air empty bed residence time (EBRT)	Usually 10–30 s, some cases as low as 2–5 s (Gabriel and Deshusses 2003a, b)
Pressure drop	1–5 cm of water column
Air temperature	10–30 °C, some cases reported as high as 60–70 °C (Cox et al. 2001; Kong et al. 2001)
Liquid recycle rate	Highly variable, from 0.01 to 10 m h^{-1}. Average values are 0. 1^{-1} m h^{-1}
pH of recycle liquid	∼ 7 for VOC control, ∼ 1–2 for H_2S control
Typical monitored parameters (on- or offline)	Temperature, pH, dissolved oxygen and conductivity of recycle liquid, liquid feed rate, trickling flow rate, pollutant inlet and outlet concentrations, pressure drop
Typical controls	Liquid feed rate, pH control, low-level liquid alarm or pump shutdown

as far as growth and biomass production is concerned. Biomass control has been investigated by several researchers (see Cox and Deshusses 2001), and will not be discussed in detail in this chapter. Deshusses and Cox (1999) proposed a cost-benefit method to decide whether it was best to design a large reactor with low-performance system, thereby avoiding plugging issues, or to design a high-performance system with high nutrient loading, and periodically remove excess biomass. Validation of the approach and more experience with full-scale biotrickling filters in the coming years should allow setting general guidelines for nutrient supply. Interestingly, if pollutant loadings are low and/or intermittent loading conditions exist, it may be difficult to reach an optimum density of bacteria on the packing to obtain sustained performance (Webster et al. 1999). This is an issue that does not exist in biofilters, as the process culture in biofilters is able to survive on the limited supply of nutrients and substrates provided by the organic packing of the biofilter.

Biotrickling filters can be operated with the gas and liquid flowing cocurrently or countercurrent. Cocurrent operation is often preferred for VOC treatment, as countercurrent operation may result in stripping of the pollutant from the liquid phase at the gas outlet port of the reactor, thereby causing low treatment. Of course, the extent of stripping depends on the concentration of the pollutant dissolved in the recycle liquid, which itself depends on numerous parameters. Countercurrent operation is often preferred for H_2S and odor treatment, as these are often mass transfer limited, and little dissolved pollutant can be found in the trickling liquid. Typical trickling velocities range from 0.01 up to $10\,\mathrm{m\,h^{-1}}$. The optimum trickling rate will depend on the nature of the process limiting the removal of the pollutant. It is clear that biotrickling filters are typically operated at trickling velocities much lower than chemical scrubbers, resulting in partial wetting of the packed bed. This is often not a problem, because it is believed that all parts of the bed will be wetted over the course of a day. However, for the treatment of acid-producing pollutants such as H_2S, the trickling liquid also serves for pH control. If the trickling rate is too low, the liquid may not be able to remove the produced acids fast enough, and the pH could be reduced to the extent of inhibiting the process culture. Another case is when significant biodegradation occurs in the sump. In such cases, increasing the trickling rate will result in greater pollutant transfer to the trickling liquid, and higher treatment rates. Another consideration is that with a high trickling rate, a thick layer of liquid covers the biofilm, and may act as a significant resistance to mass transfer, potentially reducing the rate of pollutant treatment. Overall, the above discussion illustrates that the effects of the trickling rate can be numerous and complex. Therefore, experimenting with this parameter is often required in order to fine-tune the operating conditions.

One misconception is that oxygen limitation does not occur in bioreactors for air pollution control. This stems from the fact that oxygen concentration in air is several orders of magnitude higher than the pollutant being treated. However, the high Henry's law constant of oxygen makes it possible that the penetration depth of oxygen in water or the biofilm is lower than that of the pollutant, causing anaerobic zones in the deeper parts of the biofilm close to the substratum (Fig. 7.1). This has been experimentally confirmed using an oxygen microelectrode technique, showing depletion of oxygen at a depth of about 400 μm into the biofilm of a biotrickling filter treating diethyl ether (Alonso et al. 1998). At first approximation, one can estimate whether oxygen limitation may occur by calculating (liquid) interfacial concentrations of oxygen (often equal to saturation with air, roughly $8\,mg\,l^{-1}$) and that of the pollutant, and determining from the stoichiometry of biodegradation whether there is an excess or a deficiency of oxygen. For example, one would predict that oxygen limitation should occur at gaseous concentrations of toluene above $0.7\,g\,m^{-3}$. For more hydrophilic compounds, the gaseous concentration leading to potentially oxygen-limiting conditions is lower. This suggests that monitoring of dissolved oxygen may be worthwhile for the treatment of high concentrations of hydrophilic compounds.

Biotrickling filters require inoculation, as the usual biotrickling filter packing is inert and does not contain any indigenous microbial populations. In most cases, activated sludge from wastewater treatment plants will provide an effective inoculum, although other sources can also be utilized (e.g., compost extract, specialized enriched cultures, etc.). Typically, one would add a volume of about 20–100 l of activated sludge per cubic meter of bed, or a dry weight of cells of about $0.2-1\,kg\,m^{-3}$ of bed. If the reactor is intended to be operated with a high rate of freshwater supply and purge, it is recommended to reduce the freshwater supply and purge for 2–3 days to avoid washout and promote attachment. In most cases, for pollutants relatively easy to treat, such as H_2S, toluene, ethanol, methanol, ketones, and aromatic compounds, pollutant degradation is observed within 2–3 days of startup, and optimum conditions are reached within 10–15 days. For pollutants that are more difficult to degrade, inoculation with specialized cultures may be required to reduce the startup and acclimation time.

7.3 Conversion of Chemical Scrubbers to Biotrickling Filters

Hydrogen sulfide is one of the main odors targeted for control at publicly owned treatment works (POTWs). H_2S emissions are usually controlled by chemical scrubbing, which is expensive. Until recently, it was always thought that successful biotreatment of H_2S required gas contact times ranging from 10 to 30 s (Yang and Allen 1994; Smet et al. 1998; Chung et al. 2000; Wu et al. 2001). This is much longer than the 1–2 s contact times commonly used in chemical scrubbers (Card 2001). Consequently, biotreatment was not very practical for the control of very large airstreams, due to the large footprint required for meeting treatment objectives. There would be significant economical and safety benefits if chemical scrubbers could be converted to biotrickling filters, keeping the same air contact time, as biotrickling filters are cheaper to operate than scrubbers, they can potentially treat organic odors, and they do not require the use of toxic and dangerous chemicals. In the past 2 years, several successful conversions of full-scale chemical scrubbers to biotrickling filters have been demonstrated (Koe et al. 2002; Gabriel and Deshusses 2003a, b). The very high performance of H_2S treatment that has been obtained at gas contact time sometimes as low as 1.6 s opens new possibilities for air biotreatment.

7.3.1
First Approach to the Conversion

The feasibility of retrofitting any existing chemical scrubber depends on the technical and economical viability of the conversion. However, a minimum and critical requirement to make any retrofitting viable from an economical point of view is that the scrubber shell, the packing support, and most of the wetted parts need to be reused. In essence, the shell must be strong enough to support the weight of the packing and of the biomass and, in case of H_2S-degrading biotrickling filters, all wetted parts must be resistant to acidic conditions, because of the low pH of operation (pH 1–2). From a construction point of view, those conditions are often satisfied, since typical chemical scrubber construction materials are plastic resins. Commonly, fiberglass-reinforced plastic (FPR) is used for shells and packing supports, providing strong corrosion resistance to acids and bases and a tensile strength of 150 MPa at 65 °C. Other materials such as CPVC, PVC, PP or, less frequently, metal alloys such as stainless steel or Hastelloy offer good chemical resistance.

Prior to the conversion of a chemical scrubber, several technical issues need to be clarified. Table 7.3 shows key technical questions to pose to evaluate wet chemical scrubber construction and biotrickling filter requirements. The technical issues of Table 7.3 should be considered as a starting point for the evaluation of the conversion of chemical scrubbers to biotrickling filters at POTWs. Of course, a case-by-case analysis will be necessary for answering some of the questions, because of the differences in the construction, location, and operation between wet chemical scrubbers. A general cost-benefit analysis of the conversion will be required as well. An additional factor to be considered in a first conversion is to consider the possibility of returning to chemical scrubber operation if, for whatever reason, biotrickling filtration does not meet the expectations. In such cases, the number of changes performed to the chemical scrubber will be kept to the necessary minimum.

Experience acquired by the authors in converting wet chemical scrubbers for H_2S treatment in Orange County Sanitation District (OCSD), California, indicates that in most of the cases, there are a similar number and type of changes usually required for any scrubber to be converted, independently of the type and actual construction of the scrubber. One change found to be a key to success in the conversions performed is the replacement of the packing material of the chemical scrubber. This is needed because packing used for chemical scrubbers is commonly made of plastic materials with a low specific surface area that does not allow proper attachment of biomass to the bed to sustain a high biomass density.

Table 7.3. Summary of general technical questions used to evaluate the feasibility of converting wet scrubbers, and typical answers for chemical scrubbers

Question	Specific answer for most OCSD scrubbers
Will odor/H_2S treatment in the converted scrubber meet treatment objectives?	H_2S/odor removal will require case-by-case evaluation
Is the existing packing material suitable for biotrickling filtration of targeted compounds?	Scrubber packing is commonly not suitable for biotrickling filtration, it needs to be replaced by a new packing
Is the support for the packing strong enough to handle the additional weight of the packing and biomass?	For safety, an additional support may be installed, especially for large/old scrubbers. Intermediate supports may be needed to split tall beds into shorter sections
Is the demister resistant to low pH and resistant to clogging by biomass?	Demisters in most scrubbers are resistant to low pH, but clogging of the demisters needs to be checked during operation as a biotrickling filter
Is the existing blower suitable?	There will most probably be an increase in pressure drop if the same airflow rate is maintained. Accounting for packing compaction may be necessary
Is the liquid distribution system suitable to operate at lower flow rates?	Scrubbers fitted with nozzles need replacement of the nozzles. Scrubbers with weir troughs need no modification, but testing for adequate liquid distribution is warranted
Is a nutrient source available at a reasonable distance from the scrubber?	Secondary effluent may be plumbed into existing scrubbers
Can the controls be modified to accommodate for biotrickling filter operation?	In most cases, yes, but requires further case-by-case detailed evaluation and recalibration of some probes

Depending on the type, location, and actual construction of the chemical scrubber, the extent of modification of the liquid distribution or modification of the mist eliminator or need of air duct modifications may influence the economical viability of the conversion. Note that it is particularly critical to provide a uniform water distribution, taking into account that biotrickling filter operations will require a much smaller trickling rate than the one used in a chemical scrubber. Also imperative is to provide a nutrient source for biomass growth and to control the pH of the recycle liquid. The demister must be effective in removing fine liquid droplets from the outlet airstream, although potential for mist carry-over from the biotrickling filter is low because the liquid recycle rate is much lower in biotrickling filters than in chemical scrubbers.

7.3.2
General Procedure to Convert Full-Scale Chemical Scrubbers

Analysis of different types of wet scrubbers, taking into account all key factors required for the conversion of chemical scrubbers to biotrickling filters resulted in the general conversion procedure presented in this section. The procedure consists of several steps that often apply in the conversion of a chemical scrubber, independently of whether they are end-of-pipe scrubbers or scrubbers that possess a post-treatment. Steps are summarized in Table 7.4, and discussed in depth in the next sections. The experience is based on the actual conversions of five scrubbers at the two wastewater treatment plants operated by OCSD. There may be essentially two types of possible applications for biotrickling filtration: conversion of scrubbers that have a post-treatment, so that a biotrickling filtration serves as a pretreatment, and conversion of end-of-pipe scrubbers, where usually low concentrations are treated down to regulatory requirements. The former case can usually involve higher concentrations of H_2S ranging from 10 to 250 ppm_V, while the latter case might be in the 3–15 ppm_V H_2S range or for the treatment of organic odors. An example of the application of the procedure for a pretreatment scrubber at OCSD is detailed in Table 7.4, and will serve as an example throughout the text.

The time required for a conversion is usually short, therefore limiting the downtime of the scrubber. Obviously, downtime depends on the complexity of the changes that need to be performed, but in general terms, assuming that all required materials are in place, an average of 4–6 days is required per scrubber.

7.3.2.1
Scrubber Preparation

Any conversion will require some preparatory work consisting in isolating the scrubber prior to the conversion. The inlet damper has to be closed, and it must be ensured that no foul air passes through the scrubber, since further steps will require entering the scrubber. This is particularly important in cases where H_2S is the predominant component of the foul air, and when chemical scrubbers are considered confined entry spaces (OSHA regulation #1910.46.)

Table 7.4. Generic ten-step conversion procedure for the conversion of wet chemical scrubbers, and its particular application to a trunkline scrubber at OCSD

Step	Action	Application to OCSD scrubber #10
1	Removal of unnecessary parts	Locking and tagging valves and lines only (experimental conversion)
2	Removal of old packing	Certified outside contractor was used due to confined entry space classification of the scrubber
3	Packing support reinforcement	Strengthening of the bed support plate with a 15-cm-diameter Schedule 80 CPVC pipe as a reinforcement pillar under the lower packing chamber grating
4	Modification of the liquid distribution system	Not necessary (distribution via weir trough)
5	Modification of the mist eliminator	Not necessary, mist eliminator is adequate
6	Liquid recycle pump replacement	Replacement of the 5.6-kW liquid recycle pump. Backup pump was kept in place. Re-piping with 3-cm-diameter CPVC Schedule 80 of a section of the recycle line to fit a 0.4-kW pump and an online rotameter
7	Modification of the inlet/outlet air ducts	Not necessary, no airflow/path change
8	Installation of secondary effluent supply	Not necessary, already installed
9	Installation of the new packing	Dumping new packing material in the scrubber through the upper manhole
10	Modification of the controls to accommodate biotrickling filter operation	Low pH alarm was disconnected. Liquid feed supply is modified

where secured access is required. As an option, locking and tagging valves and lines that will not be used, such as chemical feeding lines, will be enough for isolating the scrubber, but more commonly, if chemical scrubbing is not going to be continued, there is no need for most of the unnecessary parts to be kept in place. During preparation of the scrubber, an acid wash step prior to any other task may be included to remove the scale that had built up on the packing and in the pipes but, if the scrubber packing has not been acid washed previously, rinsing with water until a neutral pH is reached to flush all chemicals (bleach, caustic, and/or peroxide) out the system is usually enough.

7.3.2.2
Removal of Unnecessary Parts

A typical chemical scrubber contains five major components: a fan, recirculation pumps, caustic injection pumps, hydrogen peroxide or chlorine booster pumps, and a chlorine feed station. However, for a biotrickling filter, peroxide and chlorine booster pumps and a chlorine feed station are unnecessary. Thus, those parts should be removed, and caps should be installed to close up the open ends of the pipes. Also, in cases where the concentration of H_2S to be treated is reasonably low (below 200–300 ppm_v), the caustic injection pump is not needed, since water makeup addition is enough for controlling the pH during the biotrickling filter operation. Depending on the construction and degree of instrumentation of the chemical scrubber, ancillary equipment and buildings such as peroxide and caustic tanks, and chlorine sensors may be dismantled and/or demolished.

7.3.2.3
Removal of Old Packing

Important requirements for a good packing include a large specific area, high porosity, high chemical stability over time and structural strength, low weight, suitable surface for bacterial attachment and growth, and low cost (Cox and Deshusses 2001). Packing materials commonly used in chemical scrubbers, such as Lan-Pack, Jaeger TriPack Ceilcote Tellerettes, satisfy most of the above prerequisites, except for a large surface area to sustain a quantity of biomass large enough to obtain treatment efficiencies and capacities similar to those of a chemical scrubber. Thus, the packing material of chemical scrubbers needs to be replaced.

Due to the confined entry space designation of the scrubbers, removal of the old packing needs to be performed by OSHA-certified personnel, which will usually be an outside contractor. Removal of old packing material can be accomplished through the viewing ports of the bed. Old packing material can be disposed into containers or large waste/trash bags. Disposal of packing material can be accomplished by crushing or chipping to reduce volume, and then taking it to an appropriate landfill.

7.3.2.4
Packing Support Reinforcement

Water holdup and biomass growth are the major contributors to possible increase in the weight of the packed bed in a biotrickling filter. In the case of H_2S treatment, growth of autotrophic microorganisms is relatively low compared to heterotrophic biomass that often accumulates in a larger amount in VOC-treating systems. This is due to the higher biomass yield on organic substrates. However, the weight of the packing support in a well-colonized biotrickling filter may be lower than that of a chemical scrubber, if large amounts of limestone and other deposits have built up over time. In some conversions of chemical scrubbers, the weight of the packing material has been increased by a factor of ten from the original weight of a clean packing material. Our experience at OCSD indicates that packing supports such as multi-beam gas-injected supports commonly installed in chemical scrubbers do not need reinforcement or replacement. In any case, less heavy-duty packing supports may need a careful evaluation of their strength characteristic prior to the conversion, in particular in large-diameter towers.

7.3.2.5
Modification of the Liquid Distribution System

The liquid trickling rate for a biotrickling filter does not need to be as high as in a chemical scrubber. Depending on the liquid distribution system, reducing the liquid trickling rate may lead to an improper liquid distribution. Our experience indicates that gravity liquid distribution systems, such as parting boxes and weir troughs are suitable for biotrickling filter operation if properly aligned, but nozzles need replacement by smaller nozzles due to lowered trickling rates. Here, it should be stressed that nozzle clogging is the first cause of failure of biotrickling filters. Hence, the use of non-clogging nozzles such as pigtail-shaped nozzles is highly recommended. Furthermore, the distribution system should allow for easy nozzle inspection and cleaning. A set of five pigtail-shaped nozzles ensures complete and homogeneous wetting of the packing, and reduces any problems associated with nozzle clogging. Nozzles may be configured in a five to seven-point star pattern with a central nozzle.

7.3.2.6
Modification of the Mist Eliminator

Demisters play a key role in packed-tower chemical scrubbers and are almost always necessary, since carry-over of water may lead to operational problems due to water accumulation in the discharge air ducts. These include early deterioration of the blower parts in scrubbers where the blower is located downstream of the scrubber (induced draft), and increased head losses. Mist is usually reduced by using common types of mist eliminators, such as mesh pads or small random-packed sections. Mesh pads have the highest performance, but are also the most maintenance-intensive (Card 2001). Randomly packed mist eliminators are said not to be very effective in removing fine liquid droplets from the outlet airstream, but the reduced liquid flow in the converted scrubbers reduces potential of mist formation. No problems were found with scrubbers converted at OCSD with mist eliminators composed of a bed support plate and a demisting height of 0.45 m of 2-in. TriPack packing.

7.3.2.7
Liquid Recycle Pump Replacement

A wide variety of conditions have been experienced for trickling rates in biotrickling filters, usually ranging from 0.01 to $5\,\mathrm{m\,h^{-1}}$, i.e., notably lower than the trickling velocities usually used in chemical scrubbers (Cox and Deshusses 2001). The liquid flow rate for biotrickling filters needs to be usually reduced by factors of 5–10 of that operated in chemical scrubbers (Gabriel and Deshusses 2003a). Therefore, recirculation pumps of chemical scrubbers are too powerful and would use more energy than needed for the desired flow rate in biotrickling filters.

Together with the replacement of the packing material, replacement of the liquid recycle pump in order to ensure a proper trickling rate may be a key factor in the conversion of the chemical scrubber. The trickling rate influences the wetting of the packing, and therefore the mass transfer of the pollutants to the biofilm. Excessive trickling water may lead to operational problems such as increased pressure drop, or compaction of the packed bed if a foam packing material is used. In contrast, insufficient trickling rates cause bed drying and reduced efficiencies.

Recirculation pumps with stainless steel or polypropylene pump heads may be installed in the case of H_2S treatment. It must be noted that in places such as trunkline areas where highly loaded streams must be treated, providing a backup liquid recycle pump is essential. Note that in Table 7.4, installing a lower-capacity liquid recycle pump implied re-piping of a section of the recycle line, due to the differences in the dimensions between suction and discharge ports of both liquid recycle pumps. Re-piping should also include an inoculation line location on the pump suction side, in order to ease the introduction of activated sludge for biotrickling filter startup and, if needed, for future re-inoculation of the system.

7.3.2.8
Modification of the Inlet/Outlet Air Ducts

In some cases, the inlet or outlet air ducts may need to be modified with the installation of additional ducting, dampers, flanges, tees, etc. to satisfy the requirement of residence time and flow velocity for optimal H_2S conversion. This is the conversion step the less commonly implemented in practice.

7.3.2.9
Installation of Secondary Effluent Supply

A nutrient source must be provided in any biological system and, in the case of biotrickling filters with packing materials commonly made of synthetic materials that do not provide any nutrients, an external source of nutrients must be provided. The biotrickling filter system may include a tank for the nutrient solution, which is best stored as a concentrated solution to avoid bacterial growth in the stock solution, but at wastewater treatment plants, secondary effluent has been successfully used as a nutrient source in pilot and full-scale applications of biotrickling filters (Wu et al. 2001; Gabriel and Deshusses 2003a). Secondary effluent needs to be properly characterized for nutrients, chlorine, pH, buffering capacity, and organic content. Thus, plumbing of water where secondary effluent is not available will depend on the quality and plant configuration, and proximity of a secondary effluent or industrial water supply. It must be highlighted that it is not uncommon that secondary effluent may contain a low concentration of free chlorine, which may inhibit biomass growth and pollutant removal in the biotrickling filter. Some means to reduce chlorine feeding may be needed if chlorine is regularly present above 1 ppm, or if chlorination spikes are expected to enter the biotrickling filter.

Additionally, secondary effluent can be conveniently used to control the pH of the biotrickling filter. Continuous feed and purge of secondary effluent serves to maintain the concentration of biological products such as sulfate and protons in the case of H_2S removal. Large sump volumes help control pH fluctuation dynamics caused by changing inlet H_2S concentration. No control other than maintaining a constant secondary effluent flow was needed at OCSD where sump volume to makeup water feed ratios of 200–250 min^{-1} were succesfully used, even under highly fluctuating H_2S inlet concentrations ranging from 15 to over 100 ppm_v.

7.3.2.10 Installation of New Packing Material

Packing material has been identified as the most important factor responsible for the difference between the performances of biotrickling filters. Means to improve the overall mass transfer rate and pollutant removal in biotrickling filters include the selection of packing materials with a high specific surface area. Our own experiences indicate improved performance with polyurethane foam cubes over other types of packing, especially at high gas flow rates with low H_2S concentrations. Different polyurethane foams can be acquired for biotrickling filtration purposes. In all conversions at OCSD, 4-cm cubes were used. According to the foam supplier (EDT AG, Eckental-Eckenhaid, Germany), the foam packing is made of open-pore polyurethane foam, with 10–15 pores per linear inch (PPI), specific surface area of 600 $\text{m}^2\,\text{m}^{-3}$, density of 35 $\text{kg}\,\text{m}^{-3}$, and porosity of 0.97. Foam packing is resistant to temperatures between −40 and 100 °C. Over time, a certain decrease in the bed height due to packing compaction has been observed at OCSD. Possible means to avoid such compaction and the associated pressure drop include using intermediate bed supports, using stiffer foam or larger foam cubes e.g., to form a structured bed (only if the contact time is above 5 s, because of excessive pressure drop), and intermittent trickling (Gabriel and Deshusses 2003b). Further investigations are needed to determine the best solutions.

7.3.2.11
Modification of the Controls

Although the biotrickling filter may be equipped with sophisticated sensors and control routines to implement remedial action via feedback controllers (Webster et al. 1999), it should be stressed that sophisticated controls are not always necessary. The choice of sophisticated control or mostly manual operation depends on the application, the size of the reactor, the regulatory agency, and the available budget. In any case, a minimum of control and instrumentation is desired to ensure safe operation in biotrickling filters, even if commonly less controls are required than for usual operation of chemical scrubbers.

Chemical scrubbers commonly are installed with pressure drop indicators, a pH sensor, flow indicators for makeup water supply, pressure gauges in the liquid recycle line, and a low-level switch/alarm at the bottom of the reactor to prevent the recycle pump to run dry. Also common is a control loop for chemical dosage based on the measure of the pH sensor. In some cases, chemical addition is linked to the measure of the H_2S outlet concentration in a feedback control scheme. Less common are feed-forward control schemes based on H_2S inlet concentration readings or usage of ORP sensors. Conversion of a fully equipped chemical scrubber will need disconnecting all controllers for chemical addition, low-level pH alarms. Also, monitoring is an important part of operation and maintenance. Proper monitoring helps in troubleshooting and in optimizing the operation of the reactor. Most important is the monitoring of pollutant removal. Thus, online inlet and outlet H_2S concentration monitoring is highly recommended.

7.3.3
H_2S Treatment of Converted Chemical Scrubbers at OCSD

OCSD reviewed its odor emission and control management and, in this context, explored the possibility of converting some of its wet scrubbers (usually caustic, hydrogen peroxide, and/or bleach scrubbers) to biotrickling filters. The challenge was to investigate the technical feasibility of converting wet scrubbers to biotrickling filters in a simple, easy retrofitting operation, and to operate them at very short gas contact times of 1.6–3 s. Five full-scale wet scrubbers treating odorous air from different locations at two different facilities at OCSD were converted, and H_2S and odor removal performance was studied. This section focuses on technical aspects such as startup and performance of biotrickling filters. In summary, the removal of H_2S reached elimination capacities of 105–110 g H_2S m^{-3} h^{-1}, consistently maintaining outlet concentrations well below the regulatory limits (1 ppm_v). The systems were very robust. Also, a cost-benefit analysis of the conversion demonstrated that biotrickling filtration is much more cost effective than chemical scrubbing, indicating the high viability of the scrubber conversion (Gabriel and Deshusses 2003c). This is true for both roughing and end-of-pipe scrubbers.

7.3.3.1
Startup of Biotrickling Filters

The startup phase in biotrickling filters needs inoculation with a microorganism consortium. Whereas some bioreactors treating slowly biodegradable compounds will startup faster if inoculated with a selected consortium or a pure culture, biotrickling filters for H_2S treatment may be started directly with activated sludge from a wastewater treatment plant. The bottom reservoir of biotrickling filters converted at OCSD was filled with a mixture of secondary effluent from the treatment plant and return activated sludge. The liquid was recycled over the bed for 24 h, without supply of foul air or makeup water to promote cell attachment onto the packing. This was enough for observing H_2S removal 2–3 days after startup in all cases. Establishment of fully performing biotrickling filters depends on the inlet H_2S concentration, and thus different startup behaviors may be encountered. H_2S removal efficiency in highly or overloaded biotrickling filters (average inlet H_2S well over 20 ppm_v for gas contact times ~ 2 s) kept increasing after the first 2–3 days. In such systems, H_2S elimination capacities can reach values well over 100 $g\,m^{-3}\,h^{-1}$, about 8–10 days after startup. Medium-loaded biotrickling filters (5–15 ppm_v average inlet H_2S) may follow a startup pattern similar to the highly loaded biotrickling filters, with the difference that close to 100% removal is reached within 8–10 days after startup. Low-loaded biotrickling filters (less than 2–3 ppm_v average inlet H_2S) differ from the above because the low concentration of H_2S at the inlet of the reactor causes low biomass growth in the reactor. The time needed to reach a quasi-steady-state removal efficiency may be up to twice the time required for medium-loaded biotrickling filters. Special attention should be paid to inlet load fluctuations, due to the difficulty of the reactor to deal properly with H_2S spikes during startup. Extra nutrients and other possible measures to promote growth may be warranted. Note that the low concentrations of biodegradable compounds such as some reduced sulfur compounds (RSCs) often found in the foul air of wastewater treatment plants may lead to longer startup periods for removing these RSCs, because of the slow development of suitable microorganism populations.

It must be stressed that the above patterns do not apply for a restart after a short operation stop, commonly due to maintenance in the units. In these cases, effective treatment is recovered a few hours after the restart.

7.3.3.2
Performance of Converted Chemical Scrubbers

Evaluation of operation of biotrickling filters at OCSD revealed that the biotrickling filters are capable of treating H_2S at rates comparable to those of chemical scrubbers. The long-term performance of a medium-loaded trunkline biotrickling filter at OCSD is shown in Fig. 7.2. At loadings between 15 and 95 g H_2S $m^{-3}\,h^{-1}$, the system consistently degraded more than 97% of the influent H_2S. Above 95 g H_2S $m^{-3}\,h^{-1}$, breakthrough of H_2S occurred, whereas a quasi-zero-order degradation regime was observed at loadings above 105 g

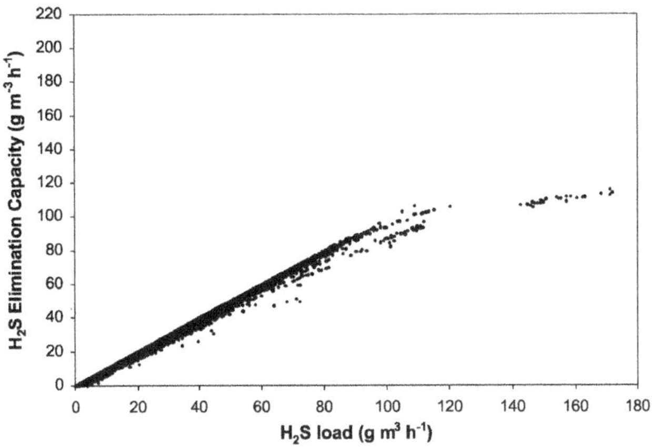

Fig. 7.2. Elimination capacity of H_2S versus H_2S loading in the biotrickling filter. Data include 8 months of continuous operation. Loading of 110 g H_2S m^{-3} h^{-1} achieved by spiking the inlet air with H_2S from a compressed gas cylinder

H_2S m^{-3} h^{-1}, with removal efficiencies gradually decreasing to below 80%. In general, removal efficiencies of 98% were commonly reached for inlet H_2S concentrations as high as 30 ppm$_v$ when the system was operated at gas contact times as low as 1.6–2.3 s. Removal efficiencies below 90% were also found at inlet concentrations below 0.2–0.5 ppm H_2S (Fig. 7.3, data shown for three converted scrubbers), possibly because of diffusion and/or kinetic limitations at such low concentrations.

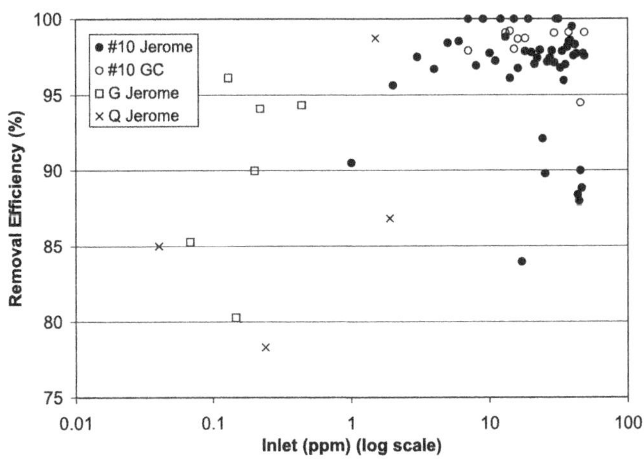

Fig. 7.3. H_2S removal in the three converted scrubbers at OCSD that treat average to low concentrations of H_2S. The log scale emphasizes the removal at low concentration. The legend (*Jerome* or *GC*) denotes the method for analysis of H_2S

Typical performance of a highly loaded trunkline biotrickling filter is shown in Fig. 7.4. The data show that lower removal efficiencies are achieved, presumably because of higher loadings, while the biotrickling filter reaches extremely high elimination capacities. This biotrickling filter, operated at 2.4 s, consistently received much higher inlet H_2S concentrations, in the range 45 to over 150 ppm_v, which probably helped developing a much larger density of H_2S-degrading microorganisms and favored faster elimination rates of H_2S.

While the focus of the OCSD project was on H_2S removal, the removal of reduced sulfur compounds (RSCs) other than H_2S, of air toxics (VOCs) and odors was also measured. Analysis of the removal of RSCs, VOCs, and odor proved to be a challenge because concentrations were always extremely low and variable, and removal was only partial. In addition, odor measurements rely on sample sniffing by odor panels and subjective judgment, and therefore it usually includes a large uncertainty. In summary, results indicate that concentrations of RSCs and VOCs were very low (ppb_v range) while odor concentrations were relatively high (values as high as 7,000 D/T were sometimes measured in the inlet air; D/T, dilution to threshold). Low concentrations and high variability probably lead to partial removal of most non-H_2S compounds. Evenso if the removal of such traces is impressive, if one considers that the air contact time is in the order of 2 s. In almost all cases, residual odor existed, which suggested that further treatment was required or that improvement of actual treatment performance was necessary. Residual odor was also observed when H_2S outlet concentrations were very low, suggesting that compounds other than H_2S, presumably RSCs and VOCs, were in great part responsible for the residual odor. On average, under standard conditions (pH $\sim$ 2, 1.6-s empty bed residence time, EBRT), RSC removal ranged between 15 and 40%, VOC removal was around 30%, while odor removal efficiencies were around

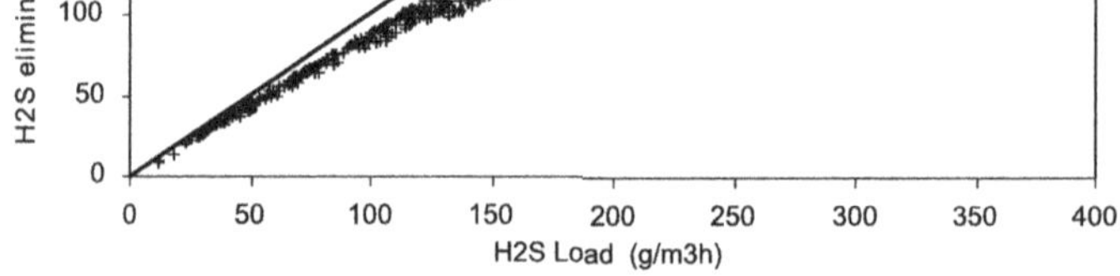

Fig. 7.4. Typical H_2S removal in biotrickling filter #I at OCSD, operated at gas contact times of $\sim$ 2.4 s. The performance shown is representative of the long-term operation of biotrickling filters exposed to high concentrations of H_2S (50 – 170 ppm). H_2S measured by electrochemical sensor

30–65% at 2,000–6,000 D/T level. Further investigations on the makeup of residual odors, and on means to improve treatment of residual odors are ongoing.

A concern with biotrickling filtration at OCSD is the pressure drop of the biotrickling filter. With the foam packing, pressure drop usually stayed at 25–30 cm of water column, which is higher compared to a clean and well-maintained chemical scrubber. The question of whether pressure drop was increased due to packing compaction at the bottom of the column was worthy of investigation. We observed qualitatively that bed compaction was proportional to the trickling rate, most probably because high tricking rates result in a higher dynamic holdup, hence a larger weight on the packing, causing more compaction. Very high trickling liquid velocities should therefore be avoided, as they cause higher pressure drops and result in no improvement of H_2S removal. An interesting observation after a period of intermittent trickling operation was that the bed expanded from the lower trickling rate, resulting in a 7–10 cm water column decrease in pressure drop. Intermittent trickling may be a means to reduce pressure drop, hence treatment costs. Another recommendation for future conversion is to avoid tall sections of bed, by splitting the bed into smaller subsections, with separated bed supports.

The biotrickling filter performance was exceptionally high compared with other biofilters or biotrickling filters, removing low concentration of H_2S even at higher gas contact times (Chung et al. 2000; Koe and Yang 2000; Cox and Deshusses 2002b). It is believed that the high performance observed at OCSD is the result of a combination of factors. First, the packing material provided a large surface area for high bacterial attachment and pollutant mass transfer. With respect to pollutant mass transfer, the air linear velocity in our systems is around 20 times higher than previously reported cases. A higher air velocity implies a more turbulent regime and higher gas–film mass transfer coefficients. This was proven to have a major impact on H_2S treatment by performing packing material removed from the OCSD biotrickling filter (Kim and Deshusses 2003). In short, the volumetric H_2S elimination capacity of single cubes of packing was determined at various air velocities in a laboratory-scale differential biotrickling filter. The elimination capacity of H_2S was found to increase as much as five- to tenfold when increasing the air velocity from $< 200\,\mathrm{m\,h^{-1}}$, i.e., the usual values in biofilters, to $6{,}300\,\mathrm{m\,h^{-1}}$, i.e., the air velocity imposed on the OCSD biotrickling filter (Kim and Deshusses 2003). Another factor contributing to the outstanding performance of the field biotrickling filter is the ample supply of micronutrients (via the secondary effluent feed) and CO_2 ($> 4{,}000\,\mathrm{ppm_v}$ in the foul air) to the process culture, such that they were never limiting the degradation of H_2S. Finally, both liquid and air were distributed homogeneously across the section of the scrubber, with no evident gas short-circuiting or extensive dry or dead zones in the packed bed. Therefore, optimum use of the biologically active section was achieved.

7.4 Conclusions

Biotrickling filtration is one of the several techniques available for biological treatment of contaminated air. It is probably one of the techniques with the greatest potential, as illustrated in this chapter with the successful conversion of chemical scrubbers to biotrickling filters for H_2S control. Biotrickling filtration is a maturing technology, and the number of full-scale biotrickling filters is rapidly increasing. At the same time, research is exploring new applications, which will open new possibilities for biotrickling filtration. Clearly, this means that there will be an increase in the use of environmentally friendly treatment techniques for air pollution control in the years to come.

References

Alonso C, Zhu X, Suidan MT, Kim BR, Kim BJ (1998) Modelling of the biodegradation process in a gas phase bioreactor – estimation of intrinsic parameters. In: Reynolds FE Jr (ed) Proc 1998 USC-TRG Conf Biofiltration. The Reynolds Group, Tustin, CA, pp 211–218

Card T (2001) Chemical odor scrubbing systems. In: Stuetz R, Frechen FB (eds) Odors in wastewater treatment: measuring, modeling and control. International Water Association, London, pp 309–344

Chung YC, Huang C, Tseng CP, Pan JR (2000) Biotreatment of H_2S and NH_3-containing waste gases by co-immobilized cells biofilter. Chemosphere 41:329–336

Cox HHJ, Deshusses MA (1998) Biological waste air treatment in biotrickling filters. Curr Opin Biotechnol 9:256–262

Cox HHJ, Deshusses MA (1999) Biomass control in waste air biotrickling filters by protozoan predation. Biotechnol Bioeng 62:216–224

Cox HHJ, Deshusses MA (2001) Biotrickling filters. In: Kennes C, Veiga MC (eds) Bioreactors for waste gas treatment. Kluwer, Dordrecht

Cox HHJ, Deshusses MA (2002a) Biotrickling filters for air pollution control. In: Bitton G (ed) The encyclopedia of environmental microbiology, vol 2. Wiley, New York, pp 782–795

Cox HHJ, Deshusses MA (2002b) Co-treatment of H_2S and toluene in a biotrickling filter. Chem Eng J 87:101–110

Cox HHJ, Nguyen TT, Deshusses MA (2000) Toluene degradation in the recycle liquid of biotrickling filters for air pollution control. Appl Microbiol Biotechnol 54:133–137

Cox HHJ, Sexton T, Shareefdeen ZM, Deshusses MA (2001) Thermophilic biotrickling filtration of ethanol vapors. Environ Sci Technol 35:2612–2619

Deshusses MA (2004) Application of immobilized cells for air pollution control. In: Nedovic V, Willaert R (eds) Cell immobilization biotechnology. Kluwer, Dordrecht (in press)

Deshusses MA, Cox HHJ (1999) A cost benefit approach to reactor sizing and nutrient supply for biotrickling filters for air pollution control. Environ Prog 18:188–196

Devinny JS, Webster TS, Torres E, Basrai S (1995) Biofiltration for removal of PCE and TCE vapors from contaminated air. Hazard Waste Hazard Mater 12:283–293

Devinny JS, Deshusses MA, Webster TS (1999) Biofiltration for air pollution control. Lewisl, Boca Raton

Gabriel D, Deshusses MA (2003a) Retrofitting existing chemical scrubbers to biotrickling filters for H_2S emission control. Proc Natl Acad Sci USA 100:6308–6312

Gabriel D, Deshusses MA (2003b) Performance of a full-scale biotrickling filter treating H_2S at a gas contact time of 1.6–2.2 seconds. Environ Prog 22:111–118

Gabriel D, Deshusses MA (2003c) Technical and economical analysis of the conversion of a full-scale scrubber to a biotrickling filter. In: Proc 2nd IWA Int Conf Odour and VOCs Measurement, Regulation and Control Techniques, 14–17 September 2003, Singapore, Pap 68

Holubar P, Andorfer C, Braun R (1999) Effects of nitrogen limitation on biofilm formation in a hydrocarbon-degrading trickle-bed filter. Appl Microbiol Biotechnol 51:536–540

Kim S, Deshusses MA (2003) Development and experimental validation of a conceptual model for biotrickling filtration of H_2S. Environ Prog 22:119–128

Koe LCC, Yang F (2000) A bioscrubber for hydrogen sulfide removal. Water Sci Technol 41:141–145

Koe L, Wu L, Loo YY, Wu Y, Chai JW, Koh YM (2002) A successful conversion of a chemical scrubber to a biotrickling filter – some experiences. In: Proc ENVIRO 2002 Conv Exhibition, 8–12 April 2002, Melbourne, Australia. International Water Association, London, Pap e21300d

Kong Z, Farhana L, Fulthorpe RR, Allen DG (2001) Treatment of volatile organic compounds in a biotrickling filter under thermophilic conditions. Environ Sci Technol 35:4347–4352

Mpanias CJ, Baltzis BC (1998) An experimental and modeling study on the removal of mono-chlorobenzene vapor in biotrickling filters. Biotechnol Bioeng 59:328–343

Oh YS, Bartha R (1997) Removal of nitrobenzene vapors by a trickling air biofilter. J Ind Microbiol Biotechnol 18:293–296

Smet E, Lens P, Van Langenhove H (1998) Treatment of waste gases contaminated with odorous sulfur compounds. Crit Rev Environ Sci Technol 28:89–117

Weber FJ, Hartmans S (1996) Prevention of clogging in a biological trickle-bed reactor removing toluene from contaminated air. Biotechnol Bioeng 50:91–97

Webster TS, Cox HHJ, Deshusses MA (1999) Resolving operational and performance problems encountered in the use of a pilot/full-scale biotrickling filter reactor. Environ Prog 18:162–172

Woertz JR, van Heiningen WNM, van Eekert MHA, Kraakman NJR, Kinney KA, van Groenestijn JW (2002) Dynamic bioreactor operation: effects of packing material and mite predation on toluene removal from off-gas. Appl Microbiol Biotechnol 58:690–694

Won YS, Cox HHJ, Walton WE, Deshusses MA (2002) An environmentally friendly method for controlling biomass in biotrickling filters for air pollution control. In: Proc Annu Meet Exhibition Air and Waste Management Association, Pittsburgh, PA, Pap no 43554, p 12

Wu L, Loo YY, Koe LCC (2001) A pilot study of a biotrickling filter for the treatment of odorous sewage air. Water Sci Technol 44:295–299

Wübker SM, Friedrich C (1996) Reduction of biomass in a bioscrubber for waste gas treatment by limited supply of phosphate and potassium ions. Appl Microbiol Biotechnol 46:475–480

Yang YH, Allen ER (1994) Biofiltration control of hydrogen sulfide. 1. Design and operational parameters. J Air Waste Manage Assoc 44:863–868

8 Bioscrubber Technology

Ajay Singh, Zarook Shareefdeen, Owen P. Ward

8.1 Introduction

Waste gases from food, animal and fish feed industries, rendering plants, chemical production, composting plants, waste handling and wastewater treatment processes, produce undesirable odors, which contribute significantly to atmospheric pollution. In wastewater treatment plants (WWTPs), odors are mainly generated during anaerobic biodegradation of sewage. Other odors associated with WWTPs come directly from industrial wastewater-containing solvents, petroleum derivatives, and volatile organic compounds (VOCs). Odors are generated by a number of chemical components, the most significant being the sulfur compounds, hydrogen sulfide (H_2S) and mercaptan (Burgess et al. 2001). Domestic sewage contains 3–6 mg l^{-1} of organic S, mainly arising from proteinaceous materials, about 4 mg l^{-1} of sulfonates arising from household detergents, and 30–60 mg inorganic sulfur as sulfates. While nitrogen-based odor compounds and the organics associated with wastewater treatment are important, H_2S is the predominant odor-causing compound, and is also corrosive to mild steel and concrete (Jensen and Webb 1995).

In order to mitigate the nuisance impact, odorous emissions generated at the open sources such as inlet works, primary sedimentation units, aeration tanks, final clarifiers, sludge processing units, and wastewater channels are being contained by covering and conveying the waste gas for treatment before discharging to the open atmosphere. Industrial waste gases have traditionally been treated by physico-chemical methods such as condensation adsorption, absorption, scrubbing, and oxidation.

Selection of a particular technology, or combination of technologies, is dependent on factors such as site characteristics, treatment objectives, waste gas airflow rates and characteristics, and concentration of odorous compound. Various technologies available for the treatment of odors and air pollution are described in Table 8.1. Physical removal is typically by adsorption or absorption. However, physical processes only transfer the VOCs to another phase (solid or liquid), which still requires treatment or disposal. Chemical removal is typically mediated by thermal or catalytic oxidation. In biological systems, the pollutants diffuse into the liquid phase where microorganisms degrade them into CO_2 and H_2O. Biological waste gas treatment

Shareefdeen/Singh (Eds.)
Biotechnology for Odor and Air Pollution Control

Table 8.1. Industrial waste gas treatment technologies

Physical	Chemical	Biological
Condensation	Chemical scrubber	Biofilter
Adsorption – activated carbon	Thermal oxidation	Biotrickling filter
Absorption	Catalytic oxidation	Bioscrubber
Membrane technology	Ozonation	Activated sludge reactor
	UV oxidation	Membrane bioreactors Rotating biological contractors

has become an attractive, new treatment alternative in the last two decades (Ottengraf and Diks 1992; Shareefdeen and Baltzis 1994; Edwards and Nirmalakhandan 1996; Kennes and Thalasso 1998; Burgess et al. 2001). Three major bioreactor designs are usually considered for the treatment of air pollutants, i.e., the biofilter, the biotrickling filter, and the bioscrubber, although other design alternatives have been proposed (Ottengraf 1987; van Groenestijn and Hesselink 1993; Kennes and Thalasso 1998; Shareefdeen et al. 2002).

8.2 Bioscrubbers

In the literature, the term 'bioscrubber' is often used to describe biotrickling filters. Some authors have distinguished two bioreactor types, the 'suspended-growth bioscrubber' and the 'fixed-film bioscrubber' (Ockeloen et al. 1996; Koe and Yang 2000). In this chapter, the bioscrubber means a 'suspended-growth bioscrubber', which is differentiated from the fixed-film biotrickling filters. A comparison of three major technologies for biological gas treatment is presented in Table 8.2.

The bioscrubber is a three-phase, fluidized-bed bioreactor system, which is quite different from a biofilter or a biotrickling filter (Fig. 8.1). Bioscrubbing consists of the physical separation or absorption of VOCs, in an aqueous phase in an absorber column, and subsequent biological treatment in a second stage in a liquid-phase bioreactor. The effluent leaving the bioreactor is recirculated to the top of the absorber unit, providing excellent gas cleaning of highly soluble pollutants (Kennes and Thalasso 1998; Humeau et al. 2000; van Groenestijn 2001).

The preferred gas–liquid contactor or absorber for this application is a packed-bed column and countercurrent flow of gas and water. The liquid-phase bioreactor is an aerated tank with a highly active microbial population and nutrient medium. To achieve a high biodegradation rate, optimized levels of oxygen and nutrients are added to the bioreactor, and the pH of the medium

Table 8.2. Comparison of major technologies for waste gas treatment

Characteristics	Biofilter	Biotrickling filter	Bioscrubber
Reactor design	Single reactor	Single reactor	Two reactors
Capital and operating cost	Low	Comparatively higher	Comparatively higher
Carrier	Organic or synthetic	Synthetic	No carrier
Area	Large area required	Compact equipment	Smaller volume of equipment
Mobile phase	Gas	Liquid	Liquid
Surface area	High	Low	Low
Process control	Limited process control	Limited process control	Good process control
Gas flow rate	100–150 $m^3 m^{-2} h^{-1}$		3,000–4,000 $m^3 m^{-2} h^{-1}$
Operation	Easy startup and operation	Relatively complicated startup procedure	Relatively complicated startup procedure
Operational stability	Channeling of airflow common	Channeling of water is common	High operational stability
Pressure drop	Medium to high	Medium to high	Low
Target compound conc.	$< 1 g m^{-3}$	$< 0.5 g m^{-3}$	$< 5 g m^{-3}$
Suitable for compounds with Henry coefficient	< 1	< 0.1	< 0.01
Nutrients	Nutrients cannot be added	Ability to add and control nutrients	Ability to add and control nutrients
Biomass	Fixed biomass	Fixed biomass	Suspended biomass
Clogging of packing	Clogging problem	Clogging problem	No clogging problem
Excess sludge	No such problems	Disposal of excess sludge required	Disposal of excess sludge required

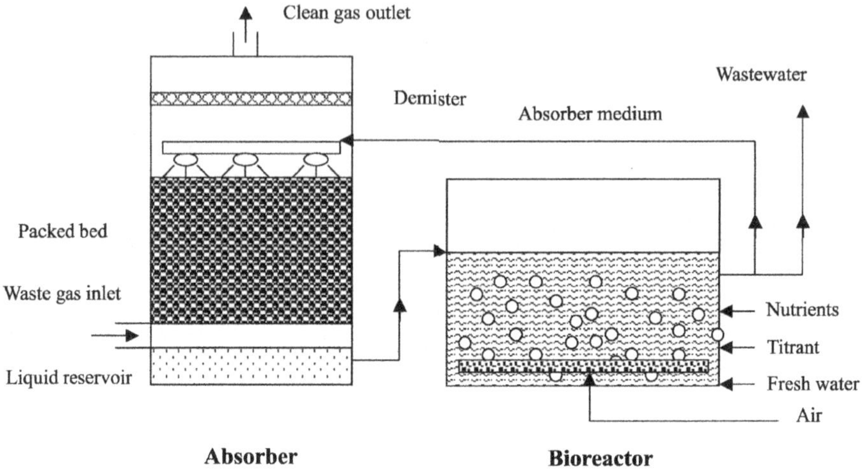

Fig. 8.1. Design of a suspended-growth bioscrubber. The waste gas is introduced into the absorber where pollutants are transferred to the water phase, which is then transferred to the bioreactor for biodegradation by an active microbial culture

is controlled in an optimum range of biological activity and/or absorption processes, using acid or alkali (Burgess et al. 2001). Suspended biomass and dissolved by-products may accumulate in the bioreactor, and a wastewater stream may be produced. To replenish any wasted or evaporated water, freshwater is added regularly.

The main advantages of bioscrubber technology are:

Smaller equipment volume, which is generally appreciated by the industry even if slightly higher capital and operating costs are involved

No clogging problems of packing material

Better process control of nutrient and pH

Suitable for elimination of water-soluble compounds such as H_2S, SO_2, alcohols, aldehydes, and fatty acids where pH control is important

Capable of handling large gas flow rates and high pollutant concentration

More reliable, predictable and stable operation

Lower concentrations of toxic byproducts generated in the reactor, due to capability of removing reaction products by washing

Good adaptation capacity of microbial biomass in response to the gas composition

The major drawbacks are considered to be:

- Requirement to dissolve the gaseous pollutants during the short residence time in the absorption column, therefore, less suitable for less water-soluble compounds
- Presumably cost-effective for pollutants with a dimensionless Henry's coefficient below 0.01
- Chances of slowest growing microorganisms being washed out
- Periods of stagnation or downtime are problematic and need to be avoided
- Excessive biomass formation may be needed to control the sludge produced in the bioreactor

Bioscrubbers were less popular than biofilters for treatment of VOCs, because most of the target VOCs in waste gas are volatile and less water-soluble. However, recent advances in bioscrubber process design and some successful applications indicate a growing interest in this technology (Le Cloirec et al. 2001). Some newer applications of bioscrubbers have been developed, for example, biological desulfurization of very large gas flow rates (up to 2×10^6 $m^3 h^{-1}$) seems to be feasible. Bioscrubbers are also considered to be one of very few examples of a technology that can be used for anaerobic waste gas treatment.

Henry's gas coefficients of some common waste gas contaminants from various industries are shown in Table 8.3.

8.3 Bioscrubber Design

In this section, the mechanism of odorous gas removal and design of the bioscrubber units and accessories are considered.

8.3.1 Mechanism for Odorous Gas Treatment by Bioscrubbers

In bioscrubbers, odorous compounds are removed by the following physical and biochemical mechanisms:

- *Absorption* – Initially odorous compounds are transferred from the gas phase into the aqueous phase. Mass transfer of the contaminant is a function of the surface area of the bubble, the contact time, and the diffusivity coefficient. Compounds such as H_2S are highly soluble in water.
- *Adsorption* – Higher molecular weight compounds with low water solubility may be physically adsorbed into the biological flocs.
- *Condensation* – Relatively warm, odorous gas condenses when it is transferred into the aqueous medium at lower temperature.

Table 8.3. Henry's coefficients and biodegradability of common waste gas contaminants

Pollutant	Henry's coefficient	Biodegradability[a]
Sulfur compounds		
Hydrogen sulfide	0.94	+ + +
Carbon disulfide	0.39	++
Nitrogen compounds		
Nitrobenzene	0.00098	+
Nitric oxide	n.a.	+
Aliphatic/aromatic hydrocarbons		
Hexane	74	++
Heptane	83.2	++
Styrene	0.11	++
Toluene	0.28	+ + +
Dichloromethane	0.093	+ + +
Chlorobenzenes	0.18	++
Oxygenated hydrocarbons		
Acetone	0.0016	+ + +
n-Butanol	0.00035	+ + +
Diethyl ether	0.028	+
Methanol	0.00019	+ + +
Methyl ethylketone	0.0024	+ + +
Methyl-*tert*-butyl ether	0.023	+
Propionaldehyde	0.0024	+ + +

[a] +, Poorly biodegradable; + + +, highly biodegradable; n.a., not available

Biodegradation or biotransformation – A high concentration of active microorganisms in the bioreactor is responsible for the significant conversion of odorous compounds, which are initially absorbed, adsorbed or condensed in the aqueous phase. Bioscrubbers can be characterized as containing heterotrophic or autotrophic microorganisms. Heterotrophic microbes must have a source of organic carbon to provide both energy and carbon for cell growth and synthesis. Autotrophic organisms get carbon from carbon dioxide in the airstream, and cellular energy for growth and respiration comes from oxidation of sulfide to either sulfate or elemental sulfur.

8.3.2 The Absorber

The primary function of an absorber unit is to provide a gas–liquid contactor under conditions favoring mass transfer of contaminants from a gas phase to an extensive area of the aqueous medium. In the case of bioscrubbers, the contactor divides the gas into small bubbles in a continuous liquid phase.

Table 8.4. Characteristics of commonly used absorbers

Absorber	Type	Typical gas velocity ($m\,s^{-1}$)	Pressure drop
Packed towers	Countercurrent	1.5–3.0	1.5–3.0 cm water m^{-1}
	Co-current	3–20	8–25 cm water m^{-1}
	Cross-flow	1.5–3.0	10 cm water total
Wet cyclone	Entrance	10–25	
	Superficial	3–5	10 cm water total
Spray tower	Countercurrent	1.0–2.5	8–25 cm water total
Venturi scrubber	Throat	40–50	20–30 cm water total

Characteristics of various types of absorbers/scrubbers are shown in Table 8.4 (Kohl and Nielsen 1997).

The packed tower absorbers are generally recommended for bioscrubbers, as other absorbers show poor elimination efficiencies for compounds with relatively poor solubility in water, i.e., compounds with Henry's coefficient around 0.01 (Kohl and Nielsen 1997). Although venturi scrubbers are best in entrainment of small mist drops ($< 10\,\mu m$), fumes ($< 1\,\mu m$) and fine dust (1–5 μm), they are less effective for transfer of VOCs. For bioscrubber operation, a countercurrent packed tower is considered better than co-current or cross-flow towers, mainly due to higher absorption efficiency, lower pressure drop, and lower associated energy costs, despite the larger volume compared to the co-current tower.

Countercurrent packed towers are comprised of the following elements (Kohl and Nielsen 1997; van Groenestijn 2001):

liquid reservoir or sump

gas inlet

packed bed filled and randomly placed with low-pressure, high-efficiency plastic packing

space with spray nozzles or other liquid distribution device

demister and a gas outlet

Conventional random packing materials are Raschig and Pall rings and saddles. Jaeger Tri-packs (Jaeger Products Inc.), Intalox Snowflake packing (Norton Chemical Process Products Corp.), Telleretes (Ceilcote Air Pollution Control) and Maspac (Clarkson Controls and Equipment Co.) are examples of some specially designed, low-pressure and high-efficiency packing material. The specific surface areas and diameters of packing material are in the range 100–300 $m^2\,m^{-3}$ and 2–8 cm, respectively.

For bioscrubber applications, some recommendations regarding the design of column and packing are as follows:

Construction material for the absorption column: the material should be resistant to corrosion caused by gases and liquids. The material for outdoor absorber should also be resistant to UV, and thermally and mechanically resistant against the temperatures and loading rates applied.

Column to packaging size ratios: a ratio of the diameter of the column and the packing material of at least 15 : 1 is recommended.

Height of packed column: due to potential deformability, plastic packing material should be limited to an unsupported height of 3–5 m.

Designing the absorber column: for the design of the absorber column (height, diameter) and water flow rate for a given gas flow, it is important to know the Henry's coefficient in order to calculate the requirement of water for a certain elimination efficiency.

Determination of absorber interphase requirement: mass transfer rate considerations give the required surface area for interphase mass transfer, and from where the packing volume can be calculated for each type of packing material.

Gas flow/pressure drop consideration: design for a gas flow rate that gives a pressure drop of less than 3 cm water m^{-1}. In the case of excessive foaming, less than 2 cm water m^{-1} is recommended.

VOC capacities: the gas-loading rate ranges between 4,000 and 10,000 $m^{-3}\,m^{-2}\,h^{-1}$ for readily soluble VOCs, which corresponds with gas velocities in the column of 1–3 $m\,s^{-1}$.

Absorber height/gas retention issues: the height of the absorbers used in bioscrubbing range from 2 to 8 m, which provides the gas retention time in the absorber as few seconds.

Water flow rate issues: for a good wetting of the packing surface, and to prevent biomass attachment (and clogging), 10–20 m^3 water m^{-2} column horizontal section h^{-1} is required.

When the polluted air stream is introduced into a countercurrent absorption column, the water leaving the column comes into contact with the influent gas. The concentration of a compound in the outgoing water cannot be higher than the concentration in equilibrium with the concentration in the gas phase. According to Henry's law:

$$H = C_g / C_l ,$$

where H is the dimensionless Henry's law coefficient ($g\,m^{-3}$ gas/$g\,m^{-3}$ water), C_g the contaminant concentration in gas phase ($g\,m^{-3}$), and C_l the contaminant concentration in liquid phase ($g\,m^{-3}$).

In order to achieve high (> 95%) elimination efficiency, the ratio between the water flow and gas flow should be at least H. The absorption factor is calculated as follows (Overcamp et al. 1993):

$$A = F_l / H F_g ,$$

where A is the dimensionless absorption factor, F_l the liquid flow rate ($m^3\,h^{-1}$), and F_g the gas flow rate ($m^3\,h^{-1}$).

For high elimination efficiency, an absorption factor of at least 1 is a prerequisite. Secondly, a sufficiently low concentration of the 'lean' absorption medium is required, which can be achieved by regeneration of the absorption medium in the bioreactor.

8.3.3 The Bioreactor

The aqueous effluent from the absorber contains the absorbed pollutant, which must be degraded in the bioreactor unit. The pollutants are converted by the active microbial culture into CO_2, H_2O, and biomass. The effluent 'lean' water is recirculated to and reused in the absorber. Most bioreactors in bioscrubbers are tanks, fitted with spargers for aeration, containing nutrient medium with a suspension of activated sludge. The properties and design of such bioreactors show great similarity to activated sludge tanks for wastewater treatment systems. The most important difference is that the hydraulic retention time is much longer, generally the same as the sludge retention time. Therefore, sludge retention facilities are not required in bioscrubbers.

In bioscrubbers, the volume of the bioreactor is normally larger than that of the absorber. The volume of the bioreactor ranges from 7 to 1,000 m^3, and the ratio between the gas flow rate ($m^3\,h^{-1}$) and bioreactor volume (m^3) ranges between 100 and 3,000, with a balanced average of 1,000 (VDI/DIN 1996). Compared with biofilters, which are generally loaded with about 100 m^3 gas m^{-3} filter bed h^{-1}, the volume of a bioscrubber is almost an order of magnitude lower than that of biofilters.

For bioscrubber applications, the following parameters need to be considered in determining the design of the bioreactor:

- Concentration and type of target compound
- Growth rate of microorganisms involved
- Required amount of biomass in bioreactor
- Specific substrate conversion rate
- Required amount of nutrient addition
- Required amount of titrant (alkali or acid) for pH control

Aeration system based on required amount of oxygen

Bioreactor volume and dilution rate

Engineering models of fixed-film bioscrubbers can be used to increase removal efficiencies by increasing the column height, decreasing the superficial gas velocity or the superficial liquid velocity, or by treating the liquid prior to recirculation to the absorber (Ockeloen et al. 1996).

8.3.4
Variations in Bioscrubber Designs

The design of a conventional suspended-growth bioscrubber, consisting of an absorption column and a fluidized-bed reactor, has been discussed above. Most existing bioscrubber designs focus on the removal of one type of compound. In order to provide more effective odor control or operational flexibility, various designs and improvements in conventional bioscrubber designs have been proposed by different groups of investigators.

8.3.4.1
Two-Stage Bioscrubber

Joyce and Sorensen (1999) have proposed a two-stage bioscrubber design. In the first-stage bioscrubber, autotrophic organisms (*Thiobacillus*) would be used for the oxidation of inorganic compound (H_2S), whereas a fixed-film biofilter (heterotrophic organisms) would be used for removal of VOCs in the second stage. Such bioreactors could be placed side-by-side for more convenient maintenance access or stacked in small footprint areas. However, more research is needed to optimize such a bioscrubber with respect to irrigation cycles, media type, nutrients, and biokinetics of both reactors.

The 'Dekoda' process from Vexamus (www.vexamus.com/dekoda.htm) is described as a high-rate bioscrubber and biofilter system for odor control. The system consists of two stages, where the bioscrubber first removes primary odor from the waste gas. In this unit, a continuously wetted medium with high surface area facilitates biological oxidation of pollutants such as hydrogen sulfide. The scrubbed air, in a second stage, is transferred to a polishing stage in a biofilter, using organic media to provide all necessary growth material for microorganisms.

The Thiopaq process ($H_2S + 1/2O_2 \rightarrow S^0 + H_2O$) is a desulfurization process for the production of elemental sulfur from H_2S-containing gas streams by sulfur-oxidizing bacteria (Buisman et al. 2000; Janssen and Buisman 2001). In this system, H_2S is first removed from the gas stream by absorption in a scrubber unit ($H_2S + OH \rightarrow HS^- + H_2O$), followed by biological oxidation of sulfide to elemental sulfur at neutral pH ($HS^- + 1/2O_2 \rightarrow S^0 + OH^-$). The bacteria deposit the elemental sulfur outside of the cell. The sulfur is separated in a sulfur separator and the percolation water is recycled to the scrubber.

8.3.4.2
Sorptive-Slurry Bioscrubber

The sorptive-slurry bioscrubber consists of a suspended-growth bioscrubber with powdered activated carbon (PAC) added to the biomass slurry (Kok 1992; Hammervold et al. 1995). Gaseous VOCs partition into the slurry in the absorber unit and adsorb onto the carbon. Adsorption on the column and any biodegradation in the absorber lowers the pollutant concentration in the liquid, which increases the concentration gradient and gas transfer. Biodegradation in the bioreactor unit continues to drive the pollutant concentration to low levels. The carbon is regenerated as the VOCs desorb and degrade, and is then recycled to the absorber. PAC also provides a surface for the growth of biofilms, and may assist in treating peak loads and in decreasing inhibition by toxic compounds. Addition of 2–5% PAC to biomass slurry is recommended to improve bioscrubber performance (Hammervold et al. 2000).

8.3.4.3
Anoxic Bioscrubber

Aerobic bioreactors are used to transform inorganic pollutants and degrade organic pollutants in conventional bioscrubbers, whereas systems combining scrubbers and upflow anaerobic sludge bed (USAB) bioreactors have been used for the degradation of perchloroethylene under anaerobic conditions (Parker et al. 1998). Anaerobic bioscrubbers have also been used to treat waste gases containing NO_X and SO_X (Janssen et al. 2000; Cetinkaya et al. 2000).

8.3.4.4
Two-Liquid Phase Bioscrubber

The concept of two-liquid phase bioscrubber stems from the limited application of conventional bioscrubbers for treatment only of pollutants that are readily soluble in water. The addition of an organic solvent to the water phase can enhance biodegradation of more hydrophobic compounds (Deziel et al. 1999), and facilitate the elimination of a range of hydrophilic and hydrophobic compounds. Besides improving bioavailability, solvents can reduce the toxicity of contaminants, and can act as a buffer system for fluctuating loads of pollutants. The addition of 10–30% water-immiscible high boiling-point solvent to the liquid phase facilitates the absorption of hydrophobic compounds from the gas phase in the absorber. Two-liquid phase bioreactor systems have been tested for the removal of alkanes, benzene, styrene, phenol, naphthalene, and pentachlorophenol. Solvents such silicon oil, paraffin oil, dibutyl phthalate, di-*n*-octyl phthalate, di-*n*-nonylphthalate, and pristane are good candidates for this application, and silicon oil has been found to be the best for two-liquid phase systems.

8.3.4.5
Airlift Bioscrubber

Edwards and Nirmalakhandan (1999) have described an airlift bioscrubber having a combined absorption/biodegradation reactor configuration for the removal of air-phase BTEX compounds. The reactor comprised two concentric tubes, with the inner tube shorter than the outer tube. As in some conventional airlift bioreactors, the inner tube serves as the downcomer, and the annular space between the two tubes works as riser (Ward 1989). Air was introduced into the reactor through the sparger, which is located near the bottom of the riser. The combined medium of air and water in the riser had a lower density than the water in the downcomer, and this resulted in a fluid circulation within the reactor. A mathematical model developed indicated that removal rates of $> 99\%$ can be achieved for benzene and toluene in airstream with concentrations $< 1{,}000$ ppmv, and the airlift bioscrubber should be operated at biomass concentrations of $2\,\mathrm{g\,l^{-1}}$ or greater for the greatest amount of operational stability.

8.4
Bioprocess Control in Bioscrubbers

8.4.1
Microbiology

The most significant advantage of a bioscrubber over a conventional biofilter and biotrickling filter is its ability to produce and sustain much greater quantities of microbial biomass in much smaller process units, while maintaining very high specific substrate utilization rates. The bioreactor can be started up by inoculation with activated sludge from wastewater treatment plants. Alternatively, an acclimated inoculum of a mixed culture can be used for the startup of the bioreactor. The acclimated inoculum can reduce the lag period in startup significantly. Adapted wastewater sludge or enriched digested sludge, with combinations of pure cultures, have commonly been used for biofilters and biotrickling filters.

Strains of *Thiobacillus* and *Hyphomicrobium* have often been identified as efficient degraders of sulfur-containing compounds (Jensen and Webb 1995; Kennes and Thalasso 1998). Heterotrophic *Xanthomonas* species have also been reported to remove H_2S from gas streams at the rate of $3.92\,\mathrm{mmol\,g^{-1}}$ dry cells $\mathrm{h^{-1}}$ (Cho et al. 1992). Immobilized cells of *Chlorobium limicola* can transform H_2S into elemental sulfur ($H_2S + 2CO_2 + 2H_2O \rightarrow 2CH_2O + H_2SO_4$) in an autotrophic reaction (Kim et al. 1990). Buisman et al. (1994) studied the kinetics of chemical and biological sulfide oxidation, and observed biological oxidation rates 7 and 75 times faster than the chemical oxidation rate at sulfide concentrations of about 150 and below $10\,\mathrm{mg\,l^{-1}}$, respectively. When provided with oxygen, bacteria consume ionic sulfide species and oxidize them to nonodorous sulfur species (Jensen and Webb 1995). It is also possible to recover sulfur for reuse from biological process ($2H_2S + O_2$ or $NO_3^- \rightarrow SO_4^{2-} + S^0 + 2H_2O$; Fox and Venkatsubbiah 1996; Sipma et al. 2003).

Various strains have been identified for the degradation of other pollutants, for example, acetone (*Rhodococcus*, *Corynebacterium*), butanol (*Pseudomonas*, *Rhodococcus*), ethanol (*Candida*, *Saccharomyces*), isopropanol (*Pseudomonas*), methanol (*Pseudomonas*), dichloromethane (*Hyphomicrobium*), methane (*Methylomonas*), MEK (*Rhodococccus*), phenol (*Pseudomonas*), 1,2-dichloroethane (*Xanthobacter*) and styrene (*Nocardia*; Jensen and Webb 1995; Kennes and Thalasso 1998; Burgess et al. 2001; Granström et al. 2002).

Since microbial growth rate is dependent on the environmental factors such as pH, temperature, ionic strength, presence of toxic compounds, and concentration of substrate, knowledge of the growth rate is required for calculations of the minimum biomass retention time and specific substrate conversion rate. The dependence of growth on limiting substrate concentration can be expressed by the Monod equation:

$$\mu = \frac{\mu_{max} C_l}{C_l + K_s} ,$$

where μ is the specific growth rate (h^{-1}), μ_{max} the maximum specific growth rate (h^{-1}), K_s the Monod constant ($g\,m^{-3}$), and C_l the substrate concentration ($g\,m^{-3}$).

In a completely mixed system, the substrate concentration C_l in the bioreactor is the same as that in the medium sent to the top of the absorber. The μ_{max} of many aerobic microbes growing on readily biodegradable organic compounds at optimum temperature (20–30 °C) and nutrient medium is generally higher than $0.1\,h^{-1}$. On the other hand, nitrifiers have μ_{max} of about $0.01\,h^{-1}$. The value of K_s is mostly around $1\,mg\,l^{-1}$.

To prevent washout of the active biomass, the dilution rate (volume fresh water added per hour per total volume of medium in the bioreactor) should not be higher than the maximum specific growth rate (Ward 1989). The required amount of biomass in the reactor can be calculated using the specific substrate conversion rate (amount of substrate converted per gram biomass per hour):

$$\text{Total rate of consumption (per g biomass dry wt.)} = \mu/Y + m ,$$

where Y is the yield (g biomass dry weight produced per gram substrate consumed), and m the maintenance coefficient.

The Michaelis-Menten equation may also be used to calculate specific substrate conversion rate:

$$V = \frac{V_{max} C_l}{C_l + K_m} ,$$

where V is the specific substrate conversion rate (g substrate g dry biomass^{-1} h^{-1}), V_{max} the maximum substrate conversion rate (g substrate g^{-1} dry biomass h^{-1}), and K_m the Michaelis-Menten constant ($g\,m^{-3}$).

Although the required amount of biomass can be calculated using the substrate conversion rate and expected pollutant-loading rate, the kinetic parameters are not always known while designing a bioreactor unit of a bioscrubber. Therefore, the bioreactor may be oversized, based on the estimated parameters to cope with fluctuating loading rates and stable performance. The concentration of biomass in a bioreactor for bioscrubber applications ranges between 1 and 15 g dry wt., and higher concentrations may lead to the clogging of the absorber packing (Schippert 1994; VDI/DIN 1996). The volumetric conversion rates are frequently 20–200 $\mathrm{g\,m^{-3}\,h^{-1}}$, with a maximum 400 $\mathrm{g\,m^{-3}\,h^{-1}}$ for readily biodegradable compounds. In practice, 0.1–0.4 g dry biomass $\mathrm{g^{-1}}$ substrate consumed with a biomass loading rate of 0.005–0.006 g substrate $\mathrm{g^{-1}}$ dry biomass has been observed.

Typical empty bed residence times (EBRT) are between 10 and 15 s for conventional bioscrubbers. However, some bioscrubbers operating with air streams solely consisting of organic compounds may have EBRTs longer than 30 s.

8.4.2 Nutrients

Generally, trace metals are present in either waste gas, water, or other raw materials. Often, phosphorus is the limiting factor, and sources of N and P are added as nutrients. Effluent water from WWTPs is a cheap and excellent source of these nutrients.

As compared to a biofilter, suspended biomass in a bioreactor is more sensitive to the periods without substrate. If the bioreactor is not operating for more than a couple of days, it is recommended to continue aeration and add substrate to maintain high microbial activity.

The addition of nutrients and titrants increases the ionic strength of the medium. Measurement of electrical conductivity gives a good indication of the ionic strength of the medium. A maximum of 5 $\mathrm{mS\,cm^{-1}}$ is recommended (VDI/DIN 1996), or 10 g dissolved inorganics $\mathrm{l^{-1}}$ (van Groenestijn 2001). In order to maintain these conditions, water may be slowly wasted and replaced with fresh water.

8.4.3 Oxygen

Oxygen may be added by sparging air into the bioreactor. From the waste gas flow rate, gas composition, waste gas COD, and the biological reaction equations, the required amount of oxygen can be calculated. A minimum concentration of 1–2 $\mathrm{mg\,l^{-1}}$ of dissolved oxygen is recommended for sufficient biological activity. It should be noted here that in bioreactors with a water height of 3 m, only about 10% of the oxygen in the used air is absorbed in the water phase. The efficiency of aeration is generally more than 1 kg O_2 transferred $\mathrm{kW\,h^{-1}}$ air blower energy consumed.

Bubble size is an important factor in gas treatment in the bioreactor. Fine bubble diffusers can provide better reduction (> 99.5%) of both odor and H_2S as compared to coarse bubble diffusers for reduction of odor (95%) and H_2S (92%; Bowker 2000).

8.4.4 pH and Temperature

pH is an essential parameter for H_2S removal. To maintain high cleaning efficiency and avoid any inhibitory effects, pH should be kept constant. Sodium hydroxide is dosed on the basis of continuous pH measurements. The pH optimum has been found to be 8.5–9.0, where a high biological activity could be maintained and an effective absorption of H_2S is obtained at the same time (Hansen and Rindel 2001). On the other hand, for water-soluble VOCs (ethanol, methanol, acetone, etc.), the pH in the bioreactor unit is maintained between 6 and 7 (DeHollander et al. 1998; Hammervold et al. 2000; Granström et al. 2002).

Despite the importance of temperature on microbial activity, no systematic work on the effect of temperature on bioscrubbing has been reported in the literature. Bioscrubber bioreactors have been operated to control ethanol emissions at 22–35 °C (Croonenberghs et al. 1994), and to remove VOCs from a printing press air at 19–25 °C (Granström et al. 2002). The efficiency of sulfide oxidation by *Thiobacillus denitrificans* dropped significantly below 16 °C, and the optimum temperature range was found to be 25–35 °C (Sublette et al. 1998).

8.4.5 Sludge Accumulation and Disposal

The sludge accumulation is generally lower than that experienced in a biological WWTP for the same organic loading rate. The sludge is also less toxic than that of WWTPs where heavy metals may have accumulated. In bioscrubber bioreactors, the sludge forms much smaller flocs as compared to wastewater treatment systems. Consequently, separation of sludge by sedimentation or floatation is difficult, and may require use of filtration or centrifugation methods. Where permitted, the sludge can be discharged in a sewer system, together with the wastewater.

Reduction of biomass accumulation in a bioscrubber by limiting the supply of phosphate and potassium ions has been suggested (Wübker and Friedrich 1996; Wübker et al. 1997). The residence time of water is 20–40 days. It is recommended that wastewater should be discharged from the tank at a rate of 0.2% of tank volume per hour. Addition of fresh water compensates for wastewater discharge and evaporation losses. A corrosion-resistant, positive displacement pump or a slow-running centrifugal pump is recommended to minimize destruction of biomass flocs.

8.5 Application of Bioscrubbers

8.5.1 Waste Gases from Wastewater Treatment Plant

Odor is produced from open water surfaces wherever anaerobic conditions exist. The five most dominant odor compounds at sewage treatment plants are H_2S, dimethyl sulfide, methyl mercaptan, NH_3, and trimethyl amine (Nurul Islam et al. 1998). Both biofilters and bioscrubbers have been applied to treat odors from wastewater treatment plants. Treatments using biofilters sometimes cause problems, with acidification resulting from the bioconversion of sulfur and nitrogen compounds. On the other hand, bioscrubbers show limited absorption efficiencies for relatively hydrophobic organo-sulfur compounds. However, developments in bioscrubber technology for the removal of odorous compounds from wastewater treatment have recently shown significant promise. Some examples of successful industrial applications of bioscrubbers are provided in Table 8.5.

Existing physico-chemical methods for the treatment of odorous sulfur compounds of wastewater treatment plants are generally expensive, due to the consumption of a large amount of chemicals and the requirement of a substantial labor input to replenish, control, and maintain chemicals and equipment for sulfur removal. Chemical treatment methods consume more energy, and could be more expansive than biological methods by 50–100% (Comas et al. 1999; Einarsen et al. 2000). Thus, biological methods have gained much attention over physico-chemical methods recently, not only due to their cost-effectiveness (low capital and operating costs), but also because of treat-

Table 8.5. Applications of bioscrubbers for waste gas treatment in various industries[a]

Industry	Waste gas compound	Pollutant conc. ($mg\,m^{-3}$)	Waste gas flow rate ($m^3\,h^{-1}$)	Elimination efficiency (%)	Reference
WWTP	H_2S	0–75	6,000	99	Hansen and Rindel (2001)
WWTP	Organo-sulfur	4,000–22,000 odor units	120,000	80	Diehl and Schafer-Treffenfeldt (1997)
Biogas	H_2S	10,000–15,000	400	> 99	Dijkman (1995)
Biogas	H_2S	2,000	40	< 99	Nishimura and Yoda (1997)
Flue gas	SO_2	1,000	6,000	> 98	Janssen et al. (2000)

[a] Activated sludge was used as an inoculum in the bioreactor

ment efficiency and environmental acceptability (by-products such as biomass and spent support material are innocuous) by both public and environmental regulators.

A bioscrubber with polishing filter for application at the Damhusaaen Wastewater Treatment Plant in Denmark has been described in detail (Hansen 1998; Hansen and Rindel 2000, 2001). The main design data for the bioscrubber are provided below. The data in parenthesis represent the variation in design parameters for designing the bioscrubber plant.

Airflow ($m^3\,h^{-1}$)	6,000 (0–6,000)
Air temperature (°C)	15 (5–30)
Moisture content (% R.H.)	80 (40–100)
Inert dust ($mg\,m^{-3}$ dry air)	< 1 (0–5)
H_2S ($mg\,m^{-3}$ dry air)	10 (0–50)
Methyl mercaptan ($mg\,m^{-3}$ dry air)	2.5 (0–5)
Dimethyl sulfide ($mg\,m^{-3}$ dry air)	2.5 (0–5)
Dimethyl disulfide ($mg\,m^{-3}$ dry air)	0.25 (0–1)

The dimensions of the bioscrubber plant are as follows:

Absorber diameter (m)	1.6
Absorber height (m)	12
Packed volume (m^3)	12
Sump volume (m^3)	7
Liquid circulation rate ($m^3\,h^{-1}$)	70
Bioreactor diameter (m)	1.9
Bioreactor volume (m^3)	11
Polishing filter volume (m^3)	0.68

The main observations on the operation and performance of the Damhusaaen bioscrubber were:

The bioscrubber cleaned 6,000 m^3 waste gas h^{-1} from the inlet structure, i.e., primary pumping station, screen and aerated grit chamber.

The inlet concentration of H_2S was up to 75 $mg\,m^{-3}$, and outlet concentration was less than 0.1 $mg\,m^{-3}$, representing an elimination efficiency of more than 99%.

The bioprocess was started up by adding activated sludge from the aeration tanks of the Damhusaaen WWTP. After adaptation to hydrogen sulfide, the oxidation rate was five to ten times higher than that of the unadapted sludge.

Conductivity of the liquid was measured regularly, and salt concentration was kept below 2% to avoid inhibition of bacteria by discontinuous discharge of wastewater.

pH was optimum between 8.5 and 9.0, which was also effective in absorbing H_2S. However, CO_2 absorption was limited.

The sulfuric acid formed was neutralized with NaOH. The consumption of NaOH was stoichiometric (2.4 kg NaOH per kg H_2S) in relation to the amount of H_2S absorbed at pH 8.5–9.0 in the scrubber/absorber.

Consumption of antifoam agent was about 1 l day^{-1}.

The consumption of adsorber in the polishing filter was low, since almost all the H_2S was removed in the scrubber section. Approximately 400–500 kg $year^{-1}$ of adsorbent consumption is expected.

Continuous dosing of P, K, and N was necessary, together with trace metals and vitamins. However, this dosing could be avoided or reduced by using wastewater effluent as feed or makeup water for the scrubber.

The pressure drop in the scrubber section was stable, indicating no risk of clogging. This could be due to slow-growing sulfide-oxidizing bacteria and continuous recirculation of scrubbing medium.

Maintenance included control and servicing of the instruments, and occasional dosing of the nutrients, mainly phosphate.

Hansen and Rindel (2001) have provided comparative operating and capital costs for a chemical scrubber and a modern biofilter. Operating cost of the bioscrubber was about one-half of the operating cost for a chemical scrubber at an inlet concentration of 10 mg H_2S m^{-3}, and one-fourth at an inlet concentration of 50 mg H_2S m^{-3}. Therefore, the higher the inlet concentration, the more advantageous the biological process is. This is mainly due to the significantly lower consumption of chemicals in the bioscrubber. Under the above conditions, the operating cost of bioscrubber and a modern biofilter would be almost identical. The construction of a bioscrubber would be similar to a two-stage chemical scrubber, and less expensive than a modern biofilter.

In a WWTP waste gas treatment application, the bioscrubber bioreactor can be connected to an existing activated sludge tank, and liquid can be exchanged (Diehl and Schafer-Treffenfeldt 1997). A bioscrubber comprising an absorber with 200 m^3 liquid h^{-1} recirculation rate treated 120,000 m^3 waste gas h^{-1} with organo-sulfur compounds. About 25% of the liquid flow rate was recirculated over a bioreactor, and 5% was exchanged with an activated sludge tank. Almost no nutrient or chemical addition was required, except for small amount of titrants to maintain pH between 6.5 and 7.5. Over 80% of the odor units was removed from the influent gas containing 4,000–22,000 odor units.

Biogas is mostly produced at a site where both anaerobic and aerobic biological reactors for wastewater treatment are present. Dijkman (1995) has described a bioscrubber for 400 m^3 biogas h^{-1} with a packed spray tower absorber and a 72-m^3 bioreactor. Bioscrubbing achieved $>$ 99% removal of H_2S

(10,000–15,000 ppmv in influent, and 20–120 ppmv in effluent gas). As compared to a chemical (caustic soda) scrubber, the bioscrubber required 90% less chemicals.

In another biogas treatment application, an aerobic reactor from the WWTP was used as the bioreactor in a bioscrubber (Nishimura and Yoda 1997). The bioscrubber treated 40 m^3 biogas h^{-1} in a 3-m^3 absorber with 13 bubble trays over which 20 m^3 activated sludge was recirculated. The H_2S concentration in the waste gas was reduced from 2,000 to 20 ppmv.

8.5.2 Aerobic and Anaerobic Gas Treatment

Bioscrubbers have been reported to remove more than 99% of H_2S from aerobic gases. In the bioreactors, sulfide is converted to sulfate. Alternatively, the adsorbed sulfide can be biologically converted to elemental sulfur. This incomplete oxidation is carried out by *Thiobacillus* under oxygen-limiting, but aerated conditions (Janssen et al. 1997). A pilot bioscrubber was tested to treat 160 m^3 sour gas h^{-1} from a refinery. The influent and effluent H_2S concentrations were 70–500 and 0 ppmv, respectively.

Pagella and De Faveri (2000) have described a laboratory-scale system using a bubble column absorber with a gas residence time of 1.5–2.3 s and a fixed-bed bioreactor where the bioreactor converts Fe(II) to Fe(III), while Fe(III) solution absorbs H_2S and converts it to elemental sulfur. The absorber liquid contains Fe(III) at a low pH (< 3) to prevent precipitation of ferric hydrates. The chemical reaction is:

$$H_2S + 2Fe^{3+} + 2OH^- \rightarrow S^0 + 2Fe^{2+} + 2H_2O\ .$$

The biological reaction carried out by *Thiobacillus ferrooxidans* is:

$$2Fe^{2+} + H_2O + 1/2O_2 \rightarrow 2Fe^{3+} + 2OH^-\ .$$

8.5.3 Treatment of Flue Gases

Due to high cost of chemical scrubbers (caustic soda, limestone) in the treatment of SO_x from flue gases, biological alternatives using bioscrubbers have been developed (Cetinkaya et al. 2000; Janssen et al. 2000). The hot waste gases are passed on absorber in form of a reverse jet wet scrubber in which particulate and SO_x are absorbed. The liquid from the absorber is transferred to an anaerobic bioreactor. In the bioreactor, an electron donor (H_2 gas for large scale and methanol or ethanol for small scale) is required to biologically convert sulfites and sulfates to sulfide. Sulfide can be further treated using an aerobic reactor. In a pilot plant design, 6,000 m^3 gas h^{-1}, with a temperature of 120 °C, was treated using a bioscrubber consisting of a 6-m-high absorber tower and a 10-m-high anaerobic internal circulation reactor. The aerobic reactor was an airlift loop reactor. The bioscrubber was loaded with 6 kg SO_2 h^{-1}, and SO_x removal efficiency was 98%.

Cetinkaya et al. (2000) have described removal of NO_X from flue gas using a bioscrubber. Since the major problem with NO_X removal is that 95% of the NO_X is NO, a compound poorly soluble in water, Fe(II)EDTA is used in the scrubber liquid to form a nitrosyl complex of $Fe(II)(EDTA)NO^{2-}$. The scrubber liquid is further treated in an anaerobic bioreactor, using a biological denitrification process in the presence of ethanol. Removal efficiency of more than 80% has been reported.

8.5.4 Treatment of Waste Gas from Fish Feed Factory

A bioscrubber with capabilities of both nitrification and denitrification has been developed for the removal of ammonia (Hvidtfeldt Rasmussen et al. 1994). The bioscrubber treated 40,000 $m^3\,h^{-1}$ waste gas from a fish feed factory. During production hours, organic compounds were absorbed in the absorber where some nitrification took place, whereas denitrification took place in the anoxic sump. During non-production period, further nitrification took place in the same sump under aerobic conditions.

Use of separate absorbers, nitrification reactors, and denitrification reactors for the removal of ammonia from waste gases at a livestock breeding facility was described by van Groenestijn et al. (1997). Based on laboratory experiments, a bioscrubber unit comprising of a 2-m^3 absorber, a 4-m^3 trickling filter (nitrification), and a 1-m^3 bioreactor (denitrification) was designed to treat 8,000 m^3 waste gas h^{-1} containing 15 mg NH_3 m^{-3}.

8.5.5 Treatment of Waste Gas Containing VOCs

As discussed above, suspended-growth bioscrubbers have potentially very useful applications in treating pollutants with relatively low Henry's law coefficients ($H < 0.01$). The common VOCs that can be efficiently treated using bioscrubbers are ethanol, methanol, acetone, methyl ethyl ketone, and methyl isobutyl ketone. A laboratory-scale bioscrubber (114-mm-diameter absorption column and 15-l bioreactor) with inlet concentration of 50–100 ppm methanol has been reported to exhibit removal efficiency of 69–81% (De-Hollander et al. 1998). At an inlet gas-phase acetone concentration of 50 ppmv in a sorptive-slurry bioscrubber, 95% removal of acetone was observed using biomass slurry containing 2% powdered activated carbon (Hammervold et al. 2000).

Croonenberg et al. (1994) described a pilot-scale bioscrubber, consisting of a scrubber with 2 m^3 of 2-in. Jaeger spherical polypropylene packing, and a 946-l tank fitted with plastic diffusers for treating waste gas containing ethanol at a brewery. The bioscrubber treated 445 scfm gas flow with average loading rates of 0.12 lb. h^{-1} and average ethanol elimination efficiency of 91.7%.

Hecht et al. (1995) developed a bubble column bioreactor for cometabolic degradation of trichloroethylene (TCE) from waste gases. In this system, phenol was added as the growth substrate for *Pseudomonas cepacia* to degrade TCE.

Granström et al. (2002) described the performance of an onsite pilot-scale bioscrubber for the treatment of waste gas from a printing press. The major VOC of the printing press air was ethanol, with smaller amounts of ethyl acetate, 1-propanol, 2-propanol, 1-methoxy-2-propanol and 3-ethoxy-1-propanol. The bioscrubber consisted of a 6.3-l scrubber column and a 120-l bioreactor unit. The typical gas stream flow was 6,300 $m^3 h^{-1}$ with temperature of 50–60 °C. Total inlet concentration of 3.4 g carbon VOC m^{-3} was treated with 20 mg carbon VOC m^{-3} with removal efficiency of 99.4%. The pH of the bioreactor unit was 6–7, and temperature varied in the range 19–25 °C.

8.6 Conclusion and Future Directions

Bioscrubbers can be characterized as environmentally friendly systems, which do not use oxidizing chemicals. They are useful in the treatment of waste gases containing water-soluble pollutants from wastewater treatment plants, waste treatment plants, composting plants, fish industry, animal feed industry, and rendering plants. Depending on the concentration and type of pollutant, bioscrubbers may also be efficiently utilized to treat odorous gases from other industries.

Although the efficient use of bioscrubbers is limited for water-soluble compounds (Henry's law coefficient $H < 0.01$), they may provide substantial advantages for wastewater gas treatment because of their smaller space requirements, high loading rates, reliable operation, process control, low risk of clogging, and low operating cost. When high concentrations of contaminants are to be treated, bioscrubbers offer more advantages than conventional biofilters, biotrickling filters, and chemical scrubbers. Due to limited process control capabilities, biofilters are quickly acidified. In chemical scrubbers, the consumption of oxidizing chemical or consumption of activated carbon is also increased in the case of high inlet contaminant concentration. Combination of bioscrubbers and polishing steps may improve the treatment efficiency of gases with mixtures of hydrophobic and hydrophilic compounds.

Bioscrubbers offer higher elimination efficiencies for wastewater gases such as H_2S, NH_3, and organic sulfur compounds. Due to the acidifying nature of these substances, however, substantial oxidation and sulfuric acid production in the scrubber may cause the pH to drop and decrease mass transfer efficiency. The capacity of the adsorption section for handling higher H_2S concentration needs to be further improved. This could be achieved by increasing the buffering capacity of the scrubbing medium.

The bioscrubber application can be more attractive than the use of other conventional biofilter or biotrickling filter at relatively high pollutant concentration ($> 0.5\,g\,m^{-3}$). A potential disadvantage is the longer startup time, and this may be overcome by using an inoculum of an acclimated culture prepared from activated sludge operation.

Currently, biological gas treatment accounts for only about 2% of the total market of world waste gas treatment. However, lower costs and more environmentally friendly biological methods, used alone or in combination with physical methods, are expected to slowly replace physico-chemical methods for industrial waste gas treatment.

References

Bowker RPG (2000) Biological odor control by diffusion into activated sludge basin. Water Sci Technol 41:127–132

Buisman CJN, Dijkman H, Prins W, Verbraak P, den Hartog H (1994) Biological (flue) gas desulfurization. Lucht 11:135–137

Buisman CJN, Sorokin DY, Kuenen JG, Janssen AJH, Robertson LA (2000) Process for purification of gasses containing hydrogen sulphide. US Patent no 6,156,205

Burgess JE, Parsons SA, Stuetz RM (2001) Development in odor control and waste gas treatment biotechnology: a review. Biotechnol Adv 19:35–63

Cetinkaya B, Sahlin RK, Abma WR, Dijkman H, Meyer SF, Kampeter SM (2000) Control FCC flu-gas emission. Hyrocarb Process 79:55–62

Cho K-S, Hirai M, Shoda M (1992) Degradation of hydrogen sulfide by *Xanthomonas*sp. strain DY44 isolated from peat. Appl Environ Microbiol 58: 1183–1189

Comas J, Balaguer M, Poch M, Rigola M (1999) Pilot plant evaluation for hydrogen sulfide biological treatment: determination of optimal conditions linking experimental and mathematical modeling. Environ Technol 20:53–59

Croonenberghs J, Varani F, Le Fevre P (1994) Use of bioscrubbing to control ethanol emissions. In: Proc Air & Waste Management Association 87th Annu Meet, 19–24 June 1994, Cincinnati, pp 1–7

DeHollander GR, Overcamp TJ, Grady CPL Jr (1998) Performance of a suspended-growth bioscrubber for the control of methanol. J Air Waste Manage 48:872–876

Deziel E, Comeau Y, Villemeur R (1999) Two-liquid phase bioreactors for enhanced degradation of hydrophobic/toxic compounds. Biodegradation 10:219–233

Diehl B, Schafer-Treffenfeldt W (1997) Biowäscheraufbauverfahrensvarianten und Betriebserfahrungen mit einem neuen Biowäscherkonzept. In: Prins WL, van Ham J (eds) Proc Int Symp Biological Waste Gas Cleaning. VDI, Duesseldorf, pp 345–352

Dijkman H (1995) Biological gas desulfurization. Med Fac Lanbouw, University Ghent 60/4b, pp 2677–2684

Edwards FG, Nirmalakhandan N (1996) Biological treatment of airstreams contaminated with VOCs: an overview. Water Sci Technol 34:565–571

Edwards FG, Nirmalakhandan N (1999) Modeling an airlift bioscrubber for removal of air-phase BTEX. J Environ Eng 125:1062–1070

Einrasen AM, Esoy A, Rasmussen A-I, Bungum S, Sveberg M (2000) Biological prevention and removal of hydrogen sulfide in sludge at Lillehammer wastewater treatment plant. Water Sci Technol 41:175–187

Fox P, Venkatsubbiah V (1996) Coupled anaerobic/aerobic treatment of high-sulfate wastewater with sulfate reduction and biological sulfide oxidation. Water Sci Technol 34:359–366

Granström T, Lindberg P, Nummela J, Jokela J, Leisola M (2002) Biodegradation of VOCs from printing press air by an on-site pilot plant bioscrubber and laboratory scale continuous yeast cultures. Biodegradation 13:155–162

Hammervold RE, Overcamp TJ, Smets BF, Grady CPL Jr (1995) Experimental study of the sorptive slurry bioscrubber for acetone emissions. In: Proc Air & Waste Management Association 88th Annu Meet, 18–23 June 1995, San Antonio, pp 1–14

Hammervold RE, Overcamp TJ, Grady CPL Jr, Smets BF (2000) A sorptive slurry bioscrubber for the control of acetone. J Air Waste Manage Assoc 50:954–960

Hansen NG (1998) Odor's biological option. Water Qual Int July/August, pp 12–14

Hansen NG, Rindel K (2000) Bioscrubbing, an effective and economic solution to odor control at wastewater treatment plants. Water Sci Technol 41:155–164

Hansen NG, Rindel K (2001) Bioscrubber for treating waste gases from wastewater treatment plants. In: Kennes C, Veiga MC (eds) Bioreactors for waste gas treatment, Kluwer, Dordrecht, pp 285–298

Hecht V, Brebbermann D, Bremer P, Deckwer W-D (1995) Cometabolic degradation of trichloroethylene in a bubble column bioscrubber. Biotechnol Bioeng 47:461–469

Humeau P, Baleo J-N, Reynaud F, Bourcier J, Le Cloirec P (2000) Flow characterization in a gas-liquid column: application to a bioscrubber for the deodorization of waste gases. Water Sci Technol 41:191–198

Hvidtfeldt Rasmussen H, Hansen NG, Rindel K (1994) Treatment of odorous nitrogen compounds in a bioscrubber comprising simultaneous nitrification and denitrification. VDI Ber 1104:491–497

Janssen AJH, Buisman CJN (2001) Process for biological removal of sulphide. US Patent no 6,221,652

Janssen AJH, de Hoop K, Buisman CJN (1997) The removal of H_2S from air at a petrochemical plant. In: Prins WL, van Ham J (eds) Proc Int Symp Biological Waste Gas Cleaning. VDI, Duesseldorf, pp 359–364

Janssen AJH, Dijkman H, Janssen G (2000) Novel biological processes for the removal of H_2S and SO_2 from the gas streams. In: Lens P, Hulshoff PL (eds) Environmental technologies to treat sulfur pollution. IWA, London

Jensen AB, Webb C (1995) Treatment of H_2S-containing gases: a review of microbiological alternatives. Enzyme Microb Technol 17:2–10

Joyce J, Sorensen H (1999) Bioscrubber design: how to improve odor-control flexibility and operational effectiveness. Water Environ Technol 11:37–44

Kennes C, Thalasso F (1998) Waste gas biotreatment technology. J Chem Technol Biotechnol 72:303–319

Kim BW, Kim IK, Chang HN (1990) Bioconversion of hydrogen sulfide by free and immobilized cells of *Chlorobium thiosulfatophilum*. Biotechnol Lett 12:381–386

Koe LCC, Yang F (2000) A bioscrubber for hydrogen sulfide removal. Water Sci Technol 41:141–145

Kohl AL, Nielsen RB (1997) Gas purification, 5th edn. Gulf, Houston

Kok HJG (1992) Bioscrubbing of air contaminated with high concentrations of hydrocarbons. In: Dragt AJ, van Ham J (eds) Biotechniques for air pollution abatement and odor control policies. Elsevier, Amsterdam, pp 77–82

Le Cloirec P, Humeau P, Ramirez-Lopez EM (2001) Biotreatment of odors: control and performances of a biofilter and a bioscrubber. Water Sci Technol 44:219–226

Nishimura S, Yoda M (1997) Removal of hydrogen sulfide from an anaerobic biogas using a bioscrubber. Water Sci Technol 36:349–356

Nurul Islam AKM, Hanaki K, Matsuo T (1998) Fate of dissolved odorous compounds in sewage treatment plants. Water Sci Technol 38:337–344

Ockeloen HF, Overcamp TJ, Grady CPL Jr (1996) Engineering model for fixed-film bioscrubbers. J Environ Eng 122:191–197

Ottengraf SPP (1987) Biological systems for waste gas elimination. Trends Biotechnol 5:132–136

Ottengraf SPP, Diks RMM (1992) Process technology of biotechniques. In: Dragt AJ, van Ham J (eds) Biotechniques for air pollution abatement and odor control policies. Elsevier, Amsterdam, pp 17–32

Overcamp TJ, Chang H-C, Grady CPL Jr (1993) An integrated theory for suspended-growth bioscrubbers. Air Waste 43:753–759

Pagella C, De Faveri DM (2000) H_2S gas treatment by iron bioprocess. Chem Eng Sci 55:2185–2194

Parker WJ, Collins J, Wells J, Kennedy K (1998) Biological treatment of air streams containing chlorinated organic compounds. In: Proc Air & Waste Management Association 91st Annu Meet, 14–18 June 1998, San Diego

Schippert E (1994) Biowaeschertechnologie. VDI Ber 1104:39–56

Shareefdeen ZM, Baltzis BC (1994) Biological removal of hydrophobic solvent vapors from airstreams. In: Galindo E, Ramirez T (eds) Advances in bioprocess engineering. Kluwer, Dordrecht, pp 397–404

Shareefdeen ZM, Herner B, Webb D, Polenek S, Wilson S (2002) Removing volatile organic compound (VOC) emissions from a printed circuit board manufacturing facility using pilot- and commercial-scale biofilters. Environ Prog 21: 196–201

Sipma J, Janssen AJH, Hulshoff Pol LW, Lettinga G (2003) Development of a novel process for the biological conversion of H_2S and methanethiol to elemental sulfur. Biotechnol Bioeng 44:1–11

Sublete KL, Kolhatkar R, Raterman K (1998) Technological aspects of the microbial treatment of sulfide-rich wastewater: a case study. Biodegradation 9:259–271

van Groenestijn JW (2001) Bioscrubbers. In: Kennes C, Veiga MC (eds) Bioreactors for waste gas treatment. Kluwer, Dordrecht, pp 133–162 van Groenestijn JW, Hesselink PGM (1993) Biotechniques for air pollution control. Biodegradation 4:283–301

van Groenestijn JW, van Harkes MP, Baartmans RFW (1997) A novel bioscrubber for the removal of ammonia from off gases. In: Prins WL, van Ham J (eds) Proc Int Symp Biological Waste Gas Cleaning. VDI, Düsseldorf, pp 305–312

VDI/DIN (1996) VDI/DIN-Handbuch Reinhaltung der Luft, vol 6. Biological waste gas purification (VDI 3478). VDI, Düsseldorf

Ward OP (1989) Fermentation biotechnology. Open University Press, Milton Keynes, UK

Wübker S-M, Friedrich CG (1996) Reduction of biomass in a bioscrubber for waste gas treatment by limited supply of phosphate and potassium ions. Appl Microbiol Biotechnol 46:475–480

Wübker S-M, Laurenzis A, Werner U, Friedrich CG (1997) Controlled biomass formation and kinetics of toluene degradation in a bioscrubber and in a reactor with a periodically moved trickle-bed. Biotechnol Bioeng 55:686–692

9 Membrane Bioreactor Technology

Mark W. Fitch

9.1 Introduction

The application of membranes that allow permeation to the control of air pollution is seemingly counterintuitive but offers many advantages over conventional biofiltration. A membrane introduces a resistance to mass transfer. However, membranes have useful properties, and the first reported use of a membrane biofilter was based on the lack of removal of contaminants with low water solubility in bioscrubbers and biotrickling filters (Bäuerle et al. 1986). Although mass transfer coefficients are lower than observed in conventional biofilters, this separation of the biological phase offers advantages, as listed in Table 9.1.

In a typical membrane filter, the gas flows through a tubular or hollow fiber membrane that is surrounded by a biofilm in an aqueous phase. Contaminants in the gas phase pass through the membrane and are consumed in the biofilm. The liquid provides nutrients to the biofilm, and thus allows for some control of the biological phase.

Despite the various advantages of membrane biofilters when compared to conventional biofilters and biotrickling filters, membrane biofilters have been explored only at bench-scale at the time of this writing. This chapter presents an overview of membrane biofilter design and operating results reported in the literature. Very limited modeling of membrane biofilters has been reported, and these models are described here as they are quite different from the models of conventional biofilters described in Chapter 10.

9.2 Membrane Bioreactor Design

Design decisions for membrane biofilters must be made on both the micro- and the macro-scale. Such decisions are driven by the two mechanisms that limit performance – mass transfer and biological activity. On the micro-scale, the membrane material and form must be appropriate for the intended use. Because of the availability of many membrane units of various configurations, the macro-scale decision on reactor configuration may be independent of this

Shareefdeen/Singh (Eds.)
Biotechnology for Odor and Air Pollution Control

Table 9.1. Comparison of membrane biofilter and conventional biofilter limitations

Issue	Conventional biofilter	Membrane biofilter
Clogging and pressure drop	Major problem. Dominant topic in biofiltration (Devinny et al. 1999)	No gas-phase clogging, consistent modest pressure drop
Channeling, flooding or foaming	Major issues depending on operation	None
Mass transfer	Direct gas-to-biofilm	Significant membrane and liquid-phase resistances
Low water solubility pollutants	Poor removal due to mass transfer	Good removal
Nutrient control	Potentially difficult	Easy due to liquid-phase separation
Moisture control	Requires humidification of gas phase	Minor or negligible issue; condensation sometimes observed on gas side of membrane
Response to shock load	Poor response, related to nutrient status (Moe and Irvine 2000)	Poor response
Maximum pollutant load	$50-300\,\mathrm{g\,m^{-3}\,h^{-1}}$, $50-2{,}000\,\mathrm{ppm_V}$	Membrane flux limit should allow for treatment of tremendous loads
Acidic byproducts	Lowered pH, occasional bed death	pH control of liquid phase easy
High temperatures	Limited to 80 °C	With external radiator loop, 150 °C gas possible
Operation	Humidification required, bed replacement at 3–5 year intervals common	Periodic addition of nutrient solution
Capital cost	Low	High–dominant design issue
Full-scale use	Many	None to date

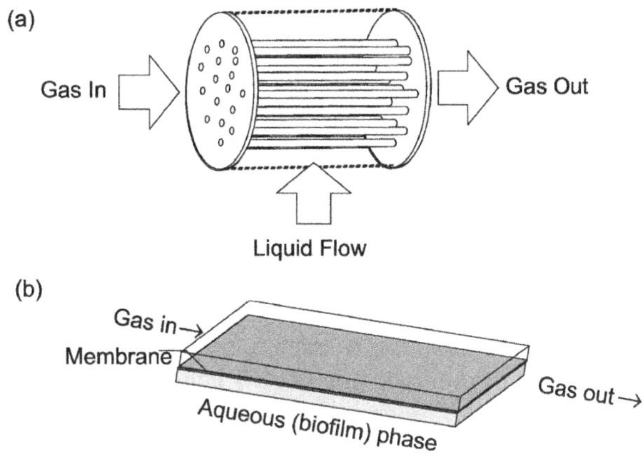

Fig. 9.1. a Hollow fiber membrane biofilter. b Flat sheet membrane biofilter

micro-scale membrane decision. Figure 9.1 shows the configuration of a typical membrane biofilter.

9.2.1 Mechanism

It is important to understand the fundamentals that drive design decisions. The membrane biofilter may be thought of as an advance in the area of pervaporation, in which volatiles transfer across a membrane between two phases. The use of a membrane to separate gas from organisms may create a large mass transfer resistance. As shown in Fig. 9.2, biofilters have high mass trans-

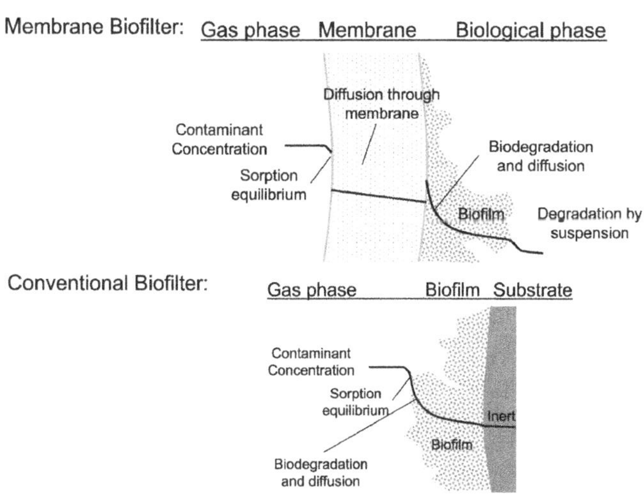

Fig. 9.2. Mass transfer and degradation in biofilters

fer rates because of direct gas–biofilm contact, but membrane biofilters have a gas–membrane–biofilm or a gas–membrane–liquid suspension mass transfer sequence that imposes a larger mass transfer resistance due to the membrane. In conventional biofilters, mass transfer is not generally considered limiting to the removal rate (Devinny et al. 1999), but in membrane biofiltration mass transfer almost certainly limits the removal rate. Therefore, minimizing mass transfer resistance is a major focus of design.

The mass transfer resistance in a membrane biofilter may be characterized as a series of resistances: gas film ($1/k_g$), membrane ($1/k_m$), and in some cases liquid film ($1/k_l$). The reported order of magnitude of these variables is $1/k_m > 1/k_l \gg 1/k_g$. The mass transfer coefficient for the gas thin film, k_g, is negligible. The liquid film mass transfer coefficient, k_l, is a function of liquid velocity and reactor design. The membrane mass transfer coefficient, k_m, is determined by the membrane material and thickness.

9.2.2 Membranes

Two types of membranes are used in membrane biofilters – dense phase and microporous. Dense-phase membranes generally have a lower permeability but do allow for some selectivity in permeation, while microporous membranes offer high rates of flux with no selectivity (Reij et al. 1998).

9.2.2.1 Dense-Phase Membranes

A dense-phase membrane has no pores for removal to occur – the contaminant must dissolve into the membrane and diffuse through the membrane. Therefore, mass transfer is promoted by using thin membranes that limit the diffusion path length. Also enhancing mass transfer is the use of membrane materials with very high gas diffusion coefficients. A high solubility of the contaminant in the membrane is desirable, although an excessively high solubility may be problematic, as described below. For high rates of gas transfer, silicone rubber has proven tremendously popular due to very high permeability.

9.2.2.2 Microporous Membranes

A microporous membrane has a highly porous structure, with typical commercial microporous membranes containing 30–85% pore space (Hartmans et al. 1992), including open pores in the surface of the membrane. The surface pores are generally submicrometer in size, which prevents organisms from passing through the pores. If the pore size distribution is not sufficiently controlled or the pore size is too large, then intrusion by organisms and organics will occur, significantly degrading mass transfer and potentially plugging the gas phase

(Hartmans et al. 1992; Attaway et al. 2002; Fitch et al. 2003). Microporous membranes are generally made of a hydrophobic material, so that water is unable to penetrate the pores except at significant trans-membrane pressures (Reij et al. 1998). Therefore, the pores remain gas-filled. Contaminants transport through these gas-filled pores much more rapidly than through dense-phase membranes because the gas presents a minimal diffusional barrier and no solubility issue exists. In theory, the mass transfer resistance of a microporous membrane is at worst one-tenth that of a silicone rubber dense-phase membrane. Operationally, this porous mass transfer may be negatively affected by operating conditions including extremes of pH and disinfection agents (Castro and Zander 1995).

9.2.2.3 Composite Membranes

Composite membranes, in which a dense-phase skin covers a porous support, have been considered (Loeb and Sourirajan 1960) as a third type of membrane. The advantage of composite membrane systems is that the dense phase can be very thin, on the order of 1–30 μm, because the porous support is responsible for structural stability. If the dense phase is 1–2 μm thick and the compound of interest is highly soluble, then a composite membrane has a mass transfer resistance similar to a polyporous membrane (de Bo 2003). However, thicker membranes or contaminants that are sparingly soluble in the dense phase impose the same issues of mass transfer resistance as any other dense-phase membrane.

9.2.2.4 Liquid- or Carrier-Mediated Membranes

Although currently no research has been done with membrane biofilters using carrier-mediated membranes, in which a carrier molecule shuttles a specific molecule across the membrane, the advantages in specificity of separation may prove sufficient to overcome issues of cost.

9.2.3 Materials

Various polymers may be used for the membrane, with choice dependent on a variety of factors. Widely adopted membrane-based separations such as dialysis and pervaporation (which is liquid/membrane/gas or vacuum) have resulted in a number of membrane units being commercially available in several configurations. Polypropylene (PP) and polysulfone, silicone rubber (PDMS), polyamide, polyethylene (PE), cellulose acetate (CA), polymethylpentene, polyimide (PI) and polytetrafluroethene (PTFE) have been used for various commercial membrane separations. PDMS, PP, PE, PTFE, and polysulfone have been reported as used for membrane biofilters (Table 9.2). Beyond commercial availability, the choice of polymer is dependent on the expected contaminant(s). Unlike most biofilter media, the membranes in membrane biofilters may be significantly affected by the contaminant. Polymers tend to swell in the presence of compatible solvents. For example, polysulfone, as shown in Table 9.2, would be expected to have a very high solubility for aromatic compounds. Indeed, benzene will act as a solvent for this polymer, resulting in swelling of the polymer due to sorption, and potentially even resulting in membrane failure due to dissolution of the polymer. A high solubility is not necessarily bad, since dense-phase permeation is described as $P = S_m D_m$ (Crank 1975), where S_m is the solubility coefficient in the polymer, and D_m is the diffusion coefficient. Thus, a high solubility is desirable, as there will be a decrease in mass transfer resistance as compared to polymers with similar diffusion coefficients but lower solubility. Changing functional groups of PDMS to enhance solubility has been shown to increase pervaporation mass transfer (Bennett et al. 1997; Yeom et al. 1999). The diffusion coefficient D_m is dependent on polymer glass transition temperature as well as contaminant properties. A good polymer for a dense-phase membrane should have a high diffusivity for the contaminants and a moderately high solubility (too high might result in solvation). In the microporous membrane, any swelling due to a highly soluble contaminant will negatively affect mass transfer due to narrowed pores, and should be avoided by choosing a polymer with low contaminant solubility. Thus, the choice of membrane material is dependent on, and must be compatible with contaminants in the gas stream.

Table 9.2. Membrane materials used for membrane biofilters

Material	Repeating unit
Polypropylene	$-CH_2-CH(CH_3)-$
Polyethylene	$-CH_2-CH_2-$
Polytetrafluorethylene	$-CFl_2-CFl_2-$
Polysulfone	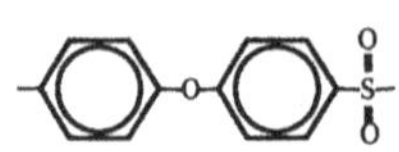
Polydimethylsiloxane (silicone rubber)	$-O-S(CH_3)_2-$

9.3 Reactor Configuration

Membrane biofilter reactor design is generally based on optimizing mass transfer. Considered as a problem of transport phenomena, mass transfer in membrane biofilters is analogous to heat transfer in heat exchangers. Therefore, membrane biofilter designs look like heat exchangers – either shell and tube (Fig. 9.1a), or flat plate (Fig. 9.1b). The flat plate design offers less surface area per volume than a shell and tube design (Reij and Hartmans 1996; Reij et al. 1997). Commercial manufacturers of membrane separation systems generally offer shell and tube designs. One issue not generally considered in the design of heat or mass transfer equipment is to support the growth of biofilm organisms. If dimensions are too small in the aqueous phase, biofilms may narrow channels and inhibit the transfer of nutrients, and in theory could become so densely grown that membranes would be compressed and air channels would be narrowed. Indeed, Reij et al. (1995) and Reij and Hartmans (1996) reported nearly 90% of tubes clogged when the tube side was used as the liquid phase for a membrane biofilter. Similarly, membrane designs such as spiral-wound systems may be unfavorable for membrane biofiltration due to narrow liquid channels. The shell and tube design is, not surprisingly, the most frequently reported configuration.

9.4 Operating Results

Although membranes are used extensively in water and wastewater treatment, membrane biofiltration has had limited development. This section summarizes the reported capabilities of membrane biofilters.

9.4.1 Loading and Elimination Capacity

Reported operating results are given in Table 9.3 for membrane biofilters reported in the literature to date. Operating results in terms of elimination capacity are comparable to other biofilters, with a wide range of values reported. Given that mass transfer limits most such systems, the flux truly limits elimination capacity. Reported flux values are roughly $100\,\mathrm{g\,m^{-2}\,h^{-1}}$ in systems using single membranes, so much higher elimination capacities can be achieved with systems using commercial designs with high specific areas (membrane area per reactor volume). In addition to good elimination capacities at gas residence times of seconds, membrane biofilters clearly are able to operate under high pollutant loads and high pollutant concentrations. At high mass loading (short residence times), membrane biofilters become mass transfer-limited rather than biologically limited, as is observed at low mass loading. This is reflected in Fig. 9.3.

Table 9.3. Membrane biofilter results from the literature

Contaminant ($g\,m^{-3}$)	Membrane	Load ($g\,m^{-3}\,h^{-1}$)	EC ($g\,m^{-3}\,h^{-1}$)	Operation (days)	Gas HRT (s)	Flux ($mg\,m^{-2}\,h^{-1}$)	Notes	Reference
Xylol (0.1–0.6)	HF silicone rubber	1.8–8.1	1.8–7.9	n.r.	2.5–14	67–302		Bäuerle et al. (1986)
Butanol (0.1–0.5)		1.8–7.3	1.3–5.4			48–207		
DCM (0.2–0.4)		2.4–11	1–4			37–151		
Toluene (0.38)	HF polyporous PP	2–31	33–1,500	~ 90	0.9–6.3	13–180	Suspension rather than biofilm	Ergas and McGrath (1997)
Toluene (0.7–3.4)	HF polyporous PP	23–121	16–42	> 20	0.9–1.8	15–35		Ergas et al. (1999)
Nitrogen oxide (0.1)	HF polyporous PP	1.3	0.1–0.9	105	1.9	3–8	20–55 °C	Min et al. (2002)
DCM (0.65)	Spiral-wound silicone rubber	14–35	12–28	11	n.r.	25–55		Freitas dos Santos et al. (1995)
Methane, TCE in liquid (264)	Silicone rubber tubing	n.a.	12	~ 120	n.a.	130	Cometabolic groundwater remediation	Clapp et al. (1999)
Dimethyl sulfide (0.036–1.81)	Flat composite silicone rubber on polysulfone	5–270	5–200	79	8–24			De Bo et al. (2002)
Toluene (0.75–1.5)	HF polyporous polysulfone, two in series	n.r.	n.r.	> 140	16 or 32	80		Parvatiyar et al. (1996)
Methanol (0.01–2.6)	Composite with silicone rubber	36	25	n.r.	n.r.	420		Resier et al. (1994)
Toluene (0.03–4.2)		58	40			670		
Hexane (0.03–2.4)		33	24			400		

n.r., Not reported or not sufficient data to calculate; n.a., not applicable (e.g., static gas phase)
[a] Some values calculated from reported data

Table 9.3. (continued)

Contaminant ($g\,m^{-3}$)	Membrane	Load ($g\,m^{-3}\,h^{-1}$)	EC ($g\,m^{-3}\,h^{-1}$)	Operation (days)	Gas HRT (s)	Flux ($mg\,m^{-2}\,h^{-1}$)	Notes	Reference
BTEX (7.7–15.4)	Composite with silicone rubber	600	360	52	4.3–15	170		Attaway et al. (2001)
BTEX (2.2–9.8)	HF polyporous PP	118	105	20	8–16	312		Attaway et al. (2002)
DCM (0.16)	Flat polyporous PP	180	80	<1	1.6–9.6	320		Hartmans et al. (1992)
Toluene (0.075)		84	29			117		
Toluene (0.004–3.18)	Flat composite, silicone rubber on PVDF	32–470	30–395	339	2–24	790		De Bo (2003)
Toluene (0–0.97)	Flat composite, silicone rubber on PVDF	0–170	0–170	20	24	n.r.		De Bo (2003)
TCE (0.04)		6	0–6					
Dimethyl sulfide (0.011–1.63)	Flat composite, silicone rubber on PVDF	5–170	5–130	20–144	4–24	n.r.		De Bo (2003)
Toluene (0.5–0.75)	HF polyporous PP	16–96[a]	32–72[a]	>8	1.8–7.2	6–60[a]	Cometabolism, various feeds, TCE toxicity	Dolasa and Ergas (2000)
TCE (0.04–0.2)		3–9[a]	0–6[a]		3.6–7.2	0–5[a]		

n.r., Not reported or not sufficient data to calculate; n.a., not applicable (e.g., static gas phase)
[a] Some values calculated from reported data

Table 9.3. (continued)

Contaminant ($g\,m^{-3}$)	Membrane	Load ($g\,m^{-3}\,h^{-1}$)	EC ($g\,m^{-3}\,h^{-1}$)	Operation (days)	Gas HRT (s)	Flux ($mg\,m^{-2}\,h^{-1}$)	Notes	Reference
Butanol (0.6–2.3)	HF polyporous polysulfone	13–26	3.8–13	105	1.6 and 2.9	13–326	Diurnal loading	Fitch et al. (2002)
TCE (0.13–0.21)	HF polyporous PP	0.03–0.1	0.01–0.06	21	96–300	0.06–0.2	Cometabolism, methanol feed to liquid suspension circulated through lumen	Pressman et al. (2000)
Benzene (0.1)	HF polyporous PP	7–60	4.8–58	100	0.17–1.4	4.2–20.4		Fitch et al. (2003)
Benzene (0.1)	Latex rubber tubing	7–28	2.5–18	40	0.55–1.4	670–2,700		Fitch et al. (2003)

n.r., Not reported or not sufficient data to calculate; n.a., not applicable (e.g., static gas phase)
[a] Some values calculated from reported data

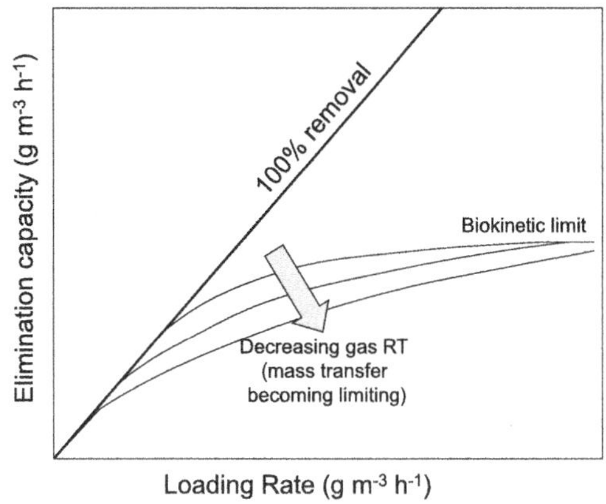

Fig. 9.3. Membrane biofilter performance trends

As with biofilters, the form of kinetic limitation (if such exists in a membrane biofilter) may be due to either the electron acceptor or donor. Both enter the active biofilm from the membrane side, so the relative ratios of diffusion coefficient and degradation stoichiometry determine limitation, as is seen even in trickling filters. De Bo (2003) noted that ammonia provided as a nutrient from the liquid side may decrease elimination capacity, perhaps because nitrifiers compete for oxygen with heterotrophs degrading organic contaminants.

9.4.2 Transient Loads and Aging

Biofilms in membrane biofilters do show some aging. For example, clogging was reported when the liquid phase was on the tube side of hollow fiber membrane systems (Reij et al. 1998). However, consistent removal has been reported for such systems in operation for at least a year. Another temporal issue is the aging of the membrane material. We have observed a decrease in the permeability of dense-phase silicone rubber systems used intermittently over two years, and apparent intrusion of organisms into microporous membranes has been reported in a number of studies (Attaway et al. 2002; Fitch et al. 2003).

Little study has been explicitly made of the response of membrane bioreactors to transient loads. Three important time-dependent conditions exist for biofilters: startup response, response to varying loads, and long-term performance. Startup generally is accomplished by inoculation with an acclimated suspension, followed by a rapid development of activity, with apparent steady-state performance after 1–2 weeks (Ergas et al. 1999). During this startup period, an initial high removal at 1–2 days is apparently followed by a decline in performance, attributed to either starvation in the liquid phase as the forming biofilm inhibits mass transfer, or to changes in the membrane due to swelling.

Transient and variable loads may result from changes in flow rate, contaminant concentration, or simply from the hourly organization of a typical working week. Membrane biofilters appear to respond well to diurnal loading based on a 40-h week (Fitch et al. 2000, 2002). Indeed, the removal at the start of a "workday" was higher than at the end of the day, which was attributed to some combination of sorption and favorable concentration gradient due to low initial concentrations in the biofilm. A number of investigators have shown that membrane biofilters respond well to changing loads, with new steady states established in a few days, but have not reported results during this transition period. For a dense-phase silicone rubber unit treating wastewater containing an aromatic and dichloromethane during alternating weeks, aromatics were found to be degraded nearly completely within hours of switching feeds, but the acclimation period was 2–3 days for dichloroethane (Ferreira Jorge and Livingston 2000a, 2000b). The authors speculated that the microbial community was readily able to use the aromatic, but required regrowth of the dichloroethane-degrading strain, which starved during aromatic feeding.

9.4.3
Biofilm Thickness

Biofilm formation on the membrane can result in enhancement or decrease in removal (Pressman et al. 1999; Nicolella et al. 2000a, b). Enhancement occurs with thin biofilms due to a larger concentration gradient across the membrane, due to the lowered substrate concentration at the membrane surface resulting from biodegradation. However, depending on the relative diffusivity and kinetics of degradation, which may be expressed as a Thiele modulus (Nicolella et al. 2000b), a thicker biofilm may result in decreased mass transfer and removal because of low rates of diffusion of the substrate within the biofilm.

9.4.4
Heat

Thermophilic biofiltration occurs at temperatures up to 62 °C (Cox et al. 2001), and many polluted airstreams are at elevated temperatures. Membrane biofilters have been reported to operate equally well regardless of temperature in the range of 20–55 °C. The separation of the organisms from the gas stream by a well-insulating membrane may allow for treatment of high-temperature gas streams if the liquid is recirculated through a radiator.

9.5
Models of Membrane Biofiltration

9.5.1
Mass Transfer

Mass transfer models for membrane systems have been extensively reported (Yang and Cussler 1986; Semmens et al. 1989; Crowder and Cussler 1998;

Oliveira et al. 2001; Cocchini et al. 2002), but very few models of coupled membrane and biofilm systems have been developed. The membrane bioreactor has unique boundary conditions – the solid support surface is also the interface for mass transfer of a growth substrate. In most biofilm models, the solid substrate is inert, so that a distinct boundary condition such as flux = 0 can be applied. For mass transfer without a biofilm, the resistance-in-series model is generally used (Coté and Lipski 1988):

$$\frac{1}{K_\mathrm{L}} = \frac{1}{k_\mathrm{g}H} + \frac{1}{k_\mathrm{m}H} + \frac{1}{k_\mathrm{l}}$$

in which the overall mass transfer coefficient, K_L, is related to the mass transfer coefficients for the gas, k_g, membrane, k_m, and liquid, k_l. Because the liquid and gas phases have differing chemical activities, the gas/liquid partitioning coefficient, generally described by the Henry's coefficient, H, is factored into mass transport considerations. Gas- and liquid-phase mass transfer coefficients have been described by dimensionless correlations analogous to heat transfer (Lévêque 1928; Yang and Cussler 1986; Coté and Lipski 1988; Pressman et al. 2000; Oliveira et al. 2001). The general form of such equations is:

$$k_\mathrm{l} = aRe^b Sc^c \left(\frac{d}{L}\right)^e \quad \text{or} \quad k_\mathrm{l} = aGr^b \left(\frac{D_\mathrm{L}}{d}\right),$$

where Re is the Reynolds number, Sc is the Schmidt number, d is the tube diameter, L is the tube length, D_L is the diffusion coefficient of the contaminant in the liquid, and a, b, c, and e are fitted values. The mass transfer resistance is thus a direct function of velocity and diffusivity, with k_l values varying in the range $1-15 \times 10^{-5}\,\mathrm{m\,s^{-1}}$ (Livingston et al. 1996).

For dense-phase membranes, k_m is described as the permeability divided by the membrane thickness, δ (Crank 1975):

$$k_\mathrm{m} = \frac{P}{\delta} = \frac{S_\mathrm{m}D_\mathrm{m}}{\delta}\,.$$

Dense-phase membranes have dominantly been silicone rubber due to high permeability. Reported membrane mass transfer coefficients (k_m) for currently available systems are in the range of $10^{-5}\,\mathrm{m\,s^{-1}}$, but thicker membranes may have much lower mass transfer coefficients (Cocchini et al. 2002). In microporous membranes, the mass transfer resistance is much lower, as it is the rate at which the contaminant diffuses through the gas-filled pores:

$$k_\mathrm{m} = \frac{D\varepsilon}{\delta\tau},$$

where D is the diffusion coefficient for the contaminant in the gas phase, ε is the porosity, and τ is the tortuosity of the pores. The membrane resistance in microporous membranes is reported to be similar to the gas-phase resistance, on the order of $0.1\,\mathrm{m\,s^{-1}}$ (Ergas and McGrath 1997).

These mass transfer resistances differ in order of magnitude. Reported liquid-film resistance values are generally orders of magnitude larger than that of the gas phase. Overall mass transfer coefficients from abiotic characterization are thus typically on the order of $10^{-5}-10^{-4}\ \mathrm{m\,s^{-1}}$, and are dominated by the liquid-phase mass transfer resistance, and in the case of dense-phase membranes, by a combination of liquid and membrane mass transfer resistance. The introduction of organisms changes things; biologically caused fouling of such membranes does occur, with reported overall mass transfer rates to the bulk fluid declining by two-thirds within 2 weeks (Ferreira Jorge and Livingston 2000a), but most membrane biofilters are operated as biofilm reactors, so that the “fouling” biofilm is intentional.

9.5.2 Biodegradation

The presence of the biofilm on the membrane surface adds to the fundamental mass balance a reaction term typically characterized by Monod kinetics. The resulting continuity equation for the biofilm on a hollow fiber is not analytically solvable:

$$D_S \frac{1}{r}\left(\frac{1}{r}\frac{\partial^2 S}{\partial r^2} + \frac{\partial S}{\partial r}\right) = -\frac{\varrho_b}{Y}\frac{\mu_{\max} S}{K_S + S}\,,$$

where D_S is the diffusion coefficient for the contaminant in the biofilm, r is the radial coordinate, S is substrate concentration (activity), ϱ_b is the biofilm density, Y is the yield on the contaminant, and $\mu_{\max}$ and K_S are the Monod kinetic parameters for the biofilm degrading the contaminant. In addition to variation with r, the concentration of the contaminant also varies with z, the axial dimension. Beyond the biofilm is the liquid phase, which may result in substantial degradation. One approach to this complex model is numerical methods, which were used with some success by Ergas et al. (1999). A sensitivity analysis indicated that the biofilm diffusion coefficient greatly affected predicted removal for such models, with decreased removal predicted if values fell below $10^{-9}\ \mathrm{m^2\,s^{-1}}$. A similar approach is to use the appropriate equations with a program that solves PDEs, such as models developed by Noguera et al. (2000) for trichloroethylene counterdiffusion through methane-grown biofilms, and by Livingston’s group (Nicolella et al. 2000b). Simpler approaches assume zero- or first-order kinetics; another approach is to assume a constant concentration across the biofilm, which allows for an integratable solution that must have the Monod-like parameters fit to the reactor (Parvatiyar et al. 1996).

The removal of nitrogen oxides (NO_x) from combustion gases has been minimally explored in membrane biofilters, but is quite feasible. The removal of nitrate from water streams by membrane bioreactors has been shown to be effective (Lee and Rittman 2000; Ergas and Reuss 2001; Ho et al. 2001, 2002; Mansell and Schroeder 2002; Min et al. 2002). NO_x has a low solubility, and treatment in a conventional biofilter requires long contact times. Membrane biofilters, well suited for low-solubility pollutants, have been demonstrated at laboratory scale to remove 50–70% of NO_X (with a liquid-phase bicarbonate feed) from a combustion gas mixture at scrubber exit gas temperatures (Min et al. 2002).

9.5.3 Model Results

Most reported models are fit to the reported results with one or more adjustable parameters, in part because biofilm thickness, activity, and diffusion coefficients are difficult to measure, and biofilms are highly heterogeneous. Sensitivity analyses on models generally indicate a mass transfer limitation, so that the membrane mass transfer resistance, the partitioning coefficient from gas to membrane or liquid, and the biofilm diffusion coefficient are reported as having the largest effect on the model results. Consistent with mass transfer limitation, kinetic parameters do not seem to affect predicted removal in the range of typical measured values. The exception to this case is thin biofilms or very little mass transfer resistance, in which case the kinetics are important and result in enhanced mass transfer (Nicolella et al. 2000b). The few models that considered dual substrate limitation for aerobic biodegradation generally show that oxygen is not rate-limiting.

9.6 Conclusions

Membrane biofilters will at some point be used at full scale because of the advantages of no gas-phase clogging, high removal of poorly soluble contaminants, minimal water requirements, and competitive elimination capacities. Currently, membrane biofilters are not used at full scale. The issues possibly limiting adoption of this technology are three: first, the cost of the membranes as compared to conventional biofilter packing, illustrated by the analysis of de Bo (2003) indicating that if microporous membranes are used and replaced every 3 years, both capital and operating costs are as much as tenfold greater than for any other common waste gas treatment method. Second, there is a lack of demonstrated multiyear performance. Third, no full-scale unit has yet been constructed. However, a simple demonstration of multiyear pilot-scale performance will overwhelm two of these three issues, and the use of inexpensive silicone rubber membranes should reduce capital costs by more than an order of magnitude.

References

Attaway H, Gooding CH, Schmidt MG (2001) Biodegradation of BTEX vapors in a silicone membrane bioreactor system. J Ind Microbiol Biotechnol 26:316–325

Attaway H, Gooding CH, Schmidt MG (2002) Comparison of microporous and nonporous membrane bioreactor systems for the treatment of BTEX in vapor streams. J Ind Microbiol Biotechnol 28:245–251

Bäuerle U, Fischer K, Bardtke D (1986) Biologische Abluftreinigung mit Hilfe eines neuartigen Permeationsreaktors. Staub Reinhalt Luft 46:233–235

Bennett M, Brisdon BJ, England R, Field RW (1997) Performance of PDMS and organofunctionalised PDMS membranes for the pervaporative recovery of organics from aqueous streams. J Membr Sci 137:63–88

Castro K, Zander AK (1995) Membrane air-stripping – effects of pretreatment. J Am Water Works Assoc 87:50–61

Clapp LW, Regan JM, Ali F, Newman JD, Park JK, Noguera DR (1999) Activity, structure, and stratification of membrane-attached methanotrophic biofilms cometabolically degrading trichloroethylene. Water Sci Technol 39:153–161

Cocchini U, Nicolella C, Livingston AG (2002) Braided silicone rubber membranes for organic extraction from aqueous solutions. I. Mass transport studies. J Membr Sci 199:85–99

Coté P, Lipski C (1988) Mass transfer limitations in pervaporation for water and wastewater treatment. In: Bakish R (ed) Proc 3rd Int Conf Pervaporation in the Chemical Industry, 19–22 September 1988, Nancy, France. Bakish Materials Corporation, Englewood Cliffs, pp 449–462

Cox HHJ, Sexton T, Shareefdeen ZM, Deshusses MA (2001) Thermophilic biotrickling filtration of ethanol vapors. Environ Sci Technol 35:2612–2619

Crank J (1975) The mathematics of diffusion. Clarendon Press, Oxford

Crowder RO, Cussler EL (1998) Mass transfer resistances in hollow fiber pervaporation. J Membr Sci 145:173–184

De Bo I (2003) Membrane biofiltartion of single-compound waste gas streams. Thesis, Universiteit Ghent, Ghent, Belgium

De Bo I, van Langenhove H, Heyman J (2002) Removal of dimethyl sulfide from waste air in a membrane bioreactor. Desalination 148:281–287

Devinny JS, Deshusses MA, Webster TS (1999) Biofiltration for air pollution control. Lewis, Boca Raton

Dolasa AR, Ergas SJ (2000) Membrane bioreactor for cometabolism of trichloroethene air emissions. J Environ Eng 126:969–973

Ergas SJ, McGrath MS (1997) Membrane bioreactor for control of volatile organic compound emissions. J Environ Eng 123:593–598

Ergas SJ, Reuss AF (2001) Hydrogenotrophic denitrification of drinking water using a hollow fibre membrane bioreactor. J Water Supply Res Technol 50:161–171

Ergas SJ, Shumway L, Fitch MW, Neemann JJ (1999) Membrane process for biological treatment of contaminated gas streams. Biotechnol Bioeng 63:431–441

Ferreira Jorge RM, Livingston AG (2000a) Microbial dynamics in an extractive membrane bioreactor exposed to an alternating sequence of organic compounds. Biotechnol Bioeng 70:313–322

Ferreira Jorge RM, Livingston AG (2000b) Biological treatment of an alternating source of organic compounds in a single tube extractive membrane bioreactor. J Chem Technol Biotechnol 75:1174–1182

Fitch MW, Sauer S, Zhang B (2000) Membrane biofilters: material choices and diurnal loading. In: Proc USC-TRG Conf Biofiltration, 19–20 October, The Reynolds Group, Los Angeles, pp 83–90

Fitch MW, England E, Zhang B (2002) 1-Butanol removal from a contaminated airstream under continuous and diurnal loading conditions. J Air Waste Manage Assoc 52:1288–1297

Fitch M, Neeman J, England E (2003) Mass transfer and benzene removal from air using latex rubber tubing and a hollow-fiber membrane module. Appl Biochem Biotechnol 104:199–214

Freitas dos Santos LM, Hommerich U, Livingston AG (1995) Dichloroethane removal from gas streams by an extractive membrane bioreactor. Biotechnol Prog 11:194–201

Hartmans S, Leenen EJTM, Voskuilen GTH (1992) Membrane bioreactor with porous hydrophobic membranes for waste-gas treatment. In: Dragt AJ, van Ham J (eds) Biotechniques for air pollution abatement and odour control policies. Elsevier, Amsterdam, pp 103–106

Ho CM, Tseng SK, Chang YJ (2001) Autotrophic denitrification via a novel membrane-attached biofilm reactor. Lett Appl Microbiol 33:201–205

Ho CM, Tseng SK, Chang YJ (2002) Simultaneous nitrification and denitrification using an autotrophic membrane-immobilized biofilm reactor. Lett Appl Microbiol 35:481–485

Lee KC, Rittmann BE (2000) A novel hollow-fibre membrane biofilm reactor for autohydrogenotrophic denitrification of drinking water. Water Sci Technol 41:219–226

Lévêque MA (1928) Les lois de transmission de chaleur par convection. Ann Mines 13:201

Livingston AG, dosSantos LMF, Pavasant P, Pistikopoulos EN, Strachan LF (1996) Detoxification of industrial wastewaters in an extractive membrane bioreactor. Water Sci Technol 33:1–8

Loeb S, Sourirajan S (1960) Sea water demineralization by means of a semipermeable membrane. Department of Engineering, University of California, Los Angeles, Rep no 60–60

Mansell BO, Schroeder ED (2002) Hydrogenotrophic denitrification in a microporous membrane bioreactor. Water Res 36:4683–4690

Min KN, Ergas SJ, Harrison JM (2002) Hollow-fiber membrane bioreactor for nitric oxide removal. Environ Eng Sci 19:575–583

Moe WM, Irvine RL (2000) Performance of periodically operated-gas phase biofilters during transient loading conditions. Water Sci Technol 41(4/5): 441–444

Nicolella C, Appendini ICG, Zhang SF, Livingston AG (2000a) Control of membrane-attached biofilms in extractive membrane bioreactors. Water Sci Technol 41:227–234

Nicolella C, Pavasant P, Livingston AG (2000b) Substrate counterdiffusion and reaction in membrane-attached biofilms: mathematical analysis of rate limiting mechanisms. Chem Eng Sci 55:1385–1398

Noguera DR, Pizarro G, Clapp LW (2000) Mathematical modeling of trichloroethylene (TCE) degradation in membrane-attached biofilms. Water Sci Technol 41:239–244

Oliveira TAC, Cocchini U, Scarpello JT, Livingston AG (2001) Pervaporation mass transfer with liquid flow in the transition regime. J Membr Sci 183:119–133

Parvatiyar MG, Govind R, Bishop DF (1996) Biodegradation of toluene in a membrane biofilter. J Membr Sci 119:17–24

Pressman JG, Georgiou G, Speitel GE (1999) Demonstration of efficient trichloroethylene biodegradation in a hollow-fiber membrane bioreactor. Biotechnol Bioeng 62:681–692

Pressman JG, Georgiou G, Speitel GE (2000) A hollow-fiber membrane bioreactor for the removal of trichloroethylene from the vapor phase. Biotechnol Bioeng 68(5):548–556

Reij MW, Hartmans S (1996) Propene removal from synthetic waste gas using a hollow-fibre membrane bioreactor. Appl Microbiol Biotechnol 45:730–736

Reij MW, de Gooijer KD, de Bont JAM, Hartmans S (1995) Membrane bioreactor with a porous hydrophobic membrane as a gas-liquid contactor for waste gas treatment. Biotechnol Bioeng 45:107–115

Reij MW, Hamann EK, Hartmans S (1997) Biofiltration of air containing low concentrations of propene using a membrane bioreactor. Biotechnol Prog 13:380–386

Reij MW, Keurentjes JTF, Hartmans S (1998) Membrane bioreactors for waste gas treatment. J Biotechnol 59:155–167

Resier M, Fischer K, Engesser KH (1994) Kombination aus Biowascher- und Biomembranverfahren zur Reinigung von Abluft und hydrophilen und hydrophoben Inhaltsstoffen. VDI Ber 1104:103

Semmens MJ, Qin R, Zander A (1989) Using a microporous hollow-fiber membrane to separate VOCs from water. J Am Water Works Assoc 81:162–167

Yang M, Cussler EL (1986) Designing hollow-fiber contactors. AIChE J 32:1910–1916

Yeom CK, Kim HK, Rhim JW (1999) Removal of trace VOCs from water through PDMS membranes and analysis of their permeation behaviors. J Appl Polym Sci 73:601–611

10 Modeling of Biofilters and Biotrickling Filters for Odor and VOC Control Applications

MARC A. DESHUSSES, ZAROOK SHAREEFDEEN

10.1 Introduction to Modeling

Biological treatment of contaminated air in biofilters and biotrickling filters involves a complex combination of physical, chemical, and biological processes. This makes it very difficult to mentally understand and integrate the subtle details of the phenomena occurring during treatment. Even simple exercises, such as to predict the behavior of biofilters and biotrickling filters under a different set of conditions, can be challenging. Fortunately, computers and mathematical models can track a multitude of complex relationships much better than the human mind. Therefore, mathematical models can be very useful tools in research and for design. Mathematical models help in the development of a fundamental understanding of the process, and to accomplish engineering tasks such as reactor design and scaleup, or process optimization.

There are many advantages to modeling and simulation, including:

The ability to gain insight into a given process.

The possibility to obtain quantitative information on a variable that is difficult or impossible to measure (e.g., concentration of contaminant in the biofilm).

The low cost and rapidity of performing "virtual experiments".

The ability to evaluate conditions that may not be possible to test.

The possible use in experimental design in order to maximize outcome and representativeness of the experimental data.

The ability to automatically perform numerous simulations to support process optimization.

At the same time, one has to acknowledge the many limitations of mathematical modeling:

Models will only be as good as the basic concept and the assumptions that were made during model development.

Shareefdeen/Singh (Eds.)
Biotechnology for Odor and Air Pollution Control

In many instances, model parameters will be unknown, and it will require significant experimentation to determine these model parameters.

The fact that simply because a model fits the experimental data does not mean that the model or the concept on which the model is based are correct.

Application of a model and extrapolation of modeling data outside the range for which the model has been validated, or application of a model for a completely different system involve risks, as uncertainties and discrepancies may be multiplied.

Models may be classified depending on their features and structures. In this chapter, the focus was placed on models that make attempts to describe the actual phenomena occurring during a biological treatment. These are often called structured or conceptual models. A number of other models, ranging from simple multi-parameter correlations to lumped parameter models, exist but these are not discussed in great detail here.

10.1.1 General Model Concepts

The process of modeling implies some kind of translation of concepts into equations that can be solved. In most cases, modeling of biofilters and biotrickling filters involves mass balances of the contaminant(s), oxygen, products resulting from contaminant degradation, and biomass generation in all possible phases in the reactor. These include the gas phase, solid phase (depending on the system), biofilm and liquid phases in the case of biotrickling filters. Depending on the model concept and assumptions, the resulting equations may include terms for the accumulation of reactive products (i.e., odors, VOCs, oxygen, nutrients, by-products) in all phases, dispersion effects in air, mass transfer between the air and biofilm phases, diffusional mass transfer in the biofilm, consumption due to biological oxidation, adsorption onto solid media, biomass growth and physiological state of the biomass, etc. Thus, the resulting model is often comprised of a complex set of partial differential equations with initial and boundary conditions, and algebraic equations, e.g., partition of the pollutant, biodegradation rates, etc. Most of the time, analytical solutions are difficult or impossible, and finite-element or finite-difference methods are used instead, with the appropriate numerical algorithms to obtain solutions to the model. Although the development of realistic mathematical models is possible, complex models often result in an increased number of parameters, values of which are hard to determine, or nonexistent in the literature. Realistic models play a significant role in advancing fundamental theory. However, they are not yet readily applicable to practical situations, due to the number of unknowns involved.

In an effort to obtain easier solutions, the complexity of the models is often substantially reduced using assumptions such as (1) equilibrium between biofilm/air interface, (2) ideal mixing in liquid phase, (3) plug flow for air phase, (4) constant biofilm thickness, biomass density or negligible biomass growth,

(5) steady-state conditions, (6) availability of excess oxygen, and (7) use of simple kinetics for biological oxidation. These assumptions make solutions somewhat simpler, and lead to valuable correlations that are used by consultants and engineers to calculate design variables, and by researchers to elaborate on specific aspects of the process. However, in some cases, simplified assumptions make models unrealistic, and may lead to errors in estimating design parameters. The objective of this chapter is to give readers a brief overview and application of modeling of biofilters and biotrickling filters.

10.1.2 Importance of Modeling in Design and Operation

Designing of large-scale biofilters and biotrickling filters requires accurate estimation of medium volume, pressure drop calculations, blower sizing, reactor dimensions, equipment specifications for pretreatment processes, water demand for humidification, and winterization requirements that are needed for operation in cold climates. Often, biofilters and biotrickling filters are scaled-up from laboratory and/or pilot-scale tests that are carried out at customers' sites. Odor measurements and analysis are expensive; hence, tests and results are based on spot readings that do not take into consideration fluctuations in the process conditions. The probability of variations in flow, concentrations levels, process changes, and future expansion make these tests alone inadequate for accurate designs that are risk-free. In a large-scale biofilter project, 5% over- or underestimation can cause significant variation in capital cost and design deficiency. Furthermore, customers require assurance that the installed systems perform as specified. For these reasons, predictive models that are validated with laboratory and pilot test data have become valuable tools in the design of equipment and operations.

10.2 A Review of Biofilter Models

10.2.1 Steady-State Models

One of the earliest and widely used steady-state biofiltration models was developed by Ottengraf and van den Oever (1983). Because of its simplicity, this model has been used by a number of other researchers (van Lith 1989; Dharmavaram 1991; Leson and Winer 1991; Ergas et al. 1993; Deshusses and Hamer 1993). The model assumes plug flow in the gas phase, and flat geometry for the biofilm. The model neglects gas film mass transfer resistance, and assumes the pollutant to be in equilibrium at the air/biofilm boundary and related by Henry's law. Pollutant transport in the biofilm is by diffusion and can be described by an effective diffusion coefficient. For the micro-kinetics of biodegradation occurring in the biofilm, the authors start with the Michaelis-Menten-type expression (growth is neglected), and consider two limiting cases of first- and zero-order kinetics. For the first-order kinetics, an exponential decrease in the pollutant concentration is expected.

10.2.1.1
First-Order Model

$$\frac{C_{\mathrm{out}}}{C_{\mathrm{in}}} = \exp\left(\frac{-EBRT\,A_{\mathrm{S}}D_{\mathrm{e}}}{m\delta}\phi \tan h\phi\right) \quad \text{where} \quad \phi = \delta\sqrt{\frac{K}{D_{\mathrm{e}}}} \tag{10.1}$$

At high pollutant concentrations, the rate expression approaches zero-order kinetics in the substrate concentration, for which case two situations are distinguished (Ottengraf and van den Oever 1983; Ottengraf 1987) – reaction limitation, and diffusion limitation. In reaction-limited process, the biofilm is fully active, and hence the conversion rate is controlled by the reaction, while in the diffusion-limited case, the biofilm is not fully active. Thus, depth of diffusion of the pollutant in the biofilm is smaller than the actual thickness.

10.2.1.2
Zero-Order Reaction-Limited Model

$$\frac{C_{\mathrm{out}}}{C_{\mathrm{in}}} = 1 - \frac{K_0 A_{\mathrm{S}}\delta\, EBRT}{C_{\mathrm{in}}}\,. \tag{10.2}$$

10.2.1.3
Zero-Order Diffusion-Limited Model

$$\frac{C_{\mathrm{out}}}{C_{\mathrm{in}}} = \left(1 - EBRT\,A_{\mathrm{S}}\sqrt{\frac{K_0 D_{\mathrm{e}}}{2\,mC_{\mathrm{in}}}}\right)^2 \tag{10.3}$$

The above model equations are based on the assumption that oxygen in the biofilm is present in excess, and hence does not affect the degradation of the contaminant. The model is also limited to the case of a single pollutant. In the case of mixed pollutants, the authors propose to use the same model in an additive sense. Based on the work of Ottengraf, described above for the case of zero-order kinetics, van Lith (1989) and van Lith et al. (1990) presented a design method for biofilters. The authors have proposed the following expression for the limiting concentration at which the process goes from diffusion-controlled to reaction rate-controlled:

$$C_{\mathrm{limit}} = \left(\frac{1}{A_{\mathrm{S}}}\sqrt{\frac{mK_0}{2D_{\mathrm{e}}}}\right)^2\,. \tag{10.4}$$

Experimental studies show that biodegradation kinetics are more complex (Chang et al. 1993; Deshusses and Hamer 1993; Shareefdeen et al. 1993; Shareefdeen and Baltzis 1994a, b; Sologar et al. 2002) than zero or first order-type kinetics, and are different when this compound is present alone or in a mixture of compounds. The kinetic interference also plays an important role in biofiltration of VOC mixtures. For example, Deshusses and Hamer (1993) discussed kinetic interference of methyl-ethyl-ketone/methyl-isobutyl-ketone mixture. Shareefdeen and Baltzis (1994a) extended Ottengraf's steady-state model (Ottengraf and van den Oever 1983) by incorporating potential limiting effects of oxygen, and have shown that the presence of oxygen plays a significant role in the biodegradation of VOCs. In a recent work, Sagastume et al. (2003) have presented a model that estimates changes in physical structure of the biofilter media over time.

10.2.2 Transient Models

In practice, transient operations due to fluctuating concentrations and process conditions are more common than steady-state ones. Thus, steady-state models, in general, are not applicable for startup performance analysis and predictions under unsteady state. Questions such as how well can a biofilter respond to peak values of concentrations and variations in volumetric flows are of paramount importance in the design and commercial application of biofilters. The fact that biofilters are most likely to operate under varying load conditions was recognized early by Ottengraf and van den Oever (1983).

Shareefdeen and Baltzis (1994b) have developed a transient model for the biofiltration of a single VOC. This model is an improvement of their steady-state model (Shareefdeen and Baltzis 1994a) for a single compound. A major feature of the model is that the authors have considered the adsorption phenomenon that plays a major role during startup and fluctuations in process conditions. By excluding plug-flow assumption, Hodge and Devinny (1995) have extended the model to incorporate axial dispersion under transient conditions. The model also describes the evolution of carbon dioxide. Deshusses et al. (1995a) have presented a model for the determination of both the transient and steady-state behavior of biofilters degrading MEK and MIBK vapors. For the kinetics of biodegradation, Michaelis-Menten-type kinetics with competition between substrates was assumed. Shareefdeen et al. (1997a) presented a general transient biofilter model for single volatile organic compound (VOC) removal, and subsequently used the model in shock loading studies that were carried out by introducing random and step changes to the inlet toluene concentration and flow rate. Shareefdeen et al. (1997b) later extended their work to VOC mixtures, and developed a comprehensive model that takes into account, among others, general interactive kinetics, multi-component adsorption phenomena, axial dispersion effects, and oxygen limitation. Devinny et al. (1999) presented a comparison of these models. Ozis et al. (2002) presented a discrete percolation model to describe biomass clogging effects on biofilter performance. Solutions of transient models do not yield simple analytical equations and require sometimes lengthy numerical computations. For keeping the chapter concise, details of model equations and modeling results are not included here but readers are encouraged to review the references for more details.

10.2.3 Critical Parameters

Solution of the model requires estimation of critical parameters. The kinetic parameters (K_0 and K_1 in the case of the model presented above) are characteristic of the process culture, the compound being degraded, and the biofilter media. In many laboratory studies, kinetic parameters were determined using shake-flask batch cultures. In pilot-scale studies or when dealing with mixtures of compounds, macro-kinetic parameters are commonly used. The determination of kinetic parameters in shake flasks is questionable as the culture conditions are markedly different than in a biofilter. Indeed, often shake-flask kinetic data do not agree with full-scale systems, especially where airstreams consist of more complex mixture of VOCs or odors. It is also possible that indigenous microorganisms in the biofilter media are different from the cultures used in the batch kinetic experiments.

Effective diffusivity (D_e) and biodegradation rate are a function of biofilm density. In the literature, there is a wealth of information about biofilm densities for three-phase systems (Shareefdeen 1994) that can be used to estimate effective diffusivity. A typical value for biofilm density is about $100\,\text{kg}\,\text{m}^{-3}$. This value implies that 90% of the biofilm is made of water. Even so, the physicochemical properties of the biofilm can be quite different from water. Unfortunately, the differences are case-dependent, and no simple correction factor can be applied. It is not easy to get a value for effective biofilm thickness (δ), but this can be estimated using the method described by Williamson and McCarty (1976) and Shareefdeen (1994). In this method, the value of effective biofilm thickness is determined as the distance in the biofilm that would first lead to depletion of either oxygen or the contaminant. Biofilm surface area per unit volume of reactor (A_S) is usually estimated by fitting the data to laboratory scale, and later used in full-scale predictions, unless a packing with a known area is used. Physical properties such as distribution coefficients (m) or partition coefficients between air and water can be easily obtained from the literature. Thus, accurate solution of the most simplistic model requires careful estimation of all these parameters. Solutions of the transient model that considers dispersion, kinetic interaction, oxygen limitation and multi-contaminants require estimation of many more parameters, and thus the solutions become cumbersome. Regardless of the complexities, only through such exercise of the fundamental science of biofiltration process can the technology be advanced, and in future these models will increasingly become part of the design.

10.3 Uses of Biofilter Models in Full-Scale Designs

10.3.1 Wastewater Treatment Applications

The major odor-causing compound found in wastewater treatment applications is hydrogen sulfide (H_2S), which has a nuisance odor characterized by rotten eggs, and very low odor threshold levels in parts per billion (ppb) levels. For these reasons, strict regulations are in effect for controlling emission of these gases at municipal facilities. In recent years, biofiltration has received increased acceptance for odor removal in the wastewater treatment industry.

In the last few years, BIOREM Technologies has installed over 60 full-scale biofilters across Canada and the USA, and over 70% of the projects are for wastewater treatment applications for removal of hydrogen sulfide. Operating conditions, concentrations, and period of fluctuations vary from plant to plant, and thus a single design or modular biofilter systems are not applicable to all cases. We have undertaken a project to develop a predictive model that can be used in the designing of biofilters, particularly for wastewater treatment applications where hydrogen sulfide is the prevalent substance.

Hydrogen sulfide removal data from laboratory (Shareefdeen et al. 2003) and commercial-scale biofilters have been collected to determine parameters for the model. Experimental data were collected at various inlet concentrations (21–63 ppm), and empty bed residence times (EBRT) of 20–48 s. The data followed closely the zero-order diffusion-limited model as described above (Eq. 10.3), with lump parameter estimated as

$$\alpha_{\text{lump}} = \left\{ A_S \sqrt{\frac{K_0 D_e}{2m}} \right\} = 0.18 \tag{10.5}$$

Equations (10.3) and (10.5) can be used to predict hydrogen sulfide discharge levels from a biofilter, provided inlet concentration of H_2S (in ppm) and EBRT (in seconds) are known. The equations were developed using BIOSORBENS proprietary inorganic media biofilters, developed by BIOREM Technologies. The model relationships, along with pretreatment (humidification), pressure drop correlations and heat effects, are incorporated into a user-friendly mathematical software model (Fig. 10.1a) from which design diagrams can be generated for estimating removal efficiencies (Fig. 10.1b) as a function of empty bed residence times and inlet concentrations.

Once the concentrations of contaminants are known or estimated, then for a given airflow volume, biofilter size, medium volume, water requirements for saturating the air and media volume can be easily calculated using the interface shown in Fig. 10.1a. The program also provides an interface for estimating heat requirements for winter operation. In the diagram shown in Fig. 10.1b, the outer circle scale gives inlet concentrations of H_2S, and inner circles give removal efficiencies at different EBRTs. The model and diagrams have been used in sizing biofilters for several large-scale applications. After installation

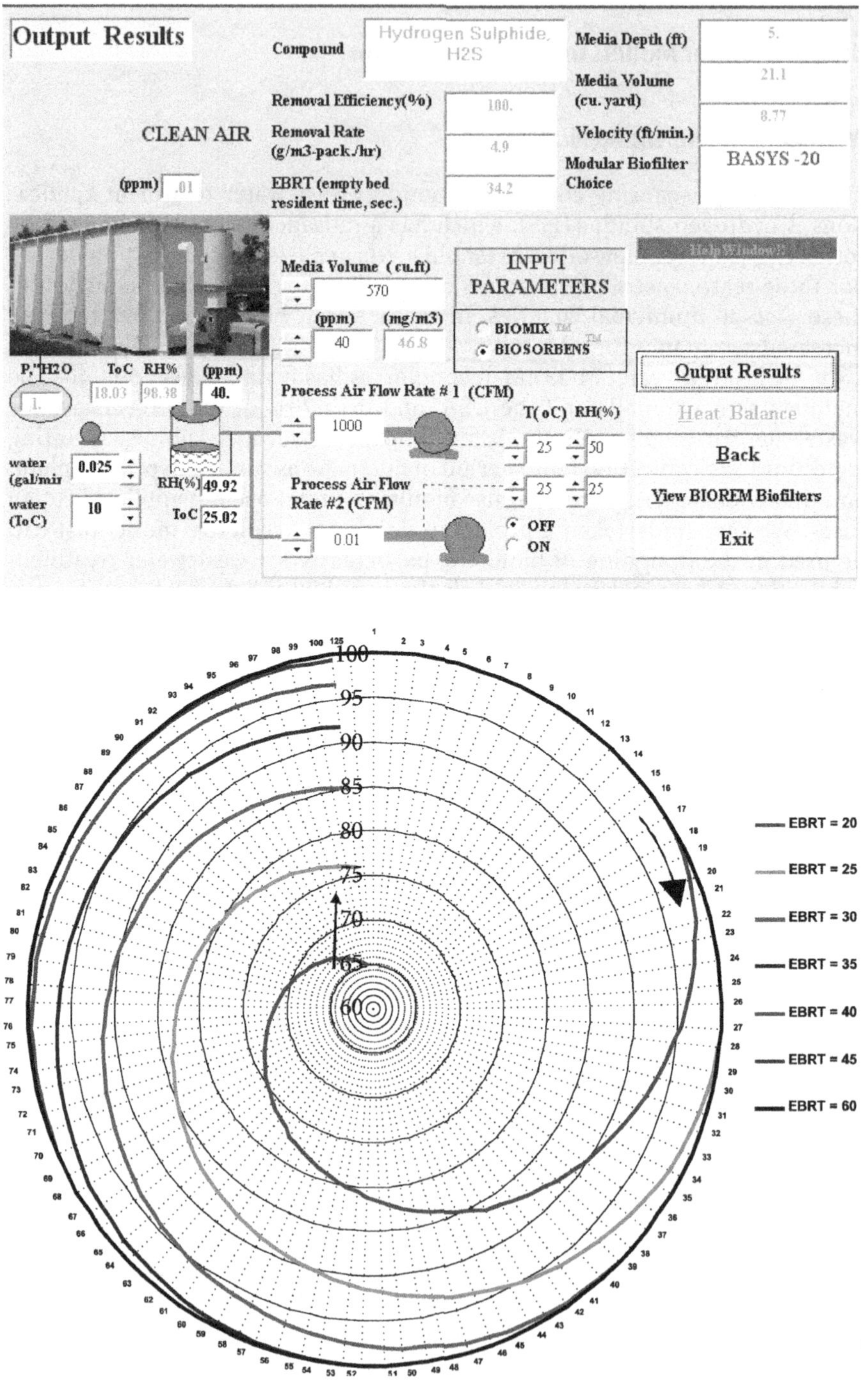

Fig. 10.1. A mathematical software model interface (a) and a design diagram (b)

of full-scale biofilters, field data have been collected and compared with the model predictions, and in most cases, the field data outperformed predicted values in terms of removal efficiency. Thus, the design based on the model significantly reduced experimentation and risks involved in larger projects.

10.3.2 Rendering Applications

Rendering applications involve recycling of animal and poultry by-product materials in order to produce tallow, grease, and protein meals. Chemical constituents emitted during several stages of process (blood processing, cooking, pressing, and drying) are sulfides, amines, ketones, aldehydes, and alcohols that are a highly odorous nuisance. Therefore, rendering plant operators are increasingly faced with problems of odor complaints from neighboring citizens, and are in need of effective systems for completely eliminating odors.

This case study is related to designing one of the world's largest synthetic media biofilter systems (Fig. 10.2) through the modeling approach discussed. The biofilter user required an average odor removal efficiency of 90% or higher, with a performance guarantee. Since several VOCs and inorganic compounds are present in the air and contribute to the odor makeup, odor concentration is treated as a single substrate and a predictive model is developed from the pilot experimentation at the site. Odor data were taken at various process conditions and empty bed residence times to estimate removal parameters. Based on the first-order model (Eq. 10.1), it was decided that a 35-s empty bed residence time was required to meet performance specification. Based on this, a biofilter has been built and commissioned in August 2003. Odor measurements were taken from the full-scale system (shown in Fig. 10.2), and

Fig. 10.2. A 250,000-cfm synthetic media biofilter system treating air emissions from a rendering plant. (Courtesy of BIOREM Technologies, ON, Canada)

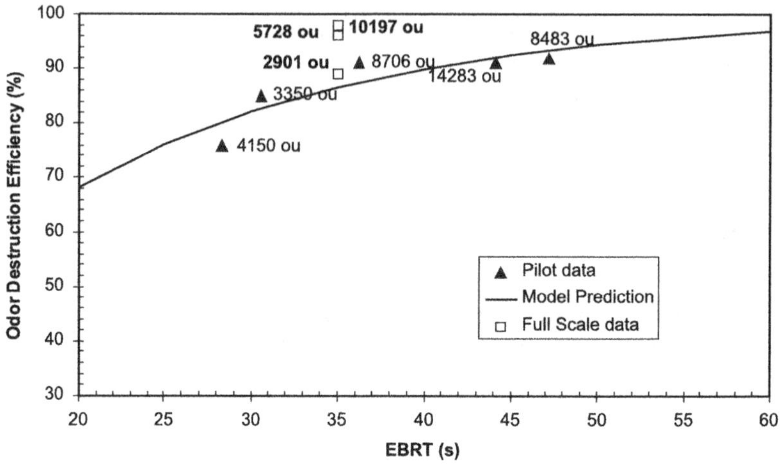

Fig. 10.3. Odor destruction efficiency as a function of empty bed residence time

compared with the pilot and model predicted data (Fig. 10.3). The full-scale system exceeded predicted performance. In full-scale applications, additional safety factors are added in the design, and often the biofilter systems operate at lower, varying concentrations and airflows than the design specifications. This may be the reason for better than predicted performance in the two cases described above.

10.4
A Review of Biotrickling Filter Models

Because biotrickling filters have a free liquid phase, modeling of biotrickling filters is fairly different from modeling biofilters. There are some additional challenges, while other factors make modeling of biotrickling filters easier. These are summarized below:

There is one more phase (the liquid phase) to model in biotrickling filters, and this will increase the number of equations and the level of complexity. Consequently, biotrickling filter models are almost exclusively solved numerically.

Biotrickling filters can be operated co-currently or countercurrently. In terms of solving the model differential equations, countercurrent operation can be a relatively stiff problem, i.e., it may be difficult for algorithms to converge to a solution, as liquid and gas flow in opposite directions. In addition, pollutant will be dissolved in the trickling liquid, which is pooled in the sump at the bottom of the biotrickling filter. Biodegradation of the pollutant in the sump may occur, although it is seldom considered. When it is not considered, this adds a very stringent constraint to the model, as top and bottom liquid concentrations should be equal. At the top of the reactor, pollutant may undergo desorption from pollutant-rich liquid. In addition to reducing the apparent performance of the biotrickling filter, this often increases computation time, and results in model instabilities.

Biotrickling filters are packed bed scrubbers in which reactive absorption is conducted. However, biotrickling filters are very often operated at liquid velocities that are at least one order of magnitude lower than packed bed scrubbers (see also Chap. 7). This means that a fraction (often unknown) of the biofilm is not wetted. Hence, the pollutant may be transferred directly from the gas to the biofilm, or first from the gas to the liquid and then from the liquid to the biofilm. Thus, multiple routes for the pollutant transfer need to be considered.

Biotrickling filters often use packing, such as plastic rings or other synthetic materials, that have a well-defined surface area. This reduces the uncertainty on the value of the interfacial area parameter.

In biofilters, the only parameter that can be experimentally determined is the gaseous concentration of pollutant. That leaves limited opportunities for model validation. In biotrickling filters, the liquid concentration of the pollutant can usually be determined relatively easily. This makes model calibration and/or validation easier.

Biotrickling filters often have visible biofilms. Because they grow on inert supports, they can be relatively easily scraped for analysis. Their thickness can be determined by direct measurement, and pollutant, pH, or dissolved oxygen can be determined using micro-sensors. This all provides a greater level of independent determinations for model validation, and may help reducing the number of fitted parameters. Biofilm growth in biotrickling filters is often significant, e.g., in VOC-degrading biotrickling filters, and increase in biofilm thickness over time may have to be considered.

Several models of biotrickling filters, and the diverse approaches that have been proposed are summarized in Table 10.1. It also illustrates that the choice of a suitable model will depend on the pollutant considered and on the treatment conditions, as those factors will affect the relative importance of biodegradation and mass transfer.

Over the past decade, more sophisticated biotrickling filter models have been proposed. It is interesting to notice that the model concepts have been heavily inspired by biofilter models, even though biofilter concepts may not always be directly transferred to biotrickling filters. There has been little influence of the numerous models that already exist for trickle bed reactors used in the chemical industry. The latter models, pioneered by Satterfield at MIT (see Goto and Smith 1975; Satterfield 1975; Satterfield et al. 1978), include a number of features that may be relevant to biotrickling filter models. These include detailed gas to liquid, or liquid to solid mass transfer considerations, diffusion limitation inside a porous carrier, partial wetting of the packing, axial

Table 10.1. Characteristics of selected biotrickling filter mathematical models

Type of reactor/pollutant	Model description	Results	Reference
Large laboratory-scale biotrickling filter packed with ceramic saddles, treating methylene chloride vapors	Steady-state model with zero-order biodegradation kinetics (no oxygen limitation). Pollutant is transferred to the liquid and then to the biofilm; there is no mass transfer limitation at the liquid–biofilm interface	Model and experimental results were in good agreement. The references cited compare experimental and modeled concentration profiles along the height for co- and countercurrent, and effect of selected parameters	Diks and Ottengraf (1991a, b)
Large laboratory-scale biotrickling filter packed with a plastic structured support, treating CS_2 vapors	Model focusing on mass transfer from the gas to the liquid, and then to the biofilm. Biodegradation lumped into a single parameter (effectiveness factor). The model reduces to two parameters that can be determined independently of biotrickling filtration experiments	Model predicts well the concentration profile along the reactor height. Detailed discussion of the sensitivity to gas–liquid mass transfer and biodegradation allowed development of a single criterion for identifying rate-limiting step	Lobo et al. (1999)

Table 10.1. (continued)

Type of reactor/pollutant	Model description	Results	Reference
Laboratory-scale biotrickling filter treating chlorobenzene alone or mixed with ortho-dichlorobenzene	Model considers wetted and non-wetted biofilm, but pollutant only transfers from the gas to the liquid, and then to the biofilm. Oxygen limitation is included. The 2001 reference cited considers cross-inhibition during biodegradation of the mixture of chlorobenzenes, although it had little effect	Model and experimental results were in good agreement. Simulations show significant stripping of the pollutant at the top of the column during countercurrent operation. Effective biofilm thickness along the height of the reactor is discussed and significant differences are shown, suggesting different limitations along the height. Oxygen limitation is predicted at high chlorobenzene concentrations	Mpanias and Baltzis (1998); Baltzis et al. (2001)
Laboratory-scale biotrickling filter treating pentane and isobutene; structured polyethylene packing	Steady-state model considering biofilm to be covered by trickling liquid (one pass). Biodegradation kinetics is lumped in one parameter	Model and experimental results suggest that the biotrickling filter had a dual limitation of biokinetics and mass transfer	Barton et al. (1998)
Large laboratory-scale biotrickling filter (2.7 × 0.4 m) with structured packing treating methylene chloride	Detailed dynamic model that includes growth of methylene chloride degraders, growth of inert biomass, effect of pH and dissolved oxygen. The substrate has to diffuse through the trickling liquid to reach the biofilm	Model appears to describe well a number of experiments, including biomass growth over time, and biomass concentration along the bed height. The reference cited suggests that the model could be used to predict the time for the biomass to plug the reactor	Okkerse et al. (1999)

Table 10.1. (continued)

Type of reactor/pollutant	Model description	Results	Reference
Laboratory-scale biotrickling filter treating diethyl ether vapors; ceramic packing, low trickling rate	Detailed conceptual model, including active and inactive fractions of biomass, growth and detachment of biomass	Model describes well the experimental data. Experiments and simulations show significant mass transfer resistance of the trickling liquid, suggesting that liquid phase should be taken into account in biotrickling filter models	Alonso et al. (1998); Zhu et al. (1998)
Laboratory-scale biotrickling filter treating diethyl ether vapors; ceramic packing, low trickling rate	Detailed conceptual model, extended to include detailed description of the reduction of interfacial area, and increase in liquid velocity with growth of biomass into the packing	Model describes effect of nitrate concentration on biomass growth, and effect of biomass on treatment performance	Alonso et al. (2001)
Laboratory-scale differential biotrickling filter; very high air velocity; treatment of H_2S	Detailed conceptual model, focusing on mass transfer aspects. Model considers wetting of packing, and both gas–liquid, and gas–liquid–biofilm transport. Mass transfer described by the correlation of Onda et al. (1968). All parameters were determined independently. Growth of biomass is neglected, as H_2S degraders are slow-growing organisms	The model only describes well selected experiments, though it provides a possible explanation for the very high performance obtained in a field biotrickling filter operated at a 2-s EBRT. Model shows possible diffusion limitation at low concentrations	Gabriel and Deshusses (2003); Kim and Deshusses (2003)

Table 10.1. (continued)

Type of reactor/pollutant	Model description	Results	Reference
Modeling study, no laboratory reactors	Chemical engineering approach to biotrickling filter model. Models use simplified equations for the flux and biodegradation in the biofilm	Extensive discussion of the model parametric sensitivity (mass transfer coefficient, Henry's law constant of the pollutant, gas and liquid mass transfer coefficients, kinetic parameters, etc.). Model shows that stripping is important with countercurrent operation, and shows the importance of providing some pollutant treatment in the liquid sump	Ockeloen et al. (1996)
Biofilter and biotrickling filter treating H_2S. Model was calibrated and validated with different biotrickling filters packed with lava rocks	Conceptual model including detailed interphase mass transfer concepts. The pollutant has to be transferred to the liquid before reaching the biofilm. Model includes biomass growth and decay. Depending on the operation, the model can be run as biofilter or as biotrickling filter	Extensive discussion of the model parametric sensitivity shows the significant resistance to mass transfer of the liquid over the biofilm. Diffusion/biodegradation kinetics were the most sensitive parameters. Gas–biofilm mass transfer coefficient is not a sensitive parameter	Li et al. (2002, 2003); Martin et al. (2002)

dispersion in the air, liquid holdup, and more. As we develop more and more complex models, it may be worth remembering some of the older research, and integrate discoveries that were made before computer simulation software existed.

10.5 Conclusions and Future Work

The recent work on biofilter and biotrickling filter modeling shows that there has been significant progress since the very first models were introduced. The improvements have resulted from a better understanding of the biofiltration/biotrickling filtration process as well as from the increase in personal computing power. Models now include far more realistic representations of actual pollutant and oxygen mass transfer, fluid dynamics, and microbiological phenomena. Similar progress has been made in the definition of model parameters, although this remains an area of intense discussion, and of possible disagreement. It is encouraging to see models being used for the design of reactors. Modeling and simulation is more than a mere academic exercise. It plays an important role in the deployment of the technology in the field.

Symbols used in biofilter models

A_S biofilm surface area

C_{in} concentration at the inlet of the biofilter

C_{out} concentration at the discharge of the biofilter

C_{limit} limiting concentration, defined in Eq. (10.4)

D_e effective diffusion in the biofilm

K first-order rate constant

K_0 zero-order rate constant

$EBRT$ empty bed residence time, or media volume/volumetric flow rate

m air/biofilm distribution coefficient

α lump kinetic parameter defined in Eq. 10.5

δ biofilm thickness

ϕ parameter defined in Eq. (10.1)

References

Alonso C, Suidan MT, Kim BR, Kim BJ (1998) Dynamic mathematical model for biodegradation of VOCs in a biofilter: biomass accumulation study. Environ Sci Technol 32:3118–3123

Alonso C, Zhu X, Suidan MT, Kim BR, Kim BJ (2001) Mathematical model of biofiltration of VOCs: effect of nitrate concentration and backwashing. J Environ Eng 127:655–664

Baltzis B, Mpanias C, Bhattacharya S (2001) Modeling the removal of VOC mixtures in biotrickling filters. Biotechnol Bioeng 72:389–401

Barton JW, Hartz SM, Klasson KT, Davison BH (1998) Microbial removal of alkanes from dilute gaseous waste streams: mathematical modeling of advanced bioreactor systems. J Chem Technol Biotechnol 72:93–98

Chang MK, Voice TC, Criddle CS (1993) Kinetics of competitive inhibition and cometabolism in the biodegradation of benzene, toluene, and p-xylene by two *Pseudomonas* isolates. Biotechnol Bioeng 41:1057–1065

Deshusses MA, Hamer G (1993) The removal of volatile ketone mixtures from air in biofilters. Bioprocess Eng 9:141–146

Deshusses MA, Hamer G, Dunn IJ (1995a) Behavior of biofilters for waste air biotreatment 1. Dynamic model development. Environ Sci Technol 29:1048–1058

Deshusses MA, Hamer G, Dunn IJ (1995b) Behavior of biofilters for waste air biotreatment 2. Experimental evaluation of a dynamic model. Environ Sci Technol 29:1059–1068

Devinny JS, Deshusses MA, Webster TS (1999) Biofiltration for air pollution control. Lewis, New York

Dharmavaram S (1991) Biofiltration: a lean emissions abatement technology. In: Proc 84th Annu Meet Air and Waste Management Association, 16–21 June 1991, Vancouver, BC, Pap no 91-103.2

Diks RMM, Ottengraf SPP (1991a) Verification studies of simplified model for the removal of dichloromethane from waste gases using a biological trickling filter, part I. Bioprocess Eng 6:93–99

Diks RMM, Ottengraf SPP (1991b) Verification studies of simplified model for the removal of dichloromethane from waste gases using a biological trickling filter, part II. Bioprocess Eng 6:131–140

Ergas SJ, Schroeder ED, Chang DPY (1993) Control of air emissions of dichloromethane, tricholoroethene, and toluene by biofiltration. In: Proc 86th Annu Meet Air and Waste Management Association, 13–18 June 1993, Denver, CO, Pap no 93-WA-52B

Gabriel D, Deshusses MA (2003) Retrofitting existing chemical scrubbers to biotrickling filters for H_2S emission control. Proc Natl Acad Sci USA 100:6308–6312

Goto S, Smith JM (1975) Trickle-bed reactors performance, Part 1. Holdup and mass transfer effects. AIChE J 21:706–720

Hodge D, Devinny J (1995) Modeling removal of air contaminants by biofiltration. J Environ Eng 121:21

Kim S, Deshusses MA (2003) Development and experimental validation of a conceptual model for biotrickling filtration of H_2S. Environ Prog 22:119–128

Leson G, Winer AM (1991) Biofiltration: an innovative air pollution control technology for VOC emissions. J Air Manage Assoc 41:1045–1054

Li H, Crittenden JC, Mihelcic J, Hautakangas H (2002) Optimization of biofiltration for odor control: Model development and parameter sensitivity. Water Environ Res 74:5–16

Li H, Mihelcic J, Crittenden JC, Anderson KA (2003) Field measurements and modeling of a two-stage biofilter that treats odorous sulfur emissions. J Environ Eng 129:684–692

Lobo R, Revah S, Viveros-Garcia T (1999) An analysis of a trickle-bed bioreactor: carbon disulfide removal. Biotechnol Bioeng 63:98–109

Martin RW Jr, Li H, Mihelcic J, Crittenden JC, Lueking DR, Hatch CR, Ball P (2002) Optimization of biofiltration for odor control: model calibration, validation, and applications. Water Environ Res 74:17–27

Mpanias C, Baltzis BC (1998) An experimental and modeling study on the removal of mono-chlorobenzene vapor in biotrickling filters. Biotechnol Bioeng 59:328–343

Ockeloen HF, Overcamp TJ, Grady CPL Jr (1996) Modeling of fixed-film bioscrubbers for the control of volatile organic compounds. J Environ Eng 122:191–197

Okkerse WJH, Ottengraf SPP, Osinga-Kuipers B, Okkerse M (1999) Biomass accumulation and clogging in biotrickling filters for waste gas treatment: evaluation of a dynamic model using dichloromethane as a model pollutant. Biotechnol Bioeng 63:418–430

Onda K, Takeuchi H, Okumoto Y (1968) Mass transfer coefficients between gas and liquid phases in packed columns. J Chem Eng Jpn 1:56–62

Ottengraf SPP (1987) Biological systems for waste gas elimination. Trends Biotechnol 5:132

Ottengraf SPP, van den Oever AHC (1983) Kinetics of organic compound removal from waste gases with a biological filter. Biotechnol Bioeng 25:3089–3102

Ozis F, Yortsos Y, Devinny JS (2002) A discrete percolation model for describing biomass clogging in biofilters. In: Proc USC-TRG Conf Biofiltration, 31 October–1 November 2002, Newport Beach, CA

Sagastume JMM, Noyola A, Revah S, Ergas S (2003) Changes in physical properties of a compost biofilter treating hydrogen sulphide. J Air Waste Manage Assoc 53:1011–1021

Satterfield CN (1975) Trickle-bed reactors. AIChE J 21:209–228

Satterfield CN, van Eek MW, Bliss GS (1978) Liquid-solid mass transfer in packed beds with downward concurrent gas-liquid flow. AIChE J 24:709–717

Shareefdeen Z (1994) Engineering analysis of a packed-bed biofilter for removal of VOC (volatile organic compound) emissions. PhD Dissertation, New Jersey Institute of Technology, Newark, New Jersey

Shareefdeen Z, Baltzis BC (1994a) Biological removal of hydrophobic solvent vapors from airstreams. In: Galindo E, Ramirez OT (eds) Advances in bioprocess engineering. Kluwer, Dordrecht, pp 397–404

Shareefdeen Z, Baltzis BC (1994b) Biofiltration of toluene vapor under steady-state and transient conditions: theory and experimental results. Chem Eng Sci 49:4347–4360

Shareefdeen Z, Baltzis BC, Oh YS, Bartha R (1993) Biofiltration of methanol vapor. Biotechnol Bioeng 41:512–524

Shareefdeen Z, Shaikh AA, Ansar Z (1997a) Development, experimental validation, and dynamic analysis of a general transient biofilter model. Chem Eng Sci 52:759–773

Shareefdeen Z, Shaikh AA, Ansar Z (1997b) Biofiltration of Volatile Organic Compound (VOC) mixtures under transient conditions. Chem Eng Sci 52:4135–4142

Shareefdeen Z, Herner B, Webb D, Wilson S (2003) Hydrogen sulphide (H_2S) removal in synthetic media biofilters. J Environ Prog 22:207–213

Sologar VC, Lu Z, Allen G (2002), Modeling the biofiltration of air emissions containing reduced sulfur compounds (RSCs) and Volatile Organic Compounds (VOCs). In: Proc USC-TRG Conf Biofiltration, 31 October–1 November 2002, Newport Beach, CA

Van Lith C (1989) Biofiltration, an essential technique in air pollution control. In: Brasser LJ, Mulder WC (eds) Man and his ecosystem. Proc 8th World Clean Air Congr. Elsevier, Amsterdam, pp 393–399

Van Lith C, David SL, Marsh R (1990) Design criteria for biofilters. Trans Inst Chem Eng 68:127–132

Williamson K, McCarty PL (1976) A model of substrate utilization by bacterial films. J Water Pollut Control Fed 48:9–24

Zhu X, Alonso C, Suidan MT, Cao H, Kim BJ, Kim BR (1998) The effect of liquid phase on VOC removal in trickle-bed biofilters. Water Sci Technol 38:315–322

Part III

Biological Reactors – Applications

11 Biofilter Design and Operation for Odor Control – The New Zealand Experience

Roger Cudmore, Peter Gostomski

11.1 Introduction

The use of biofilters is extremely widespread in New Zealand for controlling industrial and municipal odor emissions. Open-top, up-flow biofilters with a packing based on bark or soil–bark mixtures is the predominant design. The first installations were designed for odor control at wastewater treatment facilities in the 1970s (Archer and Fullerton 1992). Typical applications of biofilters in New Zealand include meat rendering plants, wastewater treatment facilities, compost production, food processing, and hide-processing plants. A combination of standard practice and present New Zealand air pollution laws have resulted in virtually no biofilters being installed for non-odor-based applications. While odor control is often considered an "easy" biofilter application, the success of this technology in New Zealand has ranged from poor to very good, and the reasons for this range of experience is generally related to different bed designs and/or level of pretreatment of air ahead of the biofilter.

Other types of bioreactors, such as biotrickling filters, are not widely used for odor control for several reasons. Firstly, there is often sufficient land to construct simple open-top, up-flow biofilters. Additionally, biotrickling filters typically require a higher level of expertise to operate. Finally, biotrickling filters are normally offered by overseas vendors as a complete package in the same manner as chemical scrubbing equipment. By comparison, the vast majority of flat bed biofilter systems that have been installed in New Zealand to date are designed by engineering consulting firms or staff engineers, and the installation phase is undertaken by construction companies, local engineering contractors or the company's own engineering staff. The labor requirements for constructing and operating these designs are minimal.

These aspects combine to make biofilters the least expensive biological odor control technology to install and operate for the New Zealand market. Although the style of biofilter typically installed in New Zealand is relatively simple to construct and operate, this more informal arrangement of design and installation requires careful management and clear contractual obligations, so that any non-performance can be clearly attributed to design, installation or operational deficiencies.

Shareefdeen/Singh (Eds.)
Biotechnology for Odor and Air Pollution Control

It is the early success of soil biofilters that has led to the almost universal use of soil-based biofilters in New Zealand. This compares to the United States, where compost-based biofilters are used more often than soil-based ones for odor applications (Solomon 1994; Easter and Okonak 2000; Iranpour et al. 2001). This success is most likely due to soil-based biofilters performing better in the open-top style preferred in New Zealand. Open-top biofilters can potentially see higher moisture variations due to solar heating and rainfall or heavy frost events, and are more easily rewetted. They also are potentially more physically stable and have a higher buffering capacity. Because the flow direction has to be upward, the acid production normally associated with odor control is more easily washed from the bottom of the bed where the highest loadings are experienced (Devinny et al. 1999; Easter and Okonak 2000).

In this chapter, the authors will review the typical biofilter design used in New Zealand and, where appropriate, compare it to practice as reported elsewhere. We will cover stream characterization, pretreatment, bed design, and operational issues. The chapter will conclude with a number of common mistakes observed in the field.

11.2 Stream Characterization

The characterization of the airstream is commonly recognized in the literature as the first important step when designing a biofilter (Devinny et al. 1999; Kennes and Veiga 2001) but often the analysis is insufficient. A number of stream characteristics should be determined, including composition/loading, temperature/relative humidity, particulates, and stream dynamics.

11.2.1 Composition

Of primary concern is the relative biodegradability of the contaminants. Most odor-causing compounds are either sulfur- or nitrogen-based and are readily degraded in biofilters, but many other compounds can also be successfully degraded in biofilters (Leson 1998; Devinny et al. 1999). If the odor were due to a chlorinated solvent, then the biodegradability of this compound would need to be confirmed before proceeding with a biofilter option. Many such toxic compounds may be treated by biofiltration, but inoculation with specifically acclimated bacteria might be necessary. To the authors' knowledge, there are no commercial biofilters treating chlorinated solvents in New Zealand at present.

Although useful, a detailed chemical analysis of the airstream components via gas chromatography is not always necessary to infer the presence of the key odorous compounds. Detailed chemical analysis can often provide too much information – for example, Luo and van Oostrom (1997) identified over

300 compounds being emitted from a meat rendering facility. The odor concentration units (OU/m^3) of the inlet and discharge streams from some biofilters have been measured by dynamic olfactometry as per the Australian/New Zealand standard AS/NZS 4323.3:2001. This standard is based on the Comité Euroṕeen de Normalisation (CEN 1995). In the authors' experience, olfactometry sometimes understates the true performance and is certainly not a simple measurement technique. It is often sufficient and more effective to test for the presence of certain key odorous compounds such as hydrogen sulfide, based on process knowledge (Shareefdeen et al. 2002).

A general recommendation is that for odorous carbon-based compounds, the concentration should be below 1,000–2,000 ppm C (Leson 1998; Berger and Peters 1999). For sulfur and nitrogen, less than 10 ppm has been recommended if acidification is to be avoided through the addition of lime or other pH adjustment techniques. However, reports of successful operation at higher concentrations are available in the literature, especially when pH is allowed to drop in biofilters (Cook et al. 1999; Singleton and Milligan 2000) or biotrickling filters (Cox and Deshusses 2002). For many applications, odor units are the only quantifiable approach to determining concentration. Concentrations are often in the order of hundreds or thousands of odor units per cubic meter (OU/m^3), with successful biofiltration at 50,000 OU/m^3 and higher.

11.2.2 Process Knowledge

Consideration of the actual process that creates the odorous airstream often provides sufficient insight to allow for an effective pretreatment and biofilter design. This requires the application of detailed knowledge about the process that generates the odorous airstream.

An example of applying process knowledge is the comparison between the emissions from direct gas-fired and indirect steam-heated driers that are used in the rendering industry. Direct-fired driers produce an exhaust airstream that is laden with fine smoke aerosols, fatty tallow particulate as well as a cocktail of organic nitrogen- and sulfur-based VOCs. The smoke is thought to result from the condensation of high-boiling point organic compounds that are generated by exposing rendered organic material to the intense radiation of the gas flame and/or direct contact of this material with the flame. The resultant stream is highly odorous and contains a mix of biodegradable organic compounds, tallow-based particulate and condensed combustion aerosols. The tallow and aerosols are slow to biodegrade, and the smoke has a propensity to cause "tar-like" deposits within the air distribution system and the biofilter bed, as well as passing through the bed and creating a visible discharge from the bed surface.

The exhaust air from indirect steam-heated driers processing the same rendered material results in a stream whose odor content has a much higher level of biodegradability and less propensity to clog the biofilter system. This ultimately requires less intense pretreatment of the exhaust air, and higher bed loading rates are permissible. Common residence times for direct- and indirect-fired driers are 2 and 1.2 min, respectively.

Another important consideration with regards to the process is its consistency in discharge characteristics. Some processes do not discharge to air at a controlled steady rate, but instead have periods of relatively high volumetric discharge. It is therefore essential to understand the dynamics of the process, and account for peak flows and associated stream characteristics. Dynamic operation is common to food industry where material is often transferred and processed in batches, but is also a feature in many industrial processes such as bitumen production (Cook et al. 1999) and wood processing (Togna et al. 1997).

11.2.3 Temperature and Relative Humidity

Many biofilters in New Zealand are treating warm, wet airstreams, so humidification is not required, but cooling and the associated dewatering are essential. Ideally, airstream temperature and relative humidity are measured so that cooling equipment can be appropriately sized. However, it is often the case that an existing manufacturing plant needs to upgrade its odor control, or that a new plant is being proposed. In this instance, it is necessary to undertake the heat and mass balance calculations so as to estimate the humid enthalpy (kJ/kg) of the exhaust airstream, and therefore establish the design cooling duty that is necessary to lower the exhaust stream temperature to the biofilter design temperature. This calculation is also useful for streams requiring humidification (Gostomski et al. 1997; van Lith et al. 1997).

11.2.4 Particulates

As mentioned for rendering driers, particulates can create significant problems in biofilters due to the clogging of the media or distribution systems. Therefore, assessing the type and size of particulate within the odorous airstream is important during the design phase. Firstly in our experience, inorganic particulates are not commonly associated with odorous airstreams where biofilters are being considered. For example, odorous metallurgy exhausts often contain inorganic dust, but these streams are best treated via physical/chemical processes. Also, many wood processing streams could have wood dust associated with them, but no biofilters for this application are operational in New Zealand at present. Instead, most problems for biofilters arise from organic-based particulates that are derived either from materials handling emissions and/or the generation of particulates within the drying systems. In order of decreasing importance with regards to the problems they cause, typical types of organic particulate are:

High boilers (e.g., smoke aerosols and tar-like compounds from condensation of incomplete combustion)

Fine, fatty particulates (e.g., food processing, rendering, wastewater treatment and milk processing)

Proteinaceous (e.g., food processing)

Having ascertained the type of particulate that is likely to be present within the airstream, appropriate pretreatment methods and bed media designs can be considered. For example, a stream that has considerable smoke particulate would be best treated by a soil–bark bed biofilter (rather than a simple bark bed) when using a relatively low design temperature (i.e., less than 30 °C) and a relatively low design air loading rate of 40 m^3 air/h per m^3 media or lower (residence time > 90 s).

It should be noted that air loading rates are specified in a range of ways, including m^3 air/h per m^2 biofiler surface as well as m^3 air/h per m^3 biofilter, whereas we always relate the loading rate to the volume of active media within the biofilter, i.e., m^3 air/h per m^3 media, which excludes the distribution system and surrounding gravel supports and coarse filtering layers. The two definitions are used commonly but can result in significantly differing effective treatment capacities for the same loading rate criterion.

11.2.5 Odor Chemistry

A final consideration that is desirable, but not always essential, for a successful biofilter design is an understanding of the predominant chemical compound that is causing the odor problem, and the approximate loading rate of this compound. For example, knowing that hydrogen sulfide is the main odorous compound and knowing its mass loading rate can help infer the pH conditions that are likely to occur at the base of the bed. With this information, we can infer suitable materials of construction, and recommended loading rates of lime to the bed if pH is going to be actively controlled.

Knowledge of the primary odor's chemical nature is also useful for successful bed design. For example, if ammonia is the primary source of odor in the untreated stream, then a soil–bark biofilter is recommended over a straight bark biofilter. Straight bark biofilters are often ineffective in the total removal of ammonia. A careful consideration of alternatives, such as a chemical scrubber, should be considered in such a case.

11.3 Pretreatment/Conditioning of Airstream

Biofilter operational problems can often be attributed to inadequate pretreatment of the odorous airstream. The biofilters are often overloaded with moisture, volatile fatty compounds and/or particulate that can all cause increase in bed backpressure after commissioning. Backpressure decreases the airflow rate, which can be especially troublesome for upstream drying or other processes, as flow rate is a key operating parameter. As the medium clogs, it potentially becomes anaerobic and odor may be generated in the biofilter. Airstreams that are dominated by building air create fewer problems, except they can be prone to rapid bed drying during dry and warm weather periods. This section provides some general guidance on how to account for these factors during design.

11.3.1 Particulates

Particulate loadings of less than 5–10 mg/m^3 are recommended to minimize clogging (Leson 1998; Berger and Peters 1999). The most common method for protecting biofilter beds against particulate clogging is to pretreat the airstream via cyclones, followed by void water spray towers. Packed towers are avoided because of the associated clogging problem. As a rule of thumb, water spray towers are best operated in countercurrent mode, and sized such that the superficial air velocity is < 0.5 m/s and for retention times of approximately 5 s. Again, as an approximate guide, a water recirculation rate of 7 l/s per m^3 air/s, and a spray nozzle pressure drop of approximately 2 bar is recommended. These tower and water recirculation design parameters will generally apply to dust that is generated from rendering and food processing plants, but could need to be varied in other applications. Finally, the makeup water rates need to be controlled to ensure that effective particulate removal is maintained. This control can be based upon turbidity or, more typically, via manually adjusted water inlet valves.

The use of water spray towers in not always a viable option because of water discharge constraints, thus the use of pulsejet bag filters may be useful. The use of this filtering technology is dependent upon the collected dust not clogging the bags. There are few examples of bag-house filters being used with biofilters in New Zealand, but they can be very efficient. Additionally, the collected dust can often be recovered as a product. For example, there are bag-house filters being used to pre-clean a hot-mix asphalt plant emission ahead of a carbon filter (carbon systems require a very clean inlet airstream to avoid clogging problems).

11.3.2 Temperature

There is some debate on the maximum inlet temperature for soil-bed or bark biofilters. A common view is that the normal maximum should be 40 °C with short-term peaks up to 45 °C. In principle, it is possible to operate biofilters at temperatures exceeding 40 °C (thermophilic range), with some reports indicating a maximum of 60 °C was acceptable for soil-based biofilters (Bohn 1992). From our experience, the most important reason for limiting the inlet temperature in many cases is to restrict the loading of fatty volatiles, tar-like compounds, and moisture into the biofilter bed. For example, water-saturated air at 60 °C holds over five times the moisture of air at 30 °C (i.e., 0.155 kg/kg dry air versus 0.0276 kg/kg dry air), and heat loss in the bed can cause a considerable amount of water to condense in the biofilter.

For hot process streams, either plate or shell and tube heat exchangers are often used as the initial cooling stage. This produces hot water that can be used in the manufacturing plant. They also have the advantage of condensing excessive water vapor and fats from the exhaust airstream, and are often referred to as condensers. The temperature difference between the inlet hot air and the cooling water typically needs to be approximately 40 °C or much higher to make the use of such condensers economically viable. For many small to medium-sized facilities, the viability of shell and tube condensers significantly increased due to the availability of secondhand units from the ever-rationalizing meat processing industry, and the centralization of milk processing plants in New Zealand.

For exhaust airstreams of moderate temperature (< 60 °C), the most cost-effective cooling device is often a water spray tower as used for dust cleaning. If these systems are used to cool high-temperature streams, then insufficient cooling can add large quantities of additional moisture to the biofilter bed due to subsequent cooling and condensation in the bed. Therefore, the recommended approach is to undertake cooling in two stages, using a condenser followed by a spray tower, to achieve the final design airstream temperature. Note that it is advisable to design for a temperature that is approximately 10 °C lower than the actual target, which accounts for some design uncertainty and future fouling effects of heat exchange surfaces.

Having defined the airstream inlet temperature, and moisture content and outlet conditions, the cooling energy and cooling water requirements can be calculated using standard humid enthalpy equations and/or charts. With the cooling energy load and inlet water and hot air conditions, standard design equations can be used by experienced process engineers to size an appropriate heat exchanger (Kern 1990).

11.3.3 Relative Humidity

Inlet airstreams to biofilters need to be essentially saturated to avoid bed drying, otherwise surface irrigation is necessary. For hot, moist exhaust streams that require pre-cooling, having sufficient humidity is never a problem – rather, it is a problem of having too much. For air extraction systems that include ambient air, artificial humidification is then often used to increase humidity levels. However, these systems are often not totally successful, as they are often undersized. The water atomization rate and efficiency often need to be very high for any hope of successful humidification. To size the humidification equipment, it is necessary to establish typical dry-day temperature and humidity levels and to calculate, via psychrometric charts or software, the additional grams of water per kg of air that is necessary to saturate the airstream. It is for this reason that many industrial applications with unsaturated airstreams operate in down-flow mode to facilitate water addition at the bed inlet to overcome insufficient humidification and subsequent bed drying (van Lith et al. 1997).

11.3.4 Bed Design

As discussed earlier, the basic design for the vast majority of New Zealand biofilters is open-top and up-flow using a soil/bark-based medium. However, it is not uncommon for the biofilters to have an open roof/cover arrangement to minimize weather effects. The typical residence time is 1–2 min but commercial installations presently being operated in New Zealand have residence times varying from 30 s to 16 min (Luo and van Oostrom 1995). The combination of larger residence times and bed depths not normally exceeding 1 m of active bed leads to relatively large biofilters compared to industrial applications in other countries. The ability to implement large footprint biofilters in New Zealand is mainly due to the industrial applications being located in "remote" areas where rigorous odor control was not originally required. Therefore, many facilities have inexpensive land to commit to relatively large biofilters now that odor control is required, due either to the complaints from new neighbors or to the renewal of emission permits. As land prices increase in New Zealand, there is a continuing push for more intensive biofilter technology using a smaller footprint.

11.3.5 Air Distribution

The common method for distributing air is via PVC pipe laterals that have a standard hole pattern drilled by the supplier. Typically, these laterals are fiber-glassed into a centrally located concrete pipe that acts as the main supply manifold. The main concrete manifold should be extended beyond the biofilter bed, and allow for external access via an inspection manifold. Furthermore, it has become common practice to extend the ends of each distribution lateral upward above the bed, so as to allow for future flushing.

The spacing and size of the distribution holes, as well as the spacing between laterals and their diameter, are key design factors that are a function of the bed geometry as well as the total airflow rate. As a general guide, good air distribution is achieved when the main manifold pressure drop and that across the full length of the lateral is less than 10% of the pressure loss across the distribution holes. These pressure losses as well as the percent of flow maldistribution across the lateral can be estimated by applying Reynolds number calculations to Fanning friction factor plots (Tilton 1997).

Biofilter beds with open plenum channels to distribute the incoming airstream are less common but have been used in a number of industrial biofilters in New Zealand. This type of distribution system is significantly more expensive to install, and does not always provide effective air distribution. However, open plenums are less prone to blockages and hole-blinding problems that can occur in pipe lateral systems, and also result in less overall backpressure being exerted upon the supply fan. However, maintenance can be difficult if geotechnical liners are used and non-biodegradable compounds are deposited (Luo and van Oostrom 1998).

11.3.6 Bed Media

The PVC pipe system is supported within a bottom layer of crushed aggregate, with the biofilter bed placed above the distribution drainage layer. It is common practice to have a thin, coarse bark layer (25–75 mm size range) approximately 100–200 mm deep immediately above the aggregate. This layer is followed by the biofilter media, which typically consists of a soil and graded bark mixture that is typically in the range of 600–1,000 mm deep. The soil contribution is typically less than, or equal to, 15% of the graded bark volume to which it is added.

As a general guide, the soil added to the graded bark mix should not increase the volume of the starting bark fraction. Therefore, the soil fits within the void space of the graded bark, with room to spare (Pinnette et al. 1995). Some biofilters in New Zealand use a straight mix of graded bark for treatment media. Because of the greater structural integrity and lower specific pressure drop of bark biofilters, these are sometimes installed to depths of several meters. However, the more common design is as described above, and results in beds that have overall heights of approximately 1.5 m, which are either constructed above the ground with a timber wall framework and impervious liner, or else fully or partially sunken into the ground and surround by earth embankments.

The key criterion for either a bark or soil-bark bed is the superficial air residence time, or sometimes given as the air loading rate per unit of bed media (m^3 air/h per m^3 media). Sometimes biofilter designers refer to a residence time for air within the bed, but this requires the effective porosity of the media to be known or estimated, which is extremely difficult to do accurately. Therefore, the empty bed volume is normally used for any calculation.

The design air loading rate should be altered to account for the media type and level of smoke within the pretreated airstream. As an approximate guide, the loading rates in Table 11.1 are recommended as a function of air source and bed type based on the authors' experience.

The recommendations are for graded bark biofilters to be approximately three times larger than soil-bark biofilters for treatment of a given stream. These recommendations are solely based upon field experience with both types of bed. However, it is reasonable to assume that the inclusion of friable topsoil into the graded bark increases the effective surface area of the media, and supports a more diverse microbial population, thereby permitting a shorter residence time for effective treatment. The downside to including soil is that soil-bark beds are more prone to pressure drop increases, and proper selection and control of the soil are very important.

11.3.7 Specification of Soil and Bark

The topsoil should have the right balance of sand, silt, and clay. This requires a soil with less than 14% clay content, and what is typically referred to as a "sandy loam" soil type and at least 2% organic carbon. The graded bark is

Table 11.1. Recommended biofilter loading rates for different air sources and different media

Air source	Media type	Recommended loading rate (m^3 air/h per m^3 media)	Residence time (s)
Building air (ambient conditions, low odor)	Soil–bark	60–150	24–60
Building air (ambient conditions, low odor)	Graded bark	20–40	90–180
Cooled process air, ≤ 40 °C, saturated, no smoke	Soil–bark	30–50	72–120
Cooled process air, 40 – 60 °C, saturated, no smoke[a]	Graded bark	10–20	180–360
Saturated process air, ≤ 40 °C, saturated, significant smoke	Soil–bark	30–40	90–120
Saturated process air with significant smoke	Graded bark	Not recommended	-

[a] Graded bark beds are more stable at higher temperatures than soil–bark mixtures

generally from *Pinus radiata* (the dominant, commercially processed tree in New Zealand), and is produced by screening to remove the fines and coarse fraction, so to produce chips between 1 and 10 mm.

It is recommended that candidate soil and bark samples that appear to be acceptable based upon a visual assessment are formally analyzed for their size distributions and composition. The laboratory test methods that have been commonly used in New Zealand for testing soil and bark are summarized as follows:

Soil: (Claydon 1989): determination of particle size in fine-grained soils – pipette method. DSIR Div. Land and Soil Sciences Technical Record LH5.

Total carbon: Leco CNS analyzer. Environment Chemistry Laboratory, Landcare Research, Palmerston North.

Bark: mass size distribution of bark. Bark oven-dried, sieved using a Fritsch Analysette 3 at 2-mm amplitude at 5 min.

The soil test results provide the mass fractions of clay, silt, and total sand size fractions. This information can be applied to soil classification triangles to indicate soil type. Soils that are high in either sand or clay should be avoided. High-clay soils provide excellent treatment capacity but make the bed too prone to structural collapse and high backpressures. Some clay loams, which are close to the sandy loam boundary, have successfully been modified by the addition of beach sand. High-sand soils with little organic content can be unstable, the sand tending to move downward into the air distribution layer.

Bark and soil test results for the final bed material provide important protection against contractors whose pre-samples of soil and bark may not reflect the less expensive substitute that is actually installed. To this end, it is always essential to specify soil and bark criteria as discussed above. On more than one occasion, the authors used these specific requirements in the contract to require contractors to reinstall biofilter beds, thus avoiding eventual problems with the client for poor biofilter performance. Typical test results for soil and graded bark samples that have been used successfully in various New Zealand biofilters are detailed in Tables 11.2 and 11.3. It can be seen that for the graded bark, the major mass fraction lies within the 3–8 mm size range. This reasonably narrow size range helps ensure a consistent and predictable bed pressure drop when mixed with the soil. Of the two soil test results above, soil A is considered to be more desirable because of the lower clay and silt composition, and even distribution of sand grain size. In fact, soil B is currently used in a 1,000-m^2 rendering plant biofilter (Taranaki By-Products, Hawera, New Zealand), following some sand addition (1 part sand to 4 parts soil), so to simulate a more typical "sandy loam" composition.

Buffering: in general, lime addition has traditionally been used for pH control, but some facilities with low loading count on the buffering capacity of the soil combined with water addition to keep the pH from dropping too low. Addition of slacked lime (CaO) to the soil–bark mix is generally recommended to help maintain the bed pH over time if the reduced sulfur loading rates are considered high. Application rates are difficult to define, and depend upon the character of the stream being treated. However, a rate of 0.5 kg/m^2 bed is sufficient for most beds. Much higher lime loadings have been successfully used to rehabilitate beds that have collapsed due to the use of high-clay soils. Compost-based biofilters have been reported to be less amenable to lime addition (Bohn 1992).

Table 11.2. Typical mass size distribution of bark

Sieve size (mm)	Weight%	Cumulative (%)
16	0.0	0.0
11.2	0.2	0.2
8	22.0	22.3
5.6	48.9	71.2
3.35	18.7	89.9
2	5.1	95.0
< 2	5.0	100.0

Table 11.3. Typical fine-earth particle size distribution of soil (%)

Sample	Coarse sand (2–0.6 mm)	Medium sand (0.6–0.2 mm)	Fine sand (0.2–0.06 mm)	Silt (0.06–0.002 mm)	Clay (< 0.002 mm)
Soil A	10	21	22	37	9
Soil B	2	5	29	50	14

Bed backpressure: the backpressure across the bed media is a strong function of the bark size, soil type, and ratios of mixing. Additionally, some caution should be used if small laboratory columns are used to estimate full-scale pressure drop (Gostomski and Liaw 2001). When using graded bark alone and loading rates of approximately 50 m^3 air/h per m^3 media, the media pressure loss is relatively low (e.g., 1–20 mm H_2O/m). When the same graded bark is premixed with soil at 15% of the bark volume, then the eventual pressure drop across the soil–bark media is more likely to be in the range of 10–80 mm H_2O/m. The large variation in pressure drop ranges that has been observed relates to the different loading rates of water, smoke, and particulate associated with different applications, and variations in the actual particle size of the bark and soil.

11.4 Operation and Monitoring

Soil-bed biofilters are relatively simple odor treatment devices. Nevertheless, there are number of key parameters that should be periodically monitored. These are summarized below.

11.4.1 General Operation and Maintenance

Moisture content, pH, temperature of the medium, and the pressure drop across the medium should be monitored and controlled within suitable limits to maximize biofilter performance.

Biofilter beds need to be tilled periodically if the backpressure across the bed begins to limit the airflow rate. Cultivation intervals can vary from 6 months to 5 years.

Complete media replacement may be needed every 5–10 years, due to low pH, crusting, salt accumulation, or loss of porosity.

11.4.2 Pressure Drop

Monitoring of bed pressure drop is recommended, as this will detect compaction and indicate when loosening or replacement of the medium is required. Monitoring will also indicate if pressure drop and therefore air delivery flow rates are being affected by the addition of too much water to the medium.

Simple water-filled U-tube manometers give an accurate and reliable measurement of static pressure head, as do spot checks of the inlet airflow rate using a Pitot tube. The installation of copper or plastic tubing into the aggregate allows the pressure drop across the biofilter medium to be established, whereas the gauge pressure within the air supply manifold indicates the pressure drop across the distribution pipes and the media combined.

High-pressure drop can adversely affect odor removal performance by limiting the flow capacity of the blower. Soil–bark beds that are 0.6–1.0 m deep typically operate at around 80–150 mm water gauge backpressure (including the distribution system).

It is usually sufficient to check bed pressure drops once a week. Continuous recording and display of inlet air pressure and/or media pressure drop is not usually undertaken, but can be justified in some cases.

11.4.3 Moisture

The bed moisture content is an important factor affecting the odor removal efficiency and pressure drop of biofilters (Gostomski and Liaw 2001; Ranasinghe and Gostomski 2003). Depending on the medium used, the recommended optimal moisture content normally ranges between 40 and 60% (wet weight basis) for soil–bark beds, but lower moisture levels can occur in simple bark beds unless heavily loaded from the process itself.

Biofilters that treat potentially unsaturated air should use a surface irrigation system to control moisture levels in the bed during extended dry periods. In such cases, only high-pressure (i.e., > 30 psi) surface misting systems should be used that provide even coverage of the biofilter bed. Inexpensive spray irrigation systems can cause waterlogged and excessively dry areas to coincide in the bed, and generate associated air channeling problems.

Moisture content changes are relatively slow in biofilters, except following heavy rainfall events. Monthly monitoring is usually sufficient. Weekly or continuous monitoring is only necessary where beds are prone to rapid drying, and require continual humidification and/or surface irrigation to maintain adequate bed moisture levels.

11.4.4 Temperature

Optimal temperature is a compromise between that giving maximum microbial activity and that maximizing the adsorption of odorous compounds onto the medium. It is recommended that biofilters are operated at temperatures less than 40 °C, if this can be practically achieved.

As temperature affects not only the activity but also the composition of microbial populations, sharp temperature fluctuations in the biofilter should be avoided.

Continual monitoring and recording of inlet air temperature are generally recommended.

11.4.5 pH

The bed pH is often recommended to be in the range of 6–8, although there are beds that have worked at much lower pH levels. Furthermore, very low pH levels may exist near the base of a bed while the higher layers operate under less acidic conditions.

The pH changes in a bed occur very slowly, and therefore 1–6 monthly sampling and analysis are usually more than adequate.

11.4.6 Emission Monitoring

Most monitoring of odor removal is informal, with operators checking performance when measuring other parameters. Additionally, complaints from neighbors are always an unfortunate indication of inadequate performance. Open-top biofilters raise challenges if more formal monitoring is required. There is no overall stream for sampling, but point samples across the surface can be easily taken to provide a composite measurement of performance. Additionally, easy access to the whole bed surface makes identification of regions of poor flow or removal efficiencies easier to identify (Bohn 1995; Shareefdeen et al. 2002).

11.4.7 Biofilter Maintenance

The following routine maintenance measures are recommended for the biofilter:

Bed surface: grass and/or weeds can be allowed to establish across the bed surface. Spraying of vegetation should be avoided, unless a herbicide is used prior to vegetation stripping and bed tilling. Biofilters have been used successfully to grow grass and flowers to increase the aesthetics of an industrial site.

Bed tilling: should the mean backpressure increase with statistical significance from one year to the next, then it is good practice to till or rotary hoe at least the first 400 mm of bed media. If this is not effective in reducing increased bed backpressure, a deep tilling of the bed media down to the coarse bark or even gravel layer could be necessary. Note that it is acceptable for small, wide-tracked diggers to work on top of bed surfaces, thus providing a means of turning and mixing the media layer. Complete replacement of the bed media should not be required for several years, and the beds generally remain effective for over 10 years and well beyond.

Bed pH: this can be maintained by annual addition and/or rotary hoeing of hydrated lime into the bed surface if irrigation is insufficient and performance has dropped. Recent publications indicate that allowing pH to drop to the range of 2–3 is acceptable, but the bark lifetime could be considerably reduced at this pH.

11.4.8 Common Failures

As long as the airstream can be forced through a biofilter bed at acceptable rates, apart from the rare cases of excessive smoke discharge, efficient odor treatment is normally assured. However, as with biofilters all over the world, the main problem that arises is high pressure drops developing over time, which results in insufficient air extraction and treatment. These pressure drop increases can be gradual or sudden. The two most common reasons for sudden increases are bed-structure collapse, or sudden water buildup within the air supply/distribution system or biofilter bed.

A sudden increase in backpressure due to water buildup is usually because of the water drainage system failing (e.g., sump pumps not working, or inadequate system for drainage of the air distribution and supply pipe work). A sudden collapse of the biofilter bed structure is invariably a result of high rainfall events or overloading of the bed with moisture. Normally, a soil–bark or bark biofilter will recover from these temporary high water loadings, as long as appropriate soil and bark mixes have been used, and the bed has adequate provision to drain water from the base of the bed at peak rates. However, there have been examples in New Zealand where biofilters with high-clay soils or inadequate drainage systems have resulted in the bed structure permanently compacting and requiring complete replacement. Typically, the source of the problem relates to inadequate consideration to water drainage requirements imposed by process water inputs, rainfall events, and even groundwater infiltration.

The combination of suitable bed materials and adequate provision for both the biofilter bed and pipe work drainage helps to eliminate the sudden pressure drop-type problems that have afflicted a number of New Zealand biofilters. This can require an analysis of groundwater levels, 15-min peak rainfall events, and mass loadings of water from the process that invariably condenses out within the biofilter bed, and these being accounted for in the design.

Whereas the sudden buildup of backpressure can normally be avoided through good design, the more common problem of gradually increasing backpressure within the bed is often unavoidable, or is at least a necessary trade-off against the capital and operating cost of the airstream pretreatment system. The common causes of gradually increasing bed backpressure include:

- the breakdown of the bed's bark and soil structure
- poor bed design and choice of media
- buildup of fatty or tar-like layers within the lower regions of the bed
- gradual blinding of air distribution holes with tars or precipitated solids

The breakdown of the bed's bark and soil structure should be a relatively slow process in most instances, and is solved by periodic bed maintenance. However, the poor choice of bed materials has been a common problem, with many beds failing to a point that was unrecoverable. The use of compost or soil within beds, while using insufficient bark or bark that is too large to provide a stable support matrix, has probably been the key source of failed beds.

Bed clogging due to fats, tars or precipitation of solids can often be rectified through annual bed tilling and remixing. Often, the cost and effort required for such actions are a necessary compromise to installing more expensive pretreatment systems. When the clogging occurs within the distribution pipe work, then the ability to access individual laterals and apply high-pressure water is essential.

11.5 Conclusions

Biofiltration is a very successful odor control technology in New Zealand. A relatively simple and inexpensive design based on an open-top, up-flow configuration through soil/bark mixtures is successful in most typical applications. With increasingly tighter emission control laws coming online in New Zealand, biofilters will undoubtedly find applications beyond odor control, similar to many other countries. There will also be a push for higher-performance designs capable of being installed in smaller spaces. Undoubtedly, biofilters will see continued implementation in New Zealand in traditional and potentially more complex designs.

References

Archer HE, Fullerton RW (1992) Odor control by biological soil filters – a review of current applications in New Zealand. In: Proc Annu Conf New Zealand Water Supply and Disposal Association, Christchurch, New Zealand, pp 1–8

Berger S, Peters D (1999) Biofiltration. Center for Waste Reduction Technologies, American Institute of Chemical Engineers, New York

Bohn H (1992) Consider biofiltration for decontaminating gases. Chem Eng Prog 88:34–40

Bohn H (1995) Spatial variability and emissions monitoring of biofilter beds. In: Proc Conf Biofiltration, 5–6 October 1995, Los Angeles, CA. The Reynolds Group, Tustin, CA, pp 237–246

CEN (1995) Odor concentration measurement by dynamic olfactometry. Comité Européen de Normalisation, CEN TC264/WG2

Claydon JJ (1989) Determination of particle size in fine-grained soils-pipette method. DSIR Div Land Soil Sci Tech Rec LH5

Cook LL, Gostomski PA, Apel WA (1999) Biofiltration of asphalt emissions: full-scale operation treating off-gases from polymer-modified asphalt production. Environ Prog 18:178–187

Cox HHJ, Deshusses MA (2002) Co-treatment of H_2S and toluene in a biotrickling filter. Chem Eng J 87:101–110

Devinny JS, Deshusses MA, Webster TD (1999) Biofiltration for air pollution control. Lewis, Boca Raton

Easter CC, Okonak CC (2000) Biofilter systems for odor control in wastewater treatment applications: an overview of issues and options. In: Proc Conf Odors and VOC Emissions 2000, 16–19 April 2000, Cincinnati, OH. Water Environment Federation, Alexandria, VA, pp 68–86

Gostomski PA, Liaw LP (2001) Air permeability of biofilter media. In: Proc Air and Waste Management Association 94th Annu Conf Exhibition, 24–28 June 2001, Orlando, FL, pp 1590–1601

Gostomski PA, Sisson JB, Cherry RS (1997) Water content dynamics in biofiltration: the role of humidity and microbial heat generation. J Air Waste Manage Assoc 47:936–944

Iranpour R, Samar P, Stenstrom MK, Clark JH, Converse BM, Schroeder ED, Cox HHJ, Deshusses MA (2001) Biological treatment of odor and VOCs in biofilters and biotrickling filters: survey of field experiences. In: Proc Air and Waste Management Association 94th Annu Conf Exhibition, 24–28 June 2001, Orlando, FL, pp 1828–1846

Kennes C, Veiga MC (2001) Conventional biofilters. In: Kennes C, Veiga MC (eds) Bioreactors for waste gas treatment. Kluwer, Dordrecht, pp 47–98

Kern DQ (1990) Process heat transfer. McGraw-Hill, New York

Leson G (1998) Biofilters in practice. In: Sikdar SK, Irvine R (eds) Bioremediation. Technomic Publ, Lancaster, PA, pp 523–556

Luo J, van Oostrom A (1995) Odor control using biofilters – a survey report. Meat Industry Research Institute of New Zealand, Hamilton, New Zealand, MIRINZ 960

Luo J, van Oostrom A (1997) Biofilters for controlling animal rendering odor – a pilot-scale study. Pure Appl Chem 69:2403–2410

Luo J, van Oostrom A (1998) Odor-control biofilters: design, operation and maintenance. In: Proc New Zealand Water Wastes Association 40th Annu Conf Expo, 23–25 September 1998, Wellington, pp 74–81

Pinnette JR, Dwinal CA, Giggey MD (1995) Porosity of biofilter media. In: Proc Conf Biofiltration, 5–6 October 1995, Los Angeles, CA. The Reynolds Group, Tustin, CA, pp 207–215

Ranasinghe MA, Gostomski PA (2003) A novel reactor for exploring the effect of water content on biofilter degradation rates. Environ Prog 22:103–109

Shareefdeen Z, Herner B, Wilson S (2002) Biofiltration of nuisance sulfur gaseous odors from a meat rendering plant. J Chem Technol Biotechnol 77:1296–1299

Singleton B, Milligan D (2000) The evolution of biofiltration and the application of third generation designs. In: Proc Conf Odors and VOC Emissions 2000, 16–19 April 2000, Cincinnati, OH. Water Environment Federation, Alexandria, VA, pp 56–67

Solomon M (1994) Soil filter beds: the west coast experience. In: Proc Conf Odor and VOC Emission Control for Municipal and Industrial Treatment Facilities, 24–27 April 1994, Jacksonville, FL. Water Environment Federation, Alexandria, VA, pp 4/61–74

Tilton JE (1997) Fluid and particle dynamics. In: Perry RH, Green DW, Maloney JO (eds) Perry's chemical engineers handbook. McGraw-Hill, New York, pp 6.1–6.54

Togna AP, Fucich WJ, Loudon RE, del Vecchio M, Barshter DW, Nadeau AJ (1997) Treatment of odorous toxic air pollutants from a Hardwood Panel Board manufacturing facility using biofiltration. In: Proc 90th Annu Meet Air and Waste Management Association, Toronto, Canada, pp 1–8

Van Lith C, Leson G, Michelsen R (1997) Evaluating design options for biofilters. J Air Waste Manage Assoc 47:37–48

12 Biological Treatment of Waste Gases Containing Inorganic Compounds

MADJID MOHSENI

12.1 Introduction

The detrimental impact of many air pollutants to human health, vegetation, human property, or the global environment is evident from various studies conducted over many years. Inorganic air pollutants are among those that require special attention because of their significant environmental impacts as well as their lack of response to many physical air pollution control techniques (e.g., filters, vapor capture, and condensation). Many of these inorganic air pollutants (such as hydrogen cyanide and hydrogen sulfide) can be associated with severe and acute health impacts following industrial accidents. However, most of the interest in air pollution and health is directed at long-term and low-concentration exposure to harmful air contaminants. With the compelling evidence on the long-term impacts of inorganic air pollutants, many researchers have focused on developing and optimizing novel and cost-effective air pollution control strategies that can meet the increasingly stringent regulations on the emissions of these chemicals (e.g., Barnes et al. 1995; Chung et al. 2000; Philip and Deshusses 2003).

12.2 Common Inorganic Air Pollutants

Anthropogenic nitrogen (N) and sulfur (S) compounds constitute two of four major classes of air pollutants of significance to human health and the environment. The other two are volatile/semi-volatile compounds and toxic metals (Krupa 2003). Two major categories of anthropogenic inorganic nitrogen and sulfur air pollutants are (1) odor-causing compounds that include a large number of sulfur- and nitrogen-containing pollutants such as H_2S, NH_3, and amines; and (2) oxides of sulfur and nitrogen that contribute to acid rain and smog formation. The primary focus of this chapter will be on the problems, issues, and promising biological treatment technologies associated with nitrogen-containing air pollutants. Biological processes for the treatments of sulfur-containing odorous compounds are covered thoroughly in other chapters and will not be discussed here.

Shareefdeen/Singh (Eds.)
Biotechnology for Odor and Air Pollution Control

12.2.1 Ammonia

Ammonia (NH_3) is a colorless, toxic, reactive, and corrosive gas with a very sharp odor (Busca and Pistarino 2003). It can be smelt when it is in the air at a level higher than 50 ppm, and dissolves easily in water. The US Occupational Safety and Health Administration (OSHA) has set a limit of 50 ppmv over an 8-h workday for ammonia vapor in ambient air (Busca and Pistarino 2003).

A considerable amount of ammonia is emitted from industrial processes such as petrochemical refining, metal manufacturing, food preparations, and composting facilities (Chung et al. 1997; Busca and Pistarino 2003). Overall, of the total anthropogenic nitrogen compound emissions, ammonia accounts for about 55%, equivalent to approximately 43 million tonnes of nitrogen (Krupa 2003). Ammonia, as an air pollutant, plays an important role in acidification (acid rain) and eutrophication. Hence, control of ammonia is essential to mitigate the environmental impact and to protect public health.

12.2.2 Amines

Amines are bad-smelling compounds and can have high toxicity (Busca and Pistarino 2003). Similar to ammonia, amines can be present in waste gases emitted from various industrial operations such as animal production facilities, food industries, and meat treatment plants (Busca and Pistarino 2003). Amine emissions from animal and livestock facilities are often correlated with those of NH_3 and consist of methyl amines, ethyl amines, dimethyl amines, etc. Potential problems associated with methylamines are speculative but possible atmospheric reactions that lead to the formation of more problematic compounds such as N_2O and HCN (Schade and Crutzen 1995).

12.2.3 Nitrogen Oxides (NO_x)

Gaseous nitrogen oxides (NO_x) are produced by a variety of human activities, such as combustion of fossil fuels and synthesis of nitrogenous fertilizers (Apel et al. 1995). NO_x also arise naturally in the atmosphere by the action of microbial biodegradation of organic nitrogenous compounds in soils (Cole 1993; de Nevers 2000). The hazardous nature of NO_x is already well established, and it is known that these chemicals adversely affect human health and possess diverse biological roles (de Nevers 2000). In addition, nitrogen oxides are of focal importance with respect to the atmospheric ozone destruction, global warming, and the formation of acid rain.

Among various nitrogen oxides, nitric oxide (NO) and nitrogen dioxide (NO_2) are the toxic and the most problematic, and constitute the major NO_x

components released during combustion processes (Barnes et al. 1995; de Nevers 2000). NO, a colorless gas, is the most common of the two, and accounts for more than 95% of the total NO_X emitted from combustion sources (Min et al. 2002). NO_2, on the other hand, is formed via photochemical reactions in the atmosphere (de Nevers 2000). NO is best considered as a new class of signaling molecules that acts primarily in the nervous and cardiovascular systems, and that diffuses freely across cell boundaries to activate nearby target cells. In addition, as a highly reactive species, NO is implicated in the degradation of neurons associated with Parkinson's disease, AIDS dementia, and strokes (Stepanov and Korpela 1997).

Reduction in NO_X emission can be achieved through a variety of measures including prevention and treatment. Preventive measures include modification of combustion process parameters such as burning temperatures and gas retention times, both being quite effective in reducing NO_X production. However, post-combustion treatment is usually required to achieve current and increasingly stringent regulatory standards set by the titles I and IV of the Clean Air Act Amendment of 1990 (CAAA 1990; Grano 1995; Barnes et al. 1995; de Nevers 2000).

12.2.4 Sulfur Oxides (SO_X)

Sulfur oxides (SO_X) and sulfur dioxide (SO_2) often exist in off-gases from the combustion of fossil fuels. Almost all the fossil fuels, such as coal, petroleum, natural gas, and even wood, contain certain amounts of sulfur that will form sulfur dioxide during the combustion process (de Nevers 2000). SO_X are among the principal causes of acid rain and adversely affect human health, livestock, and the environment (Philip and Deshusses 2003). Hence, they have been the subject of environmental regulations under Section 401 of the CAAA 1990 (de Nevers 2000).

12.3 Treatment Technologies for Inorganic Air Pollutants

A range of technologies is available for the treatment of inorganic air pollutants emitted from industrial processes, wastewater treatment plants, sludge handling and composting facilities, and food processing operations. These technologies fall into three major categories – chemical (e.g., chemical/wet scrubbing, thermal oxidation, catalytic oxidation/reduction, photocatalysis), physical (e.g., adsorption, absorption, membrane separation), and biological (e.g., biofiltration, bioscrubbing, biotrickling filtration). Table 12.1 summarizes some of the most common techniques that have been investigated for the abatement of inorganic pollutants from waste gases.

Selective catalytic oxidation is widely used for flue gas NO_X removal. A variety of reducing agents such as ammonia are used for NO_X reduction in the

Table 12.1. Treatment processes for inorganic air contaminants

Treatment process	Areas of application	Advantages	Disadvantages	Level of use
Thermal and catalytic oxidations	Odor-causing nitrogen- and sulfur-containing compounds	– Effective – Proven	– High capital and energy costs – Economical for high concentrations of recalcitrant compounds	High
Selective catalytic reduction	Nitrogen oxides	– Proven and very effective – Ability to operate at high temperature	– High capital and operating costs – Catalyst toxicity and disposal issues	Very high
Wet scrubbing	*Sulfur oxides, ammonia,* and sulfur-containing odorous compounds	– *Proven and* effective – Moderate capital and operating costs	– *High maintenance* – Not effective for all chemicals – Possibility for incomplete conversion	*Very high*
Adsorption	Odor-causing nitrogen- and sulfur-containing compounds	– Consistently high performance – Reliable	– High operating cost – No mineralization – Effective for low pollutant loads	High
Advanced oxidation (photocatalysis, electron beam, etc.)	NO_X, SO_X, ammonia	– Effective – Ambient operating conditions	– Not proven – Possibility for incomplete oxidation – High capital and operating costs	Very low

Table 12.1. (continued)

Treatment process	Areas of application	Advantages	Disadvantages	Level of use
Fixed bed biological processes (biofiltration, biotrickling filtration)	Odor-causing nitrogen- and sulfur-containing compounds, NO_X	– Ambient operating conditions – Relatively simple reactor design – Good for high flow rates – Complete mineralization and little or no chemical consumption	– Effective for low pollutant loads – Large footprint – Susceptible to microbial shock under fluctuating loads	Moderate
Bioscrubbing	Odor-causing nitrogen- and sulfur-containing compounds	– Ambient operating conditions – Relatively simple reactor design – Complete mineralization and little or no chemical consumption	– Effective for low pollutant loads – Effective for water-soluble contaminants – Two-stage treatment process	Low

presence of a catalyst. Recent technical advancements allow for these reduction processes to take place in the same reactor where the desulfurization of the flue gas with hydrated lime takes place (Stepanov and Korpela 1997). Wet lime scrubbing processes are efficient for SO_2 emission control, but incapable of removing NO_x because of the poor solubility of these pollutants in water. Hence, the application of catalytic reduction process is necessary. Poisoning of the catalysts by fly ash rich in heavy metals and disposal of toxic spent catalysts are among the major problems associated with the catalytic reduction process for NO_x removal.

Thermal combustion and catalytic oxidation present effective solutions to other inorganic contaminants such as ammonia and amines. However, each of these techniques suffers from major drawbacks, including the production of NO_x due to the high-temperature reactions involving nitrogen in the air as well as complex reactor design requirements. Adsorption, as a common physical treatment process, is generally not a viable alternative for flue gas NO_x or SO_x removals because of the cost associated with specific adsorbents, high concentrations of pollutants, and high temperature of the waste gas stream. Other inorganic air pollutants such as ammonia, on the other hand, can be removed effectively by adsorption on activated carbon or zeolite (Busca and Pistarino 2003). However, this process is often costly and requires regeneration or disposal of spent adsorbent.

Other emerging technologies that have been investigated for their potentials in removing nitrogen oxides as well as inorganic odorous compounds include photoassisted heterogeneous catalysis (photocatalysis), pulsed corona discharge, and electron beam process. Photocatalysis has been used to oxidize NO to nitric acid and nitrogen dioxide (Ibuzuki and Takeuchi 1994; Lim et al. 2000; Devahasdin et al. 2003). Although effective in removing NO from flue gas, photocatalysis has the inherent problem of producing oxidation products that require further treatment (e.g., NO_2, which should be treated via activated carbon adsorption). In addition, photocatalysis along with other technologies such as electron beam and corona discharge (Licki et al. 2003; Hu et al. 2003) are still very expensive and not economical for large-scale applications.

The selection of a particular technology or combination of technologies depends upon many parameters including site specifications, operating and maintenance capabilities, treatment objectives, pollutant loadings, airflow rates, and pollutant characteristics. Industrial air emissions have traditionally been managed through physical and/or chemical processes (Table 12.1). However, these processes often generate additional wastes and/or require significant chemical and energy inputs. In other words, these technologies do not offer forward-looking and environmentally friendly, alternative solutions to air pollution issues. Biological treatment methods, on the other hand, have not been used often and only recently received increased applications for odorous inorganic compounds such as ammonia, hydrogen sulfide, and other malodorous nitrogen- and sulfur-containing compounds (Devinny et al. 1999; Burgess et al. 2001).

12.4 Biological Technologies for Inorganic Air Pollutants

The need for cost-effective waste gas treatment technologies has led to the development of biological air pollution control processes. Although chemical and physical treatment processes remain popular, the use of biological processes for contaminated air treatment has grown dramatically because of their ability to destroy the pollutants, rather than simply transfer them from the gas phase to another phase. Over the past few decades, there have been a number of applications that utilized different types of biotreatment processes as effective and economical air pollution control strategies (Devinny et al. 1999; Burgess et al. 2001). Biological treatments are cost- and energy-efficient, and particularly well suited to treating low-concentration emissions when the pollutants are readily biodegradable (Devinny et al. 1999). In particular, biological treatments are believed to be the most economical options for treating odor-causing compounds. In all types of bioreactors for waste gas treatment, the pollutants diffuse into the liquid phase where microorganisms that are either indigenous or inoculated in the system degrade them into products such as CO_2, H_2O, inorganic metabolites, or elemental species such as nitrogen or sulfur (Burgess et al. 2001).

Three major types of bioreactors that currently dominate waste gas biotreatment are biofilters, biotrickling filters, and bioscrubbers. In addition, there are several alternative processes that have been investigated by researchers. These include algal photobioreactors, membrane bioreactors, phytoremediation systems, fluidized bed bioreactors, and activated sludge diffusers (e.g., Chung et al. 2001; Min et al. 2002; Morikawa and Erkin 2003). Although different in their design and operational aspects, all these bioreactors rely on the action of microorganisms for the removal and elimination of inorganic air pollutants. Thorough and complete understanding of complex microbial processes and biodegradation mechanisms are still the subject of many research activities (Sakano and Kerkhof 1998; Arp et al. 2002). However, it is important to briefly review some of the most basic and important elements associated with the biodegradation of nitrogen-containing inorganic air pollutants discussed in this chapter.

There will probably be no single, unanimously accepted biological mechanism for the biodegradation of inorganic air pollutants. For nitrogen-containing air pollutants, biological degradations often take place through biogeochemical nitrogen cycle or microbial nitrogen transformations (Fig. 12.1) such as nitrification, denitrification, N fixation, and NH_4^+/NO_3^- reduction (du Plessis et al. 1997; Stepanov and Korpela 1997). In particular, nitrification and denitrification are believed to be the main microbial processes involved in the biodegradation of ammonia and nitrogen oxides, respectively.

12.4.1 Biodegradation of Ammonia

The biodegradation of ammonia in bioreactors takes place through the process of nitrification, a part of the biogeochemical nitrogen cycle. As seen in

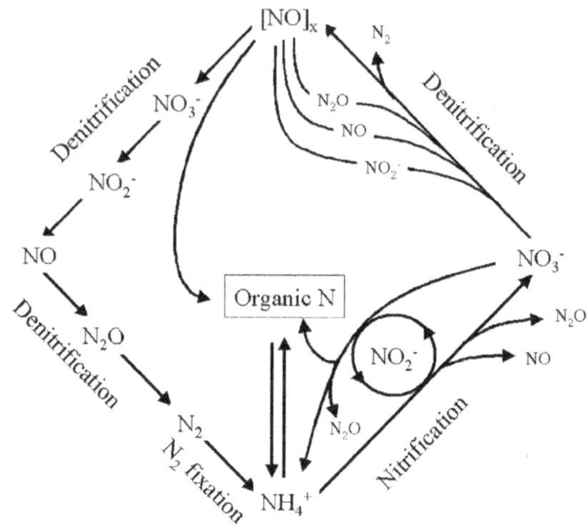

Fig. 12.1. Nitrogen cycle involving microbiological transformation reactions

Fig. 12.1, the nitrification process is a sequential transformation of NH_3 to nitrate (NO_3^-). First, ammonia-oxidizing bacteria (e.g., *Nitrosomonas* sp.) that are autotrophes derive all of their reductant required for energy and biosynthesis from the oxidation of NH_3 to nitrite (NO_2^-). These organisms also derive all of their carbon for growth from CO_2. The NO_2^- produced by ammonia-oxidizing bacteria is subsequently oxidized to NO_3^- by NO_2^--oxidizing bacteria (e.g., *Nitrobacter* sp.). Hence, the oxidation of ammonia to nitrate could be described using the following two-step process (Stepanov and Korpela 1997):

$$NH_4^+ + 0.5O_2 \rightarrow NO_2^- + 2H^+ + H_2O + \text{energy} \quad (12.1)$$

$$NO_2^- + 0.5O_2 \rightarrow NO_3^- + \text{energy}\,. \quad (12.2)$$

Ammonia oxidizers could also produce trace amounts of NO and N_2O. Also, they could produce nitrogen under suitable operating conditions (Stepanov and Korpela 1997). Overall, autotrophic nitrifiers have a high nitrification rate, but they lose their effectiveness under certain conditions such as low oxygen conditions, high ammonia concentration, and high temperature (Chung et al. 1997). The adverse impact of high NH_3 concentration during biological gas cleaning operations has been reported by a number of researchers (Hartikainen et al. 1996; Lee et al. 2001).

Degradation of ammonia can also take place by heterotrophic nitrifiers (e.g., *Alcaligenes*, *Pseudomonas*, and *Arthrobacter* sp.), which perform better under acidic environments (Chung et al. 1997). The pathway related to the nitrification of ammonia by heterotrophs involves:

$$NH_4^+ \rightarrow NH_2OH \rightarrow NO_2^- \rightarrow NO_3^-\,. \quad (12.3)$$

The process of heterotrophic nitrification is not coupled to energy generation and may not go to completion – that is, the final product of heterotrophic nitrification is often nitrite, which can supply the substrate for autotrophic nitrite oxidizers (Eq. 12.2). This offers the possibility of simultaneous auto- and heterotrophic biodegradation (Stepanov and Korpela 1997). The nitrification rate of heterotrophs is significantly lower than that of autotrophs (Chung et al. 1997), resulting in higher population size and more diversity for autotrophic organisms during ammonia remediation. Studies focusing on microbial community structures during the biological degradation of ammonia indicated that the overall diversity of heterotrophic microbial populations decreased significantly during a long-term operation of the system (Sakano and Kerkhof 1998). On the other hand, no significant decrease was observed for the diversity of the ammonia-oxidizing community, indicating the major role played by this group of organisms compared to heterotrophs.

Nitrate produced during the nitrification process can be reduced to N_2 in the subsequent step in the biogeochemical N cycle – denitrification. Although the process of denitrification was often identified as anaerobic process, it is now evident that it can take place in areas of low oxygen potential in an otherwise aerobic environment (or anoxic zones; Stepanov and Korpela 1997). Such anoxic zones are usually formed in the inner regions of biological flocs and films within the bioreactors.

12.4.2 Biodegradation of NO_x

As shown in Fig. 12.1, NO is an intermediate in the sequential reduction pathway of nitrate to N_2. Hence, it is clear that denitrifying organisms are capable of utilizing NO as an electron acceptor, oxidizing different types of organic or inorganic substances. Understanding the molecular basis of bacterial denitrification requires knowledge about the participating enzymes, their cellular organization, and identification of the necessary genes and their control mechanisms. One of the most important aspects of this process is the homology between nitric oxide reductase and heme/copper cytochrome oxidases (e.g., cytochrome bc_1, cytochromes *c*; Hendriks et al. 1998). NO reductase is a membrane protein complex containing *b*- and *c*-type cytochrome that catalyzes a key reaction of denitrification in which the N–N bond is formed (Hendriks et al. 1998):

$$2NO + 2H^+ + 2e^- \Leftrightarrow N_2O + H_2O \, . \tag{12.4}$$

The availability of reductant (mostly organic carbon compounds but inorganic compounds may be used) is also important during the denitrification process. In biological wastewater treatment, pretreated sludge provides an inexpensive carbon and energy source. For biological waste gas cleaning, organic support materials (e.g., biofilter bed media) and/or other organic pollutants in the waste gas stream could provide the necessary carbon source. This has already been demonstrated by du Plessis et al. (1997) who showed good NO removal efficiency in a toluene-treating biofilter.

The enzyme system capable of NO reduction, present in typical denitrifiers, is generally found under anaerobic growth conditions with N-oxides as electron acceptors. Stepanov and Korpela (1997) reported that for some denitrifiers the NO reductase is relatively insensitive to O_2 and therefore NO consumption by denitrification can take place under both anaerobic and aerobic conditions. Nonetheless, the rate of NO uptake under aerobic condition seems to be noticeably lower than that in the absence of oxygen (Kraemer and Conrad 1991). The adverse effects of oxygen on NO biodegradation have been demonstrated by researchers working on the biofiltration of NO in waste gases (Apel et al. 1995; Barnes et al. 1995).

Comparatively little is known about NO consumption by other microorganisms, despite the fact that nitrifying organisms have been reported to provide NO_x removal (du Plessis et al. 1997; Schroeder et al. 2000). Kraemer et al. (1990) indicated that the removal of NO and N_2O by nitrifying bacteria is mainly through reduction process rather than oxidation process. Despite these reports, it was shown the rate of NO reduction by nitrifiers is significantly slower than that by denitrifiers (Schuster and Conrad 1992; Schroeder et al. 2000).

12.5 Biofiltration

Biofilters have been used to treat malodorous gases at wastewater treatment plants since the early 1950s, and are now being used to treat a wide range of waste gases, including volatile organic compounds (VOCs; e.g., Shareefdeen and Baltzis 1994; Devinny et al. 1999; Mohseni and Allen 2000), and inorganic odorous compounds such as NH_3, H_2S, and amines (e.g., Demeestere et al. 2002; Malhautier et al. 2003). More recently, biofiltration has been investigated for the treatment of other inorganic air pollutants such as NO_x (e.g., Flanagan et al. 2002). Among nitrogen-containing inorganic pollutants, ammonia and NO_x are the most important/common contaminants and have received significant attention.

12.5.1 Biofiltration of Ammonia

There have been many studies involving the biofiltration of ammonia-contaminated waste gas streams (e.g., Chung et al. 1997; Lee et al. 2001; Demeestere et al. 2002). Table 12.2 provides a list of some of the studies conducted with biofilters containing organic- and inorganic-based packing materials. Given its high solubility and low Henry's law constant ($K_{H\,25\,°C} = 27\,\text{mol}\,\text{kg}^{-1}\,\text{bar}^{-1}$; Dean 1992), ammonia is mainly retained in biofilters by absorption into the aqueous biofilm phase formed on the solid media. As described above, nitrification by autotrophic bacteria (e.g., *Nitrosomonas* and *Nitrobacter*) generally is the main microbiological process for the degradation of ammonia in the biofilm (e.g., Sakano and Kerkhof 1998; Burgess et al. 2001; Demeestere et al. 2002). Ammonia is essentially mineralized in the form of nitrate and nitrite ions, which are retained in the biofilter material. Some denitrification and,

Table 12.2. Selected data on the successful removal of ammonia using biofiltration

Contaminant	Biofilter packing media	Microbial culture	Removal rate/efficiency	Reference
NH_3	Compost	Mixed culture (indigenous organisms)	– > 12.5 g NH_3 m^{-3} h^{-1} – Up to 87% efficiency	Demeestere et al. (2002)
NH_3	Compost + bark + peat	Mixed culture from activated sludge	0.4–1.0 g N kg^{-1} dry bed day^{-1}	Choi et al. (2003)
NH_3	Compost (yard waste)	Mixed culture (indigenous organisms)	6.8 g NH_3 m^{-3} h^{-1} for inlet loading of up to 14 g NH_3 m^{-3} h^{-1}	Lee et al. (2001)
NH_3	Ca-alginate immobilized	*Arthrobacter oxydane* CH8	1.22 g N kg^{-1} dry bed day^{-1}	Chung et al. (1997)
NH_3	Inorganic packing	*Vibrio alginolyticus* (marine bacterium)	– Up to 22.6 g N kg^{-1} dry bed day^{-1} – > 85% efficiency	Kim et al. (2000)
NH_3	Compost, oyster shells, perlite	Mixed culture (indigenous organisms)	– 10.6 g NH_3 m^{-3} h^{-1} – > 96% efficiency	Kapahi and Gross (1995)
NH_3 + reduced sulfur	Wood chips	Mixed culture (indigenous organisms)	65–93% efficiency	Sheridan et al. (2002)
NH_3 + H_2S	Co-immobilized cells	– *Thiobacillus thioparus* CH11 (for H_2S) – *Nitrosomonas europaea* (for NH_3)	> 95% efficiency for both contaminants	Chung et al. (2000)
NH_3 + H_2S	Wood chips and granular activated carbon (GAC)	– *Thiobacillus thioparus* CH11 (for H_2S) – Mixed culture of nitrifying organisms (for NH_3)	– 6–8 mg N l^{-1} h^{-1} – 45–75 mg S l^{-1} h^{-1}	Kim et al. (2002)
NH_3 + H_2S	Porous ceramic (made of diatomite and fly ash)	Mixed culture from waste-activated sludge	– > 95% for NH_3 and H_2S (inlets of 90 and 5.5 ppmv, respectively) – 1.13 kg NH_3 m^{-3} day^{-1} – 0.165 kg H_2S m^{-3} day^{-1}	Park et al. (2001)
NH_3 + H_2S	Granulated sludge	Mixed culture from sludge containing nitrifying bacteria	– 680 g H_2S m^{-3} day^{-1} – 272 g NH_3 m^{-3} day^{-1}	Malhautier et al. (2003)

hence, N_2 production may take place in the anaerobic/anoxic zones within the biofilm formed on solid packing materials.

Biofilters allow a high removal of ammonia, usually around 90–98%, at low to moderate concentrations and on a great variety of biofilter support materials, organic or inorganic (e.g., Chung et al. 1997; Kim et al. 2000; Liang et al. 2000). Chung et al. (1997) used a biofilter *Arthrobacter oxidans* CH8 immobilized with calcium alginate, and achieved ammonia removals of up to 98%, corresponding to nitrogen removal of 1.22 g N kg^{-1} bed day^{-1}. The range of concentrations tested was between 10 and 60 ppmv. The authors reported that the best temperature range for ammonia removal was 25–35 °C, with the optimum temperature being 30 °C. Kim et al. (2000) worked with a biofilter inoculated with *Vibrio alginolyticus* (marine bacterium), and obtained ammonia elimination of greater than 85% with a maximum nitrogen removal capacity of 22.6 g N kg^{-1} bed day^{-1}. Joshi et al. (2000) worked with peat-based biofilters and reported greater than 99% removal efficiency for concentrations of up to 50 ppmv. Liang et al. (2000) conducted long-term biofiltration experiments with compost-based biofilters and obtained ammonia removals of about 95%.

Despite the promising results reported in the literature, the effectiveness of biofilters toward treating waste gases with high concentrations of ammonia is debatable (Lee et al. 2001; Demeestere et al. 2002). There are some reports on high ammonia removal efficiency at very high concentrations of up to several hundred ppmv (Kim et al. 2000; Smet et al. 2000). On the other hand, Lee et al. (2001) reported that ammonia removal efficiency decreased significantly with increasing inlet ammonia concentration and/or loading rate (from 95% at 25 ppmv to 67% at 130 ppmv). Overall, the biofilter performance dropped to as low as 47% at a loading rate of 8.25 g NH_3 m^{-3} h^{-1}, despite the fact that high removal efficiencies of up to 98% were achieved at lower loading rates (Lee et al. 2001). Similar observations were made by Hartikainen et al. (1996) who noted a drastic drop in NH_3 removal efficiency at waste gas concentrations of greater than 35–60 ppmv and a loading rate of about 8 g NH_3 m^{-3} h^{-1}. These drops in biofilter performances were associated with the sensitivity of nitrifying microorganisms to high pollutant concentration in ammonia-treating biofilters (Don 1985; Demeestere et al. 2002).

Demeestere et al. (2002) related the drop in biofilter performance to the accumulation of NH_4^+/NO_x^- (i.e., NO_2^- or NO_3^-) in the bed due to high NH_3 loading. They added a carbon source to their compost-based biofilter and obtained greater than 12.5 g NH_3 m^{-3} h^{-1} removal, with up to 87% elimination capacity. The authors observed that high elimination capacity of the biofilter could be maintained as long as the NH_4^+/NO_x^- concentration in the bed medium (i.e., compost) was less than 4 g NH_4^+/NO_x^--N kg^{-1} wet compost. Above this value, osmotic effects inhibited the nitrifying activity, and the elimination capacity for ammonia decreased. Demeestere et al. (2002) reported that by adding a carbon source (methanol) and reducing NH_4^+/NO_x^- accumulated in the compost, they were able to restore the biofilter performance. In other words, the addition of methanol was a suitable technique for regenerating the compost material from osmotic inhibition caused by the high NH_3 loading (Demeestere et al. 2002).

12.5.2 Biofiltration of Mixtures of Ammonia and Hydrogen Sulfide

Waste gases emitted from some facilities such as composting and sewage treatment plants contain ammonia mixed with other odorous compounds such as hydrogen sulfide. As a result, simultaneous removal of NH_3 and H_2S has great potentials, and many investigations (Table 12.2) have focused on this subject (e.g., Amirhor et al. 1995; Chung et al. 2000; Malhautier et al. 2003). Overall, the results of these investigations showed that ammonia and hydrogen sulfide simultaneous elimination occurred in the biofilters. Amirhor et al. (1995) investigated the treatment of waste gas from composting facility, containing ammonia and reduced sulfur, and obtained elimination efficiencies of 98 and 80% for ammonia and reduced sulfur, respectively. Chung et al. (2000) reported NH_3 and H_2S removal efficiencies of greater than 95 and 90%, respectively. It was also reported that the activity of nitrifying bacteria was inhibited by high concentrations of H_2S (over 200 ppmv), but it recovered gradually after H_2S concentration decreased (Chung et al. 2000; Kim et al. 2002). In a separate study, Malhautier et al. (2003) observed similar behavior and noted that the oxidation of high amounts of H_2S would involve some environmental stress on nitrifying bacterial growth and activity, thereby reducing ammonia removal efficiency in the biofilters.

Contrary to nitrifying organisms, *Thiobacillus thioparus*, as sulfur-oxidizing bacteria, did not show inhibition with NH_3 concentration (Kim et al. 2002). However, extended operation of biofilters treating mixtures of NH_3 and H_2S resulted in reduced efficiency due to the metabolic products [elemental sulfur and $(NH_4)_2SO_4$] accumulated on the packing materials (Kim et al. 2002). Such impacts observed on the removal of either pollutant by the concentration of one contaminant create some practical implications for the biofiltration of mixed pollutants. Hence, how to optimize the operation of biofilters treating combined NH_3 and H_2S is key to the effectiveness of this technology in waste gas treatment.

12.5.3 Biofiltration of Nitrogen Oxides

Over the past several years, a number of studies focused on the application of biofiltration to the removal of NO_x from fuel combustion gases (e.g., Barnes et al. 1995; Woertz et al. 2001; Flanagan et al. 2002). Although there is no single, unanimously accepted biological mechanism for NO_x purification, it is believed that biofiltration utilizes the activity of denitrifying organisms, inoculated and/or indigenous to the bed medium, to treat NO_x-containing emissions (cf. Sect. 12.4.2). Microbial denitrifiers can utilize NO as an electron acceptor, and this requires the maintenance of anoxic conditions within a biofilter so that NO_x can be reduced to inert nitrogen gas (Barnes et al. 1995; Flanagan et al. 2002). This is despite some reports suggesting that for many denitrifiers the NO reductase may be relatively insensitive to O_2, so that NO consumption by denitrification can take place under anaerobic conditions as well (Stepanov and Korpela 1997; Okuno et al. 2000).

Most promising results on the biofiltration of NO_x involved nitric oxide (NO)-contaminated waste streams. Apel et al. (1995) used compost-based biofilters operating at elevated temperatures of 50–60 °C under oxygen-restricted environments (O_2 concentration of up to 6%). NO removal was greater than 80% for inlet concentration of more than 250 ppmv and oxygen concentrations of less than 3%. However, the removal of NO decreased by increasing the oxygen content (Apel et al. 1995; Lee et al. 1999). Apel et al. (1995) also observed that the addition of an exogenous carbon and energy source enhanced the removal of NO in the biofilters. Obtaining a relatively high NO removal under thermophilic conditions, Apel et al. (1995) concluded that biofiltration could be an effective technique for the removal of combustion gases, following scrubbers that are usually installed to reduce sulfur dioxide emissions. Barnes et al. (1995) expanded the work by Apel et al. (1995) and studied the impact of pH on the reduction of NO in biofilters. The results indicated that optimum denitrifying activity occurred at pH levels ranging between 6 and 7.

Biofilter medium is another important parameter affecting the performance of NO-removing biofilters. In particular, the selection of packing medium becomes very important when the biofilter is operating under thermophilic condition (typical of combustion flue gas), where long-term stability of the medium is an issue. Flanagan et al. (2002) examined various biofilter media (compost, perlite, and biofoam) and observed that compost provided highest NO removal efficiencies at empty bed retention time (EBRT) as low as 13 s. For longer EBRT (70–80 s), all packing materials provided high NO removals of greater than 85%. Also, perlite and biofoam biofilters showed significantly lower pressure drops than the compost biofilter. The authors also suggested that the synthetic material-based biofilters would retain bed integrity much better than the organic-based compost biofilter over long-term biofiltration operations. Given the advantages and disadvantages associated with each system, Flanagan et al. (2002) did not recommend an optimum biofilter medium. Nonetheless, they suggested that such media must be selected based on many criteria, including performance and bed integrity during long-term operations and testing.

The difficulty associated with maintaining anaerobic condition (and/or very low oxygen content) in biofilters designed for industrial applications prompted several researchers to investigate the feasibility of aerobic NO removals in biofilters (du Plessis et al. 1997; Okuno et al. 2000; Woertz et al. 2001). Du Plessis et al. (1997) used an aerobic toluene-removing biofilter and obtained up to 60 ppmv NO removal under aerobic conditions (17% v/v O_2). The authors utilized the metabolites/byproducts produced during toluene catabolism as the carbon source for denitrification, and obtained up to 97% NO removal efficiency. This was a major departure from previous reports (Shanmugasundram et al. 1993; Apel et al. 1995; Barnes et al. 1995) that found drastic decrease in NO removal even at low oxygen concentration (about 3%). Du Plessis et al. (1997) suggested that the denitrification of NO took place in the anoxic regions of the biofilm developed during the aerobic process of toluene removal. They also indicated that higher NO removal efficiency could be obtained by increasing the toluene supply to the biofilters and/or under lower oxygen concentration in the air (which is typical of combustion off-gas streams). Later, Woertz et al. (2001) used a toluene-degrading fungal biofilter and obtained significantly higher NO removal rate (93% at the inlet concentration of 250 ppmv). Okuno et al. (2000) also reported up to 80% NO removal in a soil-based biofilter, operating under aerobic conditions. However, the removal efficiency dropped significantly to about 20% after several days. Okuno et al. (2000) did not explain the cause of this major reduction.

The use of nitrifying bacteria for NO removal was also reported in the literature (Davidova et al. 1997; Hudepohl et al. 1998; Schroeder et al. 2000). Davidova et al. (1997) demonstrated that some nitrifying autotrophic bacteria were capable of oxidizing nitric oxide in the presence of atmospheric concentration of oxygen. However, the rate of oxidation was extremely low, with the biofilters requiring EBRT of greater than 10 min to obtain up to 90% NO removal at inlet concentrations of about 100 ppmv. The slow and limited growth and development of biofilm was considered responsible for the low NO removal efficiency. Schroeder et al. (2000) examined methods of enhancing the biofilm growth, including the use of NO_2 as the initial substrate to promote biofilm growth. Overall, there was little impact on the rate of NO removal, despite the fact that some temporary improvements could be achieved through reducing liquid film thickness on the biofilter packing materials (Schroeder et al. 2000). It is not clear whether such improvement was due to the adsorption of NO on the packing material or if there was indeed some enhanced biological activity.

12.6 Biotrickling Filtration

Biotrickling filters represent a viable alternative to biofilters, although they are used less frequently. Biotrickling filters utilize inert synthetic materials as packing media (see Chap. 7). Unlike biofilters that incorporate nutrients and buffer in natural organic media, biotrickling filters require the delivery of liquid nutrients and buffers to the microbial population within the biotrickling filter. One of the major advantages of biotrickling filters involving the treatment of nitrogen- and sulfur-containing contaminants is better pH control over long-term operations. In other words, acidification can be avoided due to the fact that reaction products are washed out of the media continuously. The major drawback with these systems is the problem of transferring the non-soluble pollutants such as NO_X from the contaminated air to the liquid phase (biofilm). In general, biotrickling filters are believed to be effective in the treatment of gaseous compounds with an air/water partition coefficient of less than 0.1 (Kennes and Thalasso 1998).

Biotrickling filter is potentially an attractive technology for the removal of ammonia because of its high water solubility. Ammonia will be dissolved easily in the aqueous phase where nitrification process takes place. The feed of nutrients and buffer maintains a controlled pH environment, providing high removal capacity. Sorial et al. (2001) reported on a very high ammonia removal efficiency of more than 99% at loading rates of up to 0.48 mol NH_3 m^{-3} h^{-1} and EBRT of 1 min. Smits et al. (1995) used a biotrickling filter with corrugated PVC media and obtained removals of up to 4 g NH_3 m^{-3} h^{-1}. Biotrickling filters have also provided effective removals for ammonia (as well as other odor-causing compounds) when used for odor control in large-scale applications (Kraakman et al. 1996).

One potential problem associated with the biotrickling filtration of high concentrations of water-soluble and easily biodegradable compounds such as ammonia is the excessive biofilm development on the packing materials (Kennes and Thalasso 1998). Sorial et al. (2001) observed significant biomass buildup in their biotrickling filter, and controlled this problem through periodic backwashing of the system. Filter bed backwashing with minimum fluidization twice a week for one hour proved to be an efficient means for preventing excess biomass accumulation in other systems treating various contaminants (Smith et al. 1996).

Biotrickling filters have also shown some successful applications for the removal of NO_x and SO_x (Chou and Lin 2000; Klasson and Davison 2001; Philip and Deshusses 2003). Chou and Lin (2000) obtained up to about 80% removal efficiency for NO at inlet concentrations of up to about 1,200 ppmv and EBRT of 118 s. The efficacy of the system and biofilm development was heavily dependent upon nutrient addition. In particular, the presence of carbon source was important for the activity of denitrifying organisms. Detachment of the biofilm from the filter media was reported within 2 weeks of organic carbon deficiency (Chou and Lin 2000). The importance of carbon source for bacterial activity and NO removal has also been reported in other studies (Hosseini and Moser 2000; Flanagan et al. 2002). Flanagan et al. (2002) used lactate as their carbon source in a biofoam-based biotrickling filter, and obtained up to about 40 g NO m^{-3} h^{-1} at 60% removal efficiency. The biofoam-based system provided less NO removal compared to a compost-based biofilter, which was studied in parallel and was about 50% more efficient at removing NO. However, the biotrickling filter provided significantly lower pressure drop, providing better long-term operation potentials.

Philip and Deshusses (2003) used a biotrickling filter and were able to capture SO_2 from a simulated flue gas stream with near 100% efficiency. The absorbed SO_2 was then recovered as sulfite and was treated in a post-biological treatment process. Although a two-stage process was used for SO_2 abatement, the high removal efficiency achieved in the biotrickling filter (100% removal for inlet concentrations of up to 1,000 ppmv and EBRT of 6 s) made this two-stage solution very attractive and promising (Philip and Deshusses 2003). The authors did not elaborate on the long-term effectiveness of this process and its economical competitiveness, in comparison with the common wet lime scrubbing process currently used in industry.

12.7 Bioscrubbing

Bioscrubbing has several advantages over media-based filtration (biofiltration or biotrickling filtration). The process is more easily controlled because the pH, temperature, nutrient balance, and removal of metabolic products can be altered in the liquid phase of the reactor. However, bioscrubbers rely on good water solubility of vapor-phase contaminants, as they employ the absorption of pollutants into the aqueous phase in a gas/liquid exchange system, followed by biodegradation in an aqueous-phase bioreactor (see also Chap. 8). Therefore, bioscrubbing is of interest to gaseous pollutants with a Henry's law constant or partition coefficient of less than 0.01 (Kennes and Thalasso 1998). This is of major importance because many inorganic air pollutants are water-insoluble (e.g., NO_x). This drawback, along with the fact that bioscrubbing is indeed a two-stage process, is likely among the major reasons bioscrubbers are less popular than biofilters for inorganic compounds, despite the fact that several examples of successful applications are reported in the literature for water-soluble compounds such as ammonia (Schirz 1991; Demmers 1992; van Groenestijn et al. 1997).

Van Groenestijn et al. (1997) applied a bioscrubbing system that allowed for complete removal of NH_3 from the waste gas produced by intensive stock breeding. Ammonia captured from gas stream was fully converted to nitrogen through sequential nitrification and denitrification reactors. The volumetric conversion capacity of the system was about 30 g NH_3-N m^{-3} with an elimination capacity of greater than 93% (van Groenestijn et al. 1997). Other examples of successful ammonia abatements relate to different odor-causing industries such as fish food production (Rasmussen et al. 1994), wastewater treatment plants (Dechant et al. 1999), and animal and livestock barns (Schirz 1991; Demmers 1992).

12.8 Photobiodegradation

Photosynthesis has long been known to provide CO_2 fixation and energy production. The use of photosynthetic microalgae to fix CO_2 and thereby produce energy and chemicals has been proposed in the literature (Miura et al. 1993). Microalgae have also shown the ability to provide some degrees of NO_x removal. Earlier research indicated that certain types of algae (e.g., *Nannochloris* sp.) could grow in the presence of CO_2 and NO, and assimilate some dissolved NO (Negoro et al. 1991). These findings suggested that NO and CO_2 could be simultaneously eliminated, by a microalgal culture, from flue gas that contains large amount of CO_2 and a significant quantity of NO_x generated during the combustion of fossil fuels.

Several strains of marine microalgae have been found to effectively remove NO from flue gas (Yoshihara et al. 1996; Nagase et al. 1997). Yoshihara et al. (1996) reported on some degrees of NO removals using a bubble column photobioreactor illuminated by fluorescent lamps, and inoculated with a strain of marine algae (NOA-113) isolated from seawater. For the range of NO concentrations between 100 and 300 ppmv, Yoshihara et al. (1996) obtained removal efficiencies of about 50%. Nagase et al. (1997) worked on different strains of marine algae to enhance NO removal efficiencies, and found that *Dunaliella tertiolecta* can remove NO at wide ranges of NO concentrations and gas flow rates. The bubble column photoreactor was able to provide removal efficiencies of about 65% at inlet NO concentrations ranging between 25 and 500 ppmv. The authors also observed that algal growth was not affected by the concentration of NO, and there was always some oxygen (1.6% O_2) evolved by photosynthesis (Nagase et al. 1997). The proposed mechanism of NO removal in the system suggested that NO in the gas phase was first dissolved in the aqueous phase, and then oxidized and assimilated by the algal cells.

In a more detailed study on the mechanism of NO photobiodegradation in the presence of microalgae, Nagase et al. (2001) found that upon being dissolved in the aqueous phase, NO permeates directly into the algal cells by diffusion based on the mass balance of nitrogen, and the change in nitrate and nitrite concentrations in the aqueous medium (Nagase et al. 2001). In other words, little NO is oxidized in the medium of the photoreactor before its uptake by the algal cells. Compared to nitrate, NO taken up in the algal cells is preferentially utilized as a nitrogen source for cell growth. Overall, photobiodegradation of NO using microalgae provided promising NO removals under laboratory-scale setups. It could potentially be used for removing NO from flue gas from which SO_x had been previously removed by chemical desulfurization methods (SO_x was shown to inhibit the growth of microalgae). However, it is not quite evident whether the bubble column photobioreactors investigated in laboratory-scale studies can be scaled up easily and economically for the industrial applications where they require working for high flow rates of combustion flue gas streams. Further research and studies in this area are warranted to better understand these potentially promising techniques.

12.9 Other Biological Processes

There are many other alternative biological processes investigated for the abatement of waste gases containing inorganic air pollutants. Activated sludge diffusion is a well-established process that is effectively used for the elimination of odorous air pollutants (e.g., H_2S, NH_3) emitted from wastewater treatment operations (Burgess et al. 2001). Membrane bioreactors, fluidized bed bioreactors, and phytoremediation systems are among the more innovative and recent techniques that provide promises for certain niche applications. Membrane bioreactors seem to be very effective for the removal of water-insoluble contaminants such as NO_x. Fluidized bed bioreactors, on the other hand, have great potential in treating water-soluble, odor-causing compounds such as ammonia and hydrogen sulfide. Phytoremediation processes that rely on the ability of the plants to uptake contaminants have proven effective for the removal of low concentrations of NO_X in the atmosphere (Morikawa and Erkin 2003).

12.9.1 Membrane Bioreactors

In membrane bioreactors, the gaseous pollutants are transferred from the gas to the liquid phase where they are degraded by a microbial community. Microorganisms are available in two forms – fixed film cultures (biofilms), and suspended growth cultures. The membrane forms the gas/liquid interface, and the diffusion of contaminants across this interface is driven by the concentration gradient as well as the solubility and diffusivity of the compound in the matrix (Reij et al. 1997).

Membrane bioreactors have been investigated for various air pollution control applications (Reij et al. 1997; Ergas et al. 1999). In particular, it was shown that membrane bioreactors are effective against water-insoluble compounds such as propene and nitric oxide (Reij et al. 1997; Min et al. 2002). Min et al. (2002) applied hollow fiber membrane bioreactors inoculated with nitrifying bacteria to NO-contaminated air. The treatment system was such that NO-contaminated gas passed through the microporous membrane where NO diffused through the membrane pores and partitioned into the nitrifying biofilm. Nitrifying bacteria in the biofilm were able to oxidize NO to NO_3^- with a maximum removal efficiency of 74% at a liquid velocity of 1.5 $cm\,s^{-1}$ (Min et al. 2002). Changes in temperature from 20 to 55 °C did not have any impact on the removal of NO, which stayed nearly constant at about 70% efficiency.

12.9.2 Fluidized/Spouted Bed Bioreactors

Three-phase fluidized and/or spouted bed bioreactors present an alternative to immobilized biological reactors (e.g., biofilters) for certain contaminants with high water solubility. In particular, fluidized bed bioreactors could provide better interphase mixing and mass transfer at lower pressure drop in the system (Wright and Raper 1998; Chung et al. 2001). Wright and Raper (1998) used a three-phase fluidized bed bioreactor and obtained a significantly high ammonia removal capacity. At a loading rate of up to 650 g NH_3 $m^{-3}\,h^{-1}$, the removal efficiency was around 40%. The ammonia loading rate reported by Wright and Raper (1998) was significantly higher (more than one order of magnitude) than those reported in biofilter applications. This is likely due to the way volume of the reaction zone was calculated, and also because of the actual differences that exist between reactor volumes associated with solid-based media bioreactors (e.g., biofilters) and fluidized bed systems that are based on liquid and/or solid–liquid media.

In another study, Chung et al. (2001) showed that fluidized bed bioreactors could be used for the treatment of ammonia, hydrogen sulfide, or mixtures of these two compounds. When treated separately, fluidized bed bioreactors provided more than 95% elimination capacity for either contaminant. The removal rates were about 70 g S m^{-3} h^{-1} and 9.0 g N m^{-3} h^{-1} for H_2S and NH_3, respectively. When applied for mixtures of H_2S and NH_3, both compounds showed some inhibitory effects on H_2S-degrading organisms (*Pseudomonas putida*) at high inlet concentrations. The negative effect of H_2S was far greater than that by ammonia. These observations are contrary to what was reported by Chung et al. (2000) and Malhautier et al. (2003), who showed little impact on the activity of H_2S-degrading bacteria by the presence of ammonia in the system.

12.9.3 Phytoremediation

Phytoremediation utilizes plants, alone or together with their associated microorganisms, to degrade, contain, or stabilize various environmental contaminants, including those associated with waste gases (Morikawa and Erkin 2003). Although this technology is in its infancy in terms of large-scale air pollution and odor control applications, phytoremediation has some potential advantages: (1) it is an aesthetically pleasing and solar energy-driven remediation technique; (2) there is minimal environmental disruption; (3) it can be applied to a wide range of contaminants, including those with low water solubility and little biodegradability; and (4) it can be inexpensive (about 60–80% less expensive than conventional physical and chemical treatments; Schnoor et al. 1995). Chemicals that can enter into the plant have approximately log K_{ow} values of between 1 and 3.5 (Schnoor et al. 1995). Major drawbacks of phytoremediation are associated with its very slow process and the fact that there is little information on the fate of contaminants upon uptake (Morikawa and Erkin 2003).

Many inorganic compounds such as nitrogen oxides and sulfur oxides can be taken up by plants and transformed or degraded. Plants will assimilate the nitrogen in NO_2 to organic compounds, including amino acids (Morikawa and Erkin 2003). Some specific "NO_2-philic plants" have been genetically engineered in order to grow with atmospheric NO_2 as their sole nitrogen source (Morikawa and Erkin 2003). Other research activities involve the production of transgenic plants capable of decontaminating atmospheric nitrogen dioxide (Erkin et al. 2003) and the development of gas–gas converting plants that convert nitrogen dioxide to nitrogen gas (Morikawa and Erkin 2003).

12.10 Conclusions and Further Research Needs

Laboratory and pilot-scale studies have shown that biological processes can be successfully employed for the abatement of inorganic contaminants from waste gas streams. In particular, low concentrations of odorous compounds such as ammonia, hydrogen sulfide, and amines have been effectively biodegraded in large-scale applications involving biofilters. Long-term performance of industrial-scale bioreactors for removing mixtures of inorganic odorous compounds, although not widely investigated, may be a potential problem due to the buildup of biodegradation metabolic products (e.g., elemental sulfur). Another potential limitation, for the removal of mixed pollutants, arises from the negative impacts of one compound on the activity of the microbial community degrading the other contaminants. These outstanding issues and other potential questions warrant more long-term and full-scale results to allow for proper design and optimization of operating parameters responsible for the effectiveness of biological foul air and waste gas treatment. Also, there is a need for further study of microbial communities during the biodegradation of mixtures of inorganic odorous contaminants to better understand the dynamics of microbial community development and their responses to various changes taking place during the biological gas cleaning process.

Biological degradation of NO_X is in its early stages of developments. There have been several studies indicating the effective removal of NO_X using biofilters, biotrickling filters, and photobioreactors. Also, some promising results were obtained on the removal of SO_2 using biotrickling filters (Philip and Deshusses 2003). Microorganisms were shown to have high NO_X-reducing potentials, especially with respect to NO. Most of the NO removal took place as a result of the reducing activity of denitrifiers in bioreactors and under oxygen-restricted environment. However, the extent to which oxygen plays a role in the reduction of NO is not yet well established. In addition, some degrees of NO removals by nitrifying microorganisms have been proposed and reported in the literature. Although the growth of organisms on NO, as the terminal electron acceptor, may be a common understanding among researchers, many more outstanding questions remain with respect to the microbiology of the biological processes during NO_X removal in gas phase.

Overall, despite the promising results from laboratory-scale investigations, none of the biological NO_X removal techniques has been implemented in the pilot and/or large scale. This is particularly important because the overall performance and economic assessment of biological NO_X removal depend entirely on the combined factors of long-term stability of microbial culture and support media, consistent contaminant removal capacity, and operating parameters (e.g., gas retention time and pressure drop). The results obtained by researchers so far present a general guideline toward more research in this area, which requires more long-term investigations and optimization as well as fundamental studies before biological treatments can become a technically and economically feasible alternative to the current chemical NO_X treatment processes (e.g., selective catalytic reduction).

Another important issue related to the biological degradation of inorganic air pollutants is the lack of theoretical modeling and optimization in the literature. Modeling of any biological process is fundamentally important because appropriate models allow for developing realistic bioreactor designs, reducing extensive experimental time and costs, and predicting reactor performance. Although there are very many modeling and optimization studies on the biofiltration of VOCs (e.g., Shareefdeen and Baltzis 1994; Shareefdeen et al. 1997), little attempt has been made to model the performance of the biofilters (or bioreactors) designed to degrade inorganic air pollutants. This is likely because some applications (e.g., the removal of NO_X) have not received widespread attention from the research community and, more importantly, from industry. However, there has been little modeling work for a compound such as ammonia, which represents a success for the application of biofilters to nitrogen-containing air pollutants, despite the fact that its industrial application has been evident – and hence, the need for predictive models and optimization tools. There are some kinetic analyses of ammonia oxidation (e.g., Choi et al. 2003) and some examinations of ammonia-degrading microbial communities (e.g., Sakano and Kerkhof 1998). Few fundamental studies focused on identifying and obtaining the parameters required for predictive model developments during the biodegradation of ammonia in waste gases. Studies on the biofiltration of ammonia have mainly focused on the overall removal efficiency of the biofilters, with few cases monitoring and examining ammonia-oxidizing microbial cultures. Hence, many questions remain to be answered before models can be developed to allow for predicting the performance of a given bioreactor for removing specific compounds of interest (e.g., NO_X or ammonia) from contaminated streams.

References

Amirhor P, Kuter GA, Andrade MD (1995) Biofilters and biosolids. Water Environ Technol 3:44–48

Apel WA, Barnes JM, Barrett KB (1995) Biofiltration of nitrogen oxides from fuel combustion gas streams. In: Proc 88th Air Waste Management Association Conf, 18–23 June 1995, San Antonio, TX, Pap TP9C.04

Arp DJ, Sayavedra-Soto LA, Hommes NG (2002) Molecular biology and biochemistry of ammonia oxidation by *Nitrosomonas europaea*. Arch Microbiol 178:250–255

Barnes JM, Apel WA, Barrett KB (1995) Removal of nitrogen oxides from gas streams using biofiltration. J Hazard Mater 41:315–326

Burgess JE, Parsons S, Stuetz RM (2001) Developments in odour control and waste gas treatment biotechnology: a review. Biotechnol Adv 19:35–63

Busca G, Pistarino C (2003) Abatement of ammonia and amines from waste gases: a summary. J Loss Prevent Proc Ind 16:157–163

Choi JH, Kim YH, Joo DJ, Choi SJ, Ha TW, Lee DH, Park IH, Jeong YS (2003) Removal of ammonia by biofilters: a study with flow modified system and kinetics. J Air Waste Manage Assoc 53:92–101

Chou M-S, Lin J-H (2000) Biotrickling filtration of nitric oxide. J Air Waste Manage Assoc 50:502–508

Chung YC, Huang C, Tseng CP (1997) Biotreatment of ammonia from air by an immobilized *Arthrobacter oxydans* CH8 biofilter. Biotechnol Prog 13:794–798

Chung YC, Huang C, Tseng CP, Pan JR (2000) Biotreatment of H_2S and NH_3 containing waste gases by co-immobilized cells biofilters. Chemosphere 41:329–336

Chung YC, Huang C, Liu CH, Bai H (2001) Biotreatment of hydrogen sulfide and ammonia containing waste gases by fluidized bed bioreactor. J Air Waste Manage Assoc 51:163–172

Cole J (1993) Controlling environmental nitrogen through microbial metabolism. Trends Biotechnol 11:368–372

Davidova YB, Schroeder ED, Chang DPY (1997) Biofiltration of nitric oxide. In: Proc 90th Air Waste Management Association Conf, 8–13 June 1997, Toronto, Canada, Pap WP71B.05

Dean JA (1992) Lange's handbook of chemistry. McGraw-Hill, Boston

Dechant D, Ball P, Hatch C (1999) Full scale validation of emerging bioscrubber technology for odor control. In: Proc 72nd WEFTEC Annu Conf Expo, 9–13 October 1999, New Orleans, LA, pp 2965–2975

Demeestere K, Van Langenhove H, Smet E (2002) Regeneration of a compost biofilter degrading high loads of ammonia by addition of gaseous methanol. J Air Waste Manage Assoc 52(7):796–804

Demmers TGM (1992) Ammonia removal from air with bioscrubbers. Process Technol 2:37–41

De Nevers N (2000) Air pollution control engineering. McGraw-Hill, Boston

Devahasdin S, Fan C, Li K, Chen DH (2003) TiO_2 photocatalytic oxidation of nitric oxide: transient behavior and reaction kinetics. J Photochem Photobiol A Chem 156:161–170

Devinny JS, Deshusses MA, Webster TS (1999) Biofiltration for air pollution control. Lewis, New York

Don JA (1985) The rapid development of biofiltration for the purification of diversified waste gas streams. VDI Ber 561:63–73

Du Plessis CA, Kinney KA, Schroeder ED, Chang DPY, Scow KM (1997) Denitrification and nitric oxide reduction in an aerobic toluene-treating biofilter. Biotechnol Bioeng 58:418–415

Ergas SJ, Shumway L, Fitch MW, Neemann JJ (1999) Membrane process for biological treatment of contaminated gas streams. Biotechnol Bioeng 63:431–441

Erkin OC, Takahashi M, Morikawa H (2003) Development of a regeneration and transformation system for *Raphiolepis umbellate* L. "Sharinbai" plants by using particle bombardment. Plant Biotechnol 20:145–152

Flanagan WP, Apel WA, Barnes JM, Lee BD (2002) Development of gas phase bioreactors for the removal of nitrogen oxides from synthetic flue gas streams. Fuel 81:1953–1961

Grano D (1995) Clean Air Act requirements: effect on emissions of NOx from stationary sources. In: Ozkan US, Agarwald SK, Marcelin G (eds) Reduction of nitrogen oxide emissions. American Chemical Society, Washington, DC, pp 14–31

Hartikainen T, Ruuskanen J, Banhatalo M, Martikainen PJ (1996) Removal of ammonia from air by a peat biofilter. Environ Technol 17:45–53

Hendriks J, Gohlke U, Saraste M (1998) From NO to OO: nitric oxide and dioxygen in bacterial respiration. J Bioenerg Biomembr 30(1):15–24
Hosseni M, Moser A (2000) Liquid film effects on the elimination of CO and NO in trickle bed biofilm reactor. Chem Biochem Eng Q 14:35–37
Hu X, Zhang J-J, Mukhnahallipatna S, Hamann J, Biggs MJ, Agarwal P (2003) Transformations and destruction of nitrogen oxides – NO, NO_2 and N_2O – in a pulsed corona discharge reactor. Fuel 82:1675–1684
Hudepohl NJ, Schroeder ED, Chang DPY (1998) Oxidation of nitric oxide in a biofilter. In: Proc 91st Air Waste Management Association Conf, 14–18 June 1998, San Diego, CA, Pap TA20B.06
Ibusuki T, Takachi K (1994) Removal of low concentration of nitrogen oxides through photoassisted heterogeneous catalysis. J Mol Catal 88:93–102
Joshi JA, Hogan JA, Cowan RM, Strom PF, Finstein MS (2000) Biological removal of gaseous ammonia in biofilters: space travel and earth-based application. J Air Waste Manage Assoc 50:1647–1654
Kapahi R, Gross M (1995) Biofiltration for VOC and ammonia emissions control. BioCycle 36:87–90
Kennes C, Thalasso F (1998) Waste gas biotreatment technology. J Chem Technol Biotechnol 72:303–19
Kim NJ, Sugabo Y, Hirai M, Shado M (2000) Removal of a high load of ammonia gas by a marine bacterium, *Vibrio anginolyticus*. J Biosci Bioeng 90:410–415
Kim H, Kim YJ, Chung JS, Xie Q (2002) Long term operation of a biofilter for simultaneous removal of H_2S and NH_3. J Air Waste Manage Assoc 52:1389–1398
Klasson KT, Davison BH (2001) Effect of temperature on biofiltration of nitric oxide. Appl Biochem Biotechnol 91/93:205–211
Kraakman B, Oosting R, Koers B, Urlings L (1996) Biological odor removal with a biotrickling filter-biofilter combination. In: Proc 89th Air Waste Management Association Conf, 23–28 June 2001, Nashville, TN, Pap RA87B.04
Kraemer M, Conrad R (1991) Influence of oxygen on production and consumption of nitric oxide in soil. Biol Fertil Soils 11:38–42
Kraemer M, Baumgartner M, Bender M, Conrad R (1990) Consumption of NO by methanotrophic bacteria in pure culture and in soil (heterotrophic nitrification). FEMS Microb Ecol 73:345–350
Krupa SV (2003) Effects of atmospheric ammonia (NH_3) on terrestrial vegetation: a review. Environ Pollut 124:179–221
Lee BD, Apel WA, Smith W (1999) Effect of oxygen on thermophilic denitrifying populations in biofilters treating nitric oxide containing off-gas streams. In: Proc 92nd Air Waste Management Association Conf, 20–24 June 1999, St. Louis, MO, pp 1996–2011
Lee DH, Lau AK, Pinder KL (2001) Development and performance of an alternative biofilter system. J Air Waste Manage Assoc 51:78–85
Liang Y, Quan X, Chen J, Chung JS, Sung JY, Chen S, Xue D, Zhao Y (2000) Long term results of ammonia removal and transformation by biofiltration. J Hazard Mater B80:259–269

Licki J, Chemielewski AG, Iller E, Zimek Z, Mazurek J, Sobolewski L (2003) Electron beam flue gas treatment for multicomponent air pollution control. Appl Energ 75:145–154

Lim TH, Jeong SM, Kim SD, Gyenis J (2000) Photocatalytic decomposition of NO by TiO_2 particles. J Photochem Photobiol A Chem 134:209–217

Malhautier L, Gracian C, Roux JC, Fanlo JL, Cloirec PL (2003) Biological treatment process of air loaded with an ammonia and hydrogen sulfide mixture. Chemosphere 50:145–153

Min K-N, Ergas SJ, Harrison JM (2002) Hollow-fiber membrane bioreactor for nitric oxide removal. Environ Eng Sci 19:575–583

Miura Y, Yamada W, Hirata K, Miyamoto K, Kiyohara M (1993) Stimulation of hydrogen production in algal cells grown under high CO_2 concentration and low temperature. Appl Biochem Biotechnol 39/40:753–761

Mohseni M, Allen DG (2000) Biofiltration of mixtures of hydrophilic and hydrophobic volatile organic compounds. Chem Eng Sci 55:1545–1558

Morikawa H, Erkin OC (2003) Basic processes in phytoremediation and some applications to air pollution control. Chemosphere 52:1553–1558

Nagase H, Yoshihara K-I, Eguchi K, Yokota Y, Matsui R, Hirata K, Miyamato K (1997) Characteristics of biological NO_x removal from flue gas in a *Dunaliella tertiolecta* system. J Ferment Bioeng 83:461–465

Nagase H, Yoshihara K-I, Eguchi K, Okamato Y, Murasaki S, Yamashita R, Hirata K, Miyamato K (2001) Uptake pathway and continuous removal of nitric oxide from flue gas using microalgae. Biochem Eng J 7:241–246

Nascimento DM, Hudepohl DM, Schroeder ED, Chang DPY (2000) Bio-oxidation of nitric oxide in a nitrifying, aerobic filter. In: Proc 93rd Air Waste Management Association Conf, 18–22 June 2000, Salt Lake City, UT, pp 2134–2143

Negoro M, Shioji N, Miyamato K, Miura Y (1991) Growth of microalgae in high CO_2 gas and effect of SO_x and NO_x. Appl Biochem Biotechnol 28/29:877–886

Okuno K, Hirari M, Sugiyama M, Haruta K, Shoda M (2000) Microbial removal of nitrogen monoxide (NO) under aerobic conditions. Biotechnol Lett 22:77–79

Park SJ, Nam SI, Choi ES (2001) Removal of odor emitted from composting facilities using a porous ceramic biofilter. Water Sci Technol 44:301–308

Philip L, Deshusses MA (2003) Sulfur dioxide treatment from flue gases using a biotrickling filter-bioreactor system. Environ Sci Technol 37(9):1978–1982

Rasmussen HH, Rindel KI, Kruger AS, Soeborg D (1994) Biological air cleaning processes exemplified by applications in wastewater treatment and fish industry. In: Proc Conf Organic Compound Emission Control for Municipal and Industrial Wastewater Treatment Facilities, 24–27 April 1994, Jacksonville, FL. Water Environ Fed, pp 2/13–24

Reij MW, Hammann EK, Hartmans S (1997) Biofiltration of air containing low concentrations of propene using a membrane bioreactor. Biotechnol Prog 13:380–386

Sakano Y, Kerkhof L (1998) Assessment of changes in microbial community structure during operation of an ammonia biofilter with molecular tools. Appl Environ Microbiol 64:4877–4882

Schade G, Crutzen PJ (1995) Emission of aliphatic amines from animal husbandry and their reactions: potential source of N_2O and HCN. J Atmosph Chem 22:319–346

Schirz S (1991) Practical application of bioscrubbing technique to reduce odor and ammonia. In: Nielsen VC, Voorburg JH, L'Hermite P (eds) Odour and ammonia emission from livestock farming. Elsevier, London, pp 82–91

Schnoor JL, Licht LA, McCutcheon SC, Wolfe NL, Carrieira LH (1995) Phytoremediation of organic and nutrient contaminants. Environ Sci Technol 29:318A-323A

Schroeder ED, Eweis JB, Chang DPY, Veir JK (2000) Biodegradation of recalcitrant components of organic mixtures. Water Air Soil Pollut 123(1/4):133–146

Schuster M, Conrad R (1992) Metabolism of nitric oxide during nitrification and denitrification in soil at different incubation conditions. FEMS Microbiol Ecol 101:133–143

Shanmugasundram R, Lee CM, Sublette KL (1993) Reduction of nitric oxide by denitrifying bacteria. Appl Biochem Bioetechnol 39/40:727–737

Shareefdeen Z, Baltzis BC (1994) Biofiltration of toluene vapor under steady state and transient conditions: theory and experimental results. Chem Eng Sci 49:4347–4360

Shareefdeen ZM, Shaikh AA, Ansar Z (1997) Development, experimental validation and dynamic analysis of a general transient biofilter model. Chem Eng Sci 52:759–773

Sheridan BA, Curran TP, Dodd VA (2002) Assessment of the influence of media particle size on the biofiltration of odorous exhaust ventilation air from a piggery facility. Bioresour Technol 84:129–143

Smet E, van Langenhov H, Maes K (2000) Abatement of high concentrated ammonia loaded wasted gases in compost biofilter. Water Air Soil Pollut 119:177–190

Smith FL, Sorial GA, Suidan MT, Breen AW, Biswas P, Brenner, RC (1996) Development of two biomass control strategies for extended, stable operation of highly efficient biofilters with high toluene loadings. Environ Sci Technol 30:1744–1751

Smits MCJ, Hoek AP, Osinga B, Ottengraf SPP, Wijngaard MH (1995) Removal of ammonia and odor from composting facility waste gas with a bio-trickling filter. Meded Fac Landbouw Univ Gent 60:2313–2320

Sorial GA, Smith FL, Suidan MT, Brenner RC (2001) Removal of ammonia from contaminated air by trickle bed air biofilters. J Air Waste Manage Assoc 51:756–765

Stepanov AL, Korpela TK (1997) Microbial basis for the biotechnological removal of nitrogen from flue gases. Biotechnol Appl Biochem 25:97–104

Van Groenestijn JW, Harkes MP, Baartmans RFW (199) A novel bioscrubber for the removal of ammonia from off-gases. In: Prins WL, van Ham J (eds) Biological waste gas cleaning. VDI, Düsseldorf, pp 305–312

Woertz JR, Kinney KA, Szaniszlo PJ (2001) A fungal vapor-phase bioreactor for the removal of nitric oxide from waste gas streams. J Air Waste Manage Assoc 51:895–902

Wright PC, Raper JA (1998) Investigation into the viability of a liquid film three phase spouted bed biofilter. J Chem Technol Biotechnol 73:281–291
Yoshihara K-I, Nagase H, Eguchi K, Hirata K, Miyamato K (1996) Biological elimination of nitric oxide and carbon dioxide from flue gas by marine microalgae NOA-113 cultivated in a long tubular photobioreactor. J Ferment Bioeng 82:351–354

13 Biological Treatment of Waste Gases Containing Volatile Organic Compounds

Pierre Le Cloirec, Yves Andrès, Claire Gérente, Pascaline Pré

13.1 Introduction

In recent years, a strong technological base has been developed for air quality control. The driving force has been public awareness of the need for human health and environmental protection. Consequently, environmental legislation and regulations have been initiated by governments to control air quality. Removing air pollutants, and especially volatile organic compounds (VOCs), from industrial polluted gaseous emissions is achieved by different ways:

- Transfer and/or concentration of compounds in a liquid phase (condensation, absorption) or onto a solid (adsorption).
- Destruction of molecules by thermal oxidation, catalytic treatments, flaring, or degradation of compounds by microorganisms.

Among these technologies useful for air treatment, processes involving the biodegradation of pollutants present in the gaseous phase are widely used. They may be applied to both water and air purification (Godish 1997; Liu and Liptak 1997; Heinsohn and Kabel 1999). Thus, these biological systems can remove a large number of molecules, especially volatile organic compounds and odorous molecules. They are generally utilized for the treatment of industrial gaseous emissions. The applications are possible for a concentration of pollutant above $1\,g\,m^{-3}$ to diluted air (some mg or $\mu g\,m^{-3}$). The installation designs cater for an airflow of between 1,000 to 100,000 $m^3\,h^{-1}$, or even more for some systems (Le Cloirec 1998). The principles, processes and modeling of biological systems for air pollution control have been described previously in this book.

The objective of this chapter is to present practical approaches to air treatment, especially removal of volatile organic compounds, by biological methods. Specific points and process examples are developed:

- examples and data on the biodegradation of VOCs;
- applications of biological processes: general operating conditions and case studies: biofilters, biotrickling filters and bioscrubbers;
- treatment of by-products generated by VOC biotreatments.

Shareefdeen/Singh (Eds.)
Biotechnology for Odor and Air Pollution Control

13.2 Biodegradation of Volatile Organic Compounds

For a better understanding of the use of biological treatments in the reduction of waste gases containing volatile organic and odorous compounds, some information on biodegradation capacities under aerobic conditions is provided in this chapter. Biotechnological processes are possible because microorganisms are able to use organic or inorganic molecules as nutritional sources. This latter utilization can follow two major paths – catabolism or biotransformation. However, the resulting products may be accumulated in the medium up to a toxic level for a microbial species before being mineralized by another one. This biodegradation sequence is frequently found for the degradation of complex substances. Moreover, microorganisms present the capacity to grow rapidly, with a large flexibility in the regulation of their metabolism. They are also able to settle in new areas, to resist extreme environmental conditions and, associated with other organisms, to develop a better metabolic diversity.

13.2.1 Microbial Growth

Microbial growth is a complex process involving numerous biochemical reactions that can be divided into two parts. The first one corresponds to the synthesis of the cell constituents and metabolites (anabolism). The second part is the breakdown of the cell constituents and metabolites (catabolism). Ultimately, these biosynthetic reactions result in cell development and division. Some substrates need to be broken down enzymatically into smaller molecules before their utilization in a general metabolic pathway like the tricarboxylic acid cycle. A number of parameters influence the survival and activity of microorganisms in any environment. However, all of them have some common elementary requirements like water, a source of carbon, nitrogen, mineral compounds, etc. In many cases, these nutrients are present in their environment. The occurrence and abundance of microorganisms in an environment is determined not only by the available substrate sources but also by various physical and chemical factors. These include oxygen availability, nutrient availability, temperature, pH, salinity, and water activity. Inhibition of biodegradation can be caused by the limitation imposed by any one of these factors. Moreover, these parameters have a regulatory effect on the microbial population development.

13.2.1.1 Elementary Needs

Microorganisms can be divided into a number of groups based on their source of energy, and their source of carbon, used for the synthesis of all the organic molecules making up their cell components. The two major sources of energy are light energy (phototrophs) and chemical energy (chemotrophs). Moreover, two sources of carbon can be utilized – organic material (heterotrophs) and

CO_2 (autotrophs). Some chemoheterotrophs use an amazingly large number of organic compounds. For example, certain members of the genus *Pseudomonas* can derive carbon and/or energy from more than 90 different organic substances, including unusual molecules such as naphthalene. Conversely, some microorganisms degrade only a single compound, like *Bacillus fastidious* that uses only urea as a source of both carbon and energy (Nester et al. 1995).

In addition to carbon, organisms require all the other elements found in cell constituents ($C_5H_7O_2NP_{0.5}$), primarily nitrogen, sulfur, and phosphorus. These have to be supplied in a form that can be utilized. Most microorganisms can use inorganic salts as a source of each of these elements. Nitrogen found in amino acids and in nucleic acids can be provided from NH_4^+, NO_3^-, and N_2 in some cases, or from the organic form. Sulfur is mostly present in the amino acids and supplied by organic sulfur compounds. Inorganic phosphorus (PO_4^{3-} species) is an important element present in membrane phospholipids, nucleic acids, and ATP. Many other mineral elements, like magnesium, potassium, calcium, iron, zinc, etc., are required for the function and the activity of numerous enzymes as well as for the stabilization of the cytoplasmic membrane and nucleic acids.

13.2.1.2
Physical and Chemical Factors

Various physical and chemical factors influence the development of microorganisms. Oxygen is very significant in determining the extent and rate of pollutant biodegradation. Often in air pollution treatment, only the aerobic conditions are described. The aerobes require oxygen due to their metabolic pathways. Some facultative microorganisms use oxygen in metabolic processes if it is available, but they can also grow in its absence. Biodegradation is much faster under aerobic conditions as compared to anaerobic conditions.

Temperatures for microorganism range between 5 and 75 °C, with an optimum at 20–40 °C. Each species has a well-defined upper and lower limit at which growth stops. The enzymatic reaction rate is approximately doubled for each 10 °C rise in temperature, and thus the cells develop more rapidly.

Bacterial species prefer a neutral environment for their development. However, numerous bacteria live and multiply in acidic (pH 5) to basic (pH 8) media. Yeast and fungi grow most rapidly under acidic conditions. In many cases, microbial metabolism leads to a pH modification of the environment by the production of acidic or basic by-products.

It is well known that microbial life is highly dependent on water and on variations in ionic strength in the environment. Optimal conditions for the activity of aerobic microorganisms present on a solid support occur between 38 and 81% of the solid pore space (also referred to as the water percentage saturation) because, in this range, water and oxygen availability are maximized. At higher water contents, the slow rate of oxygen diffusion through the water is limiting, and thereby decreases the aerobic activity. At lower contents, the water availability becomes limiting.

13.2.2 Microorganisms and Pollutants

Most organic molecules, regardless of their complexity, can be degraded by at least one species of microorganism or by a consortium. The biodegradation of organic contaminants lead to their transformation into energy and biomass, their complete oxidation or mineralization, or their partial degradation with the formation of stable by-products. The factors influencing biodegradation are related to the nature of the molecule, environmental conditions, and the microbial population. Lebeault (1993) proposed some molecular factors that are important for biotreatment purposes: (1) the size, molecular weight, polymerization, and substitution rate, (2) the balance between the solid/liquid/gas phases, (3) the bioavailability in relation to the solubility of the pollutants, (4) the toxicity of the pollutants and by-products, and (5) the molecular concentration.

The capacity for microorganisms to use pollutant substances as nutritional substrates results from their activity in the natural carbon, nitrogen, and sulfur cycles. Many of the chemicals can be readily degraded because of their structural similarity to naturally occurring substrates.

The efficiency of the biodegradation of volatile organic compounds depends, then, on the chemical nature and structure of the substances (Table 13.1). The metabolic pathway and the enzymatic systems involved in the biodegradation process are different according to the molecular structure: aliphatic (saturated or unsaturated), cyclic, heterocyclic or aromatic (substituted or not), chlorinated compounds. The general strategy for microorganisms is to convert the organic contaminant into an intermediary product of the cellular central metabolism. In the case of chlorinated compounds, biodegradation begins by the rupture of the carbon–chloride bond using specific enzymes. The resulting by-product is hydrochloric acid. The limiting step in the use of halogenated xenobiotic compounds as growth substrates is the capacity of dehalogenase to recognize and transform these substances. The resistance to biodegradation of these molecules is correlated

Table 13.1. Proposed relative aerobic biodegradability of various classes of chemical compounds. (Adapted from Fass et al. 1999)

In order of decreasing biodegradability
Aliphatic compounds
Light aromatic solvents
Chlorinated solvents
Cyclic hydrocarbons
Polyaromatic hydrocarbons
Chlorinated polycyclic compounds
Pesticides – herbicides
Complex molecules

with the deficiency of working enzymes. Most of the substituted alkanes or aromatics with a high number of chlorine atoms are not directly dehalogenated under aerobic conditions, but an intermediary anaerobic step is required.

Due to their structural similarity to fatty acids and plant paraffins, which are ubiquitous in nature, microorganisms in the environment use *n*-alkanes as a sole source of carbon and energy. As a result, alkanes are usually considered the most readily biodegradable hydrocarbons. The general strategy is to transform them into fatty acids. Their biodegradation occurs with a high demand for biological oxygen. The more common pathway is the direct incorporation of one oxygen atom onto one of the end carbons by a monooxygenase enzyme. The result is the formation of a primary alcohol. Alternatively, a dioxygenase enzyme can incorporate two oxygen atoms into the alkane to form a hydroperoxide. The result of both pathways is the production of a primary fatty acid, which can follow the common pathway used in the cell (β-oxidation). The biodegradation of alkenes, containing one or more double bonds, follows the same pathway as alkanes. The initial step can be the attack at the double bond, which can yield a primary or secondary alcohol or an epoxide (Maier 2000).

A wide variety of bacteria and fungi can carry out aromatic transformations, either partial or complete, under a range of environmental conditions. In aerobic media, the most common initial transformation is a hydroxylation that involves the incorporation of molecular oxygen and leads to the formation of a catechol. This catechol ring is cleaved by a second dioxygenase, either between the two hydroxyl groups (ortho pathway) or next to one of the hydroxyl groups (meta pathway), and this continues until final degradation (Fig. 13.1). Table 13.2 reports some values for organic compound biodegradation using biological processes.

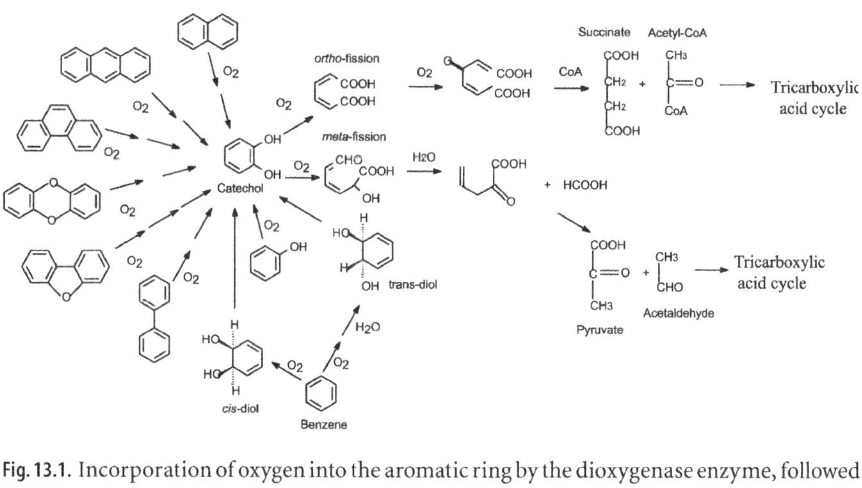

Fig. 13.1. Incorporation of oxygen into the aromatic ring by the dioxygenase enzyme, followed by the meta or ortho ring cleavage

Table 13.2. Microbial biodegradation capacities of organic compounds by various microbial species. (Adapted from Kennes and Thalasso 1998)

Compound	Microorganism	Biodegradation capacity (g substrate m^{-3} h^{-1})
Ethanol	Mixed culture	53–219
Methanol	Consortium of eight bacterial species	112.8
Dichloromethane	*Hyphomicrobium* sp.	200
Ethylacetate	Mixed culture	175
Propionaldehyde	*Pseudomonas fluorescens*	500
Toluene, ethylbenzene, xylene	Mixed culture	70
Toluene + ethylbenzene + xylene	Mixed culture	75–80
Phenol	*Pseudomonas putida*	124–700
Styrene	Mixed culture	70–80
Toluene	Mixed culture	35
Ethanol : toluene	*Bacillus* sp. + *Pseudomonas* sp.	120 : 70
Methanol + styrene	Mixed culture	110

13.3 Applications of Biological Processes

13.3.1 General Operating Conditions

In Section 13.2, it was shown that a large number of VOCs present in the air are degraded by microorganisms. Some conditions are required for optimal bacterial growth and complete pollutant degradation, such as equilibrated nutrients, water and oligo-elements. Moreover, the processes have to achieve both pollutant air–liquid transfer and biodegradation. Depending on the water-phase mobility and microorganisms that are supported or not on a solid phase, three processes are defined – biofilters, biotrickling filters, and bioscrubbers (Table 13.3). Figure 13.2 illustrates the different processes used for air treatment. Some operating conditions are similar for all three bioreactor processes:

- *Water requirement* – water is necessary for the good growth of microorganisms, and to obtain an efficient performance of the biological system. However, excessive water can lead to bypasses in biofilters or flooding in bioscrubbers and trickling filters.
- *Temperature* – biological systems work at temperatures ranging between 10 and 40 °C due to microorganism growth. In a biofilter, the exothermal biodegradation reactions increase the filter temperature by 2–10 °C. Thus,

Table 13.3. Classification of biological processes used for air treatment

Biomass	Aqueous phase	
	Mobile	Immobile
Activatedsludge	Bioscrubber	–
Attached biomass	Biotrickling filter	Biofilter

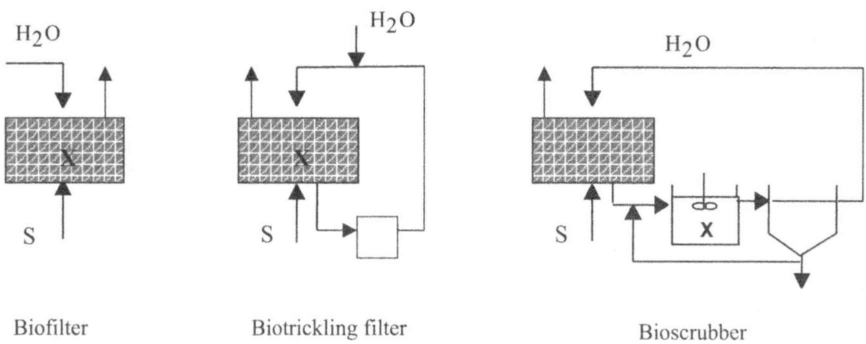

Fig. 13.2. Schematic presentation of biological treatment processes for air. X Biomass, S substrate, H_2O aqueous phase

a heat exchanger is sometimes required to decrease the inlet air temperature, or the liquid rate is also a good parameter to adjust the temperature in a bioscrubber or in a trickling filter.

- *pH variation and control* – to favor biomass growth, a pH ranging from 5 to 9 is generally chosen. The stability of this parameter promotes the performance of the treatment. The carbon dioxide production due to the total bio-oxidation of VOCs acidifies the aqueous solution. In this case, a pH correction (lime) or the utilization of a buffer solution are sometimes necessary.

- *Dust removal in the inlet air* in order to avoid rapid clogging of the filter, dust has to be removed. A number of processes are available – filtration, gas washing, and centrifugal separators (Coulson and Richardson 1997).

13.3.2 Biofilters

Biofiltration technology for air treatment was initially developed in western countries in the 1960s. Currently, a large number of biofilters are being used for VOC and odor reduction in industrial waste gases (Le Cloirec 1998; Devigny et al. 1999; Kennes and Veiga 2001).

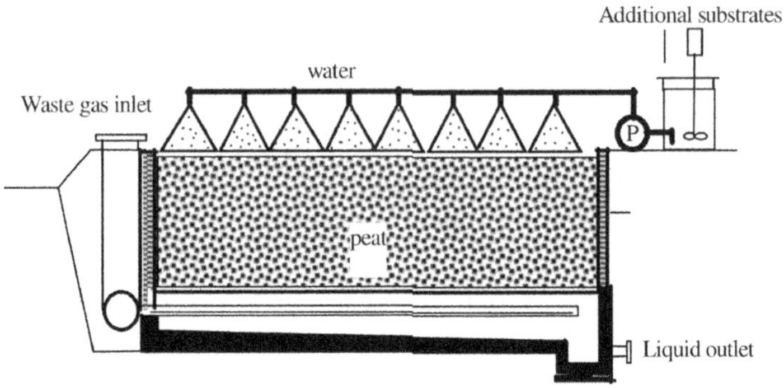

Fig. 13.3. Schematic presentation of an open biofilter

The biofilter is the most basic system for air treatment. The reactor is either open or closed, as shown in Figs. 13.3 and 13.4. Air loaded with pollutants flows through a wet packing material that supports microorganisms. Pollutants are transferred into the aqueous liquid film coating the solid support, and are degraded by bacteria. Substrate equilibrium is necessary to get a good biomass growth and a significant VOC removal. Water containing additional salts (ammonium, phosphates, potassium) has a counter- (closed biofilters) or co-current flow (open biofilter). Due to the large volumes required, the packing medium is generally very cheap. For example, peat, soil, or wood pellets can be used as bacterial supports.

Some practical data related to the design and operating conditions for VOC treatment are reported in Table 13.4. The depth (about 2 m) is limited in order to get a low pressure drop. The residence time is high compared to other air treatment processes, due to the low degradation kinetics. Thus, the reactor volume increases for less biodegradable compounds

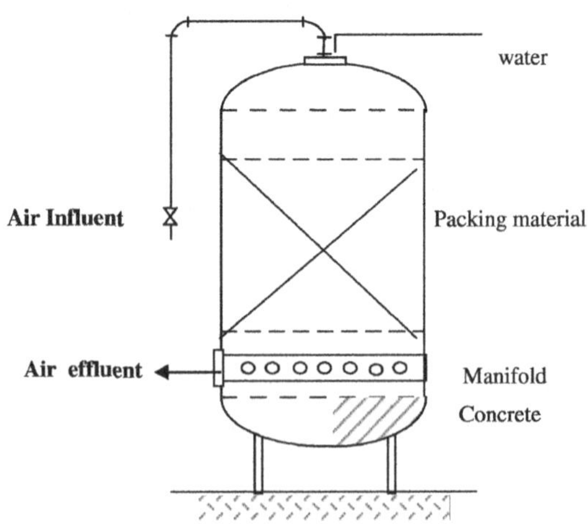

Fig. 13.4. Schematic presentation of a co-current closed biofilter

Table 13.4. Some operating conditions for biofilters used for VOC removal in air

Parameter	Value	Remarks
Filter depth (m)	0.5–2.5	Optimization between the residence time and the pressure drop
Life time (year)	2–5	Utilization of inorganic packing material increases the lifetime
Pressure drop (m H_2O)	0.1–1	Variation as a function of support compressing and/or clogging
Bed porosity	0.5–0.9	Possible bypass
Specific surface area ($m^2\,m^{-3}$)	300–1,000	High values increase the mass transfer coefficient
Residence time (s)	15–90	Dependent on degradation kinetics: alcohols > ketones > linear alkanes > aromatics
Velocity of air ($m\,h^{-1}$)	100–500	Weak values due to low kinetics of degradation
Air temperature (°C)	10–40	Temperature range required for microorganism growth
Air humidity (%)	60–100	100% level is interesting
Water level in the filter (%)	40–60	Higher levels give anaerobic death zones and a transfer limitation in the filter
pH	5–9	
Acclimation time (day)	10–30	Utilization of an inoculum (diluted activated sludge)
Pollutant concentration range ($mg\,m^{-3}$)	10–1,000	Possible inhibitions at higher concentrations
Performances (%)	90–99.9	Depending on the VOC molecules

(complex molecules, chlorinated compounds). The humidity (40–60%) of the support is an important parameter to minimize channeling and to obtain an efficient performance. Bacteria are sensitive to pH, moisture, and air temperature.

Devinny et al. (1999) described a large number of biofilters applied to VOC removal in industrial emissions. An example of biofilter application in the removal of ethanol and a mixture of ethanol, dichloromethane (DCM), methyl ethyl ketone (MEK) and toluene, at concentrations ranging from 50 to 300 $mg\,m^{-3}$, is provided here. A heterogeneous material, such as wood bark (10/25 or 25/40), was used as packing. The design data of the biofilter and the characteristics of wood barks are given in Table 13.5. First, the pressure drop was determined in clean, wet, bacteria-colonized materials. The results are

Table 13.5. Some characteristics of the biofilter and wood bark used as packing material

Parameters	Value
Filter depth (m)	1.1
Velocity of air ($\mathrm{m\,h^{-1}}$)	
For pressure drop studies	100–2,200
For VOC removal studies	100–1,300
Residence time (s)	2–40
Air temperature (°C)	20
Air relative humidity (%)	60
Wood bark size (mm)	10–20
Bed density ($\mathrm{kg\,m^{-3}}$)	258
Bed porosity	0.53
Specific surface area ($\mathrm{m^2\,m^{-3}}$)	320
Water level in the filter (wt%)	45

given in Fig. 13.5 for particle Reynolds numbers ranging from 31 to 1,460. Due to the heterogeneity of the material, it is difficult to model the linear pressure drop ($\Delta P/H$) as a function of the empty bed velocity (U_0) with the classical Ergun equation. Therefore, a specific relation was used (Comiti and Renaud 1989; Le Cloirec et al. 2001):

$$\frac{\Delta P}{HU_0} = M^* U_0 + N^* , \tag{13.1}$$

where M^* ($\mathrm{kPa\,s^{-2}\,m^{-3}}$) and N^* ($\mathrm{kPa\,s^{-1}\,m^{-2}}$) are experimental parameters defined by the following equations:

$$M^* = \left\{0.0413\left(1-\left(1-\frac{d_\mathrm{p}}{c}\right)^2\right) + 0.0968\left(1-\frac{d_\mathrm{p}}{c}\right)^2\right\}\tau^3 \varrho a_\mathrm{vd}\frac{(1-\varepsilon)}{\varepsilon^3} \tag{13.2}$$

$$N^* = 2\mu\tau^2 a_\mathrm{vd}^2\left[1+\frac{4}{a_\mathrm{vd}c(1-\varepsilon)}\right]^2\frac{(1-\varepsilon)^2}{\varepsilon^3} , \tag{13.3}$$

where d_p is particle diameter (m), c is bioreactor diameter (m), ϱ is air density ($\mathrm{kg\,m^{-3}}$), ε is bed porosity (dimensionless), μ is air viscosity (Pa s), a_vd is the specific surface in contact with the fluid ($\mathrm{m^2\,m^{-3}}$), and τ is tortuosity (dimensionless).

The determination of a_vd and τ gives $1.8 < \tau < 2.9$ and $350 < a_\mathrm{vd} < 950$, depending on the wood bark size. These values are high compared to those determined in filters. It is found that the moisture content (10–45%) of the packing material has no strong effect on these two parameters.

The performance of the process to degrade ethanol alone is given in Fig. 13.6. A good efficiency is found up to a volumetric loading rate around 150 $\mathrm{g\,m^{-3}\,h^{-1}}$.

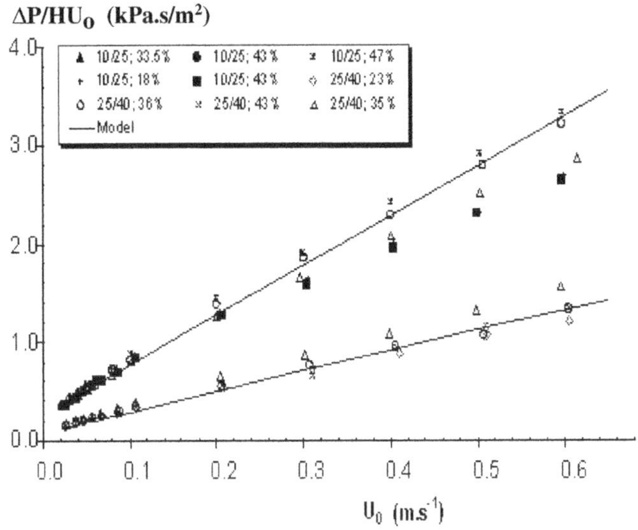

Fig. 13.5. Pressure drop in a wood bark biofilter as a function of the empty bed velocity, the size of the material, and the humidity in the process

A similar approach was carried out for a mixture of VOCs as described previously. After 84 days of working, the biofilter efficiency for each compound at air velocities of about 100 m h^{-1} is given in Table 13.6. It should be noted that the degradation is a function of the molecular structure and, in fact, of the biodegradability of each solvent.

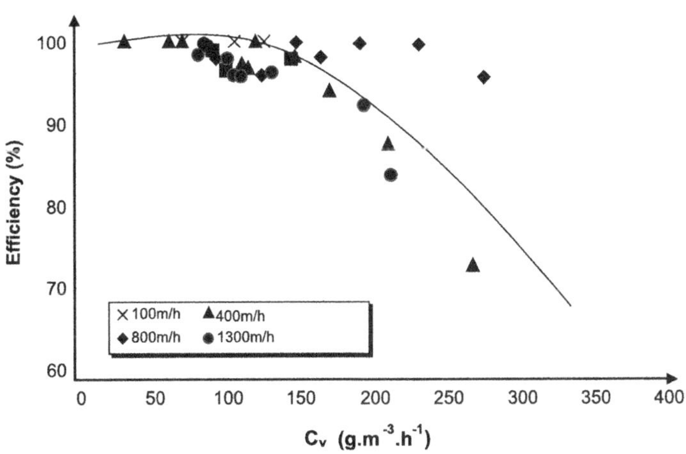

Fig. 13.6. Performance (EE) of the biofilter as a function of the volume load

Table 13.6. Removal efficiencies measured in a biofilter after 84 days for a mixture of VOCs

Parameter[a]	Value
Ethanol removal (%)	100
Methyl ethyl ketone removal (%)	60–100
Dichloromethane removal (%)	40–60
Toluene removal (%)	40–60

[a] $U_0 = 100\,\mathrm{m\,h^{-1}}$, residence time = 37 s, C_v EtOH = $35\,\mathrm{g\,m^{-3}\,h^{-1}}$, C_v MEK = $8\,\mathrm{g\,m^{-3}\,h^{-1}}$, C_v DCM = $10\,\mathrm{g\,m^{-3}\,h^{-1}}$, C_v toluene = $7\,\mathrm{g\,m^{-3}\,h^{-1}}$

13.3.3 Biotrickling Filters

In biotrickling filters, contaminated air flows through a packed bed of inert material, and the pollutant is transferred into the water and degraded by microorganisms present in the packing material (Fig. 13.2). Table 13.7 gives a range of general operating conditions.

Cox and Deshusses (2001) gave examples of pollutant removal obtained on a large number of biotrickling filters. However, in order to illustrate the industrial utilization of this biotechnology, a specific example is described here. In an industrial site producing electronic boards with their components, a gaseous emission of about $3{,}000\,\mathrm{m^3\,h^{-1}}$ loaded with isopropanol at a concentration ranging from 150 to $350\,\mathrm{mg\,m^{-3}}$ was, in a first step, treated with a trickling biofilter. The air empty bed velocity was $330\,\mathrm{m\,h^{-1}}$. The pH was continuously controlled. To avoid clogging the filter, pressurized air was injected in a countercurrent. The system worked 16 h per day with an efficiency of more than 90%. In a second step, due to some modifications in the production, the flow rose from 3,000 to $6{,}000\,\mathrm{m^3\,h^{-1}}$, the isopropanol concentration was $220\,\mathrm{mg\,m^{-3}}$, and the working time reduced from 16 to $8\,\mathrm{h\,day^{-1}}$. The removal efficiency ranged between 69 and 90%, with an average performance of about 80%.

The investment was about 229,000 Euro (EUR), and the modification cost of the system was EUR 12,000, i.e., EUR 40 $1{,}000\,\mathrm{m^{-3}}$ of treated air. The operating cost was EUR 12 $\mathrm{day^{-1}}$, i.e., EUR 0.75 $\mathrm{year^{-1}\,m^{-3}}$, including EUR 2.30 for nutrients, EUR 1.15 for pH control, EUR 2.0 for water, and the remainder for electricity. The personal cost was estimated at about EUR 5,000 $\mathrm{year^{-1}}$.

13.3.4 Bioscrubbers

VOC removal with a bioscrubber is performed by an association of different stages of treatment – a pollutant–liquid transfer column, an activated sludge basin, and possibly a settling tank. A packed column or an empty column

Table 13.7. General operating conditions applied in biotrickling filters

Parameter	Value	Remarks
Liquid velocity (U_L; m h^{-1})	0.05–20	Compromise between the mass transfer in water, the flooding point, and the biomass loss
Liquid holdup (%)	< 5	
Air velocity (U_G; m h^{-1})	100–1,000	Limited by the clogging due to microorganism growth
Air residence time (s)	< 60	Residence time lower than for biofilters
Bed porosity (dimensionless)	05–0.95	High value avoids clogging
Specific surface area (S; m^2 m^{-3})	100–400	Lower value compared to biofilters
Filter depth (H; m)	10–15	Light packing materials allow high columns
pH	5–9	Easy to control
Temperature (°C)	10–40	Mesophilic microorganisms. Treatments with temperatures of 50–70 °C could be efficient with thermophilic populations
Lag time (day)	5–210	Function of the biodegradability of the molecule

with a spray injector are the principal types of gas absorption equipment. The operating conditions are close to those used in the biological trickling filters in terms of mass transfer in the column (Table 13.8). Gas and liquid velocities are functions of the solubility (Henry constant) of the pollutant. The absorbed VOC is degraded in the bioreactor and, after sedimentation, the solution is recycled in the column.

In some cases, where the solubility of the VOC is very low, a non-biodegradable and non-miscible organic liquid is used to absorb the pollutant. A continuously stirred reactor is then used to transfer the pollutant slowly from the organic liquid to an aqueous biological solution where the molecules are biodegraded. Glycol and heavy oil have been tested in this type of system (Cotte et al. 1995; Roustan 2003).

Bioscrubbers are used for air treatment in different industrial activities, as shown in Table 13.9. This process is applied to a large range of flow rates,

Table 13.8. Examples of VOC removal in biotrickling filters

Pollutant	Biodegradability	Henry's coefficient	Lag period (day)	Epuration capacity ($g\,m^{-3}\,h^{-1}$)	References
Aliphatics					
Hexane	++	74	50	7.5	Plaggemeier et al. (1997)
Heptane	++	83.2		24	Schlinder and Friedl (1995)
Aromatics					
Styrene	++	0.11	15	32	Pol et al. (1998)
Toluene	+ + +	0.28	10	80	Cox and Deshusses (1999)
Oxygenated compounds					
Propionaldehyde	+ + +	0.0024		300	Kirchner et al. (1992)
Acetone			15	450	Morchadi et al. (1994)
	+ + +	0.0016		500	
Methyl-ethyl ketone	+ + +	0.0024		40	Kirchner et al. (1992)
Methanol	+ + +	0.00019	5	100	Chou and Huang (1997)
n-Butanol	+ + +	0.00035		100	Allen et al. (2000)
Diethyl ether	+	0.028	15	60	Heinze and Friedrich (1997)
Methyl tertio-butyl ether	+	0.023	210	45	Fortin and Deshusses (1999)
Chlorinated compounds	+ + +	0.093	7	200	Hartman and Tramper (1991)
Dichloromethane				157	Diks and Ottengraf (1991)
			15	150	Morchadi et al. (1994)
	++	0.18		300	Oh and Bartha (1994)
Chlorobenzenes				60	Mpaniasc and Baltziz (1998)

Table 13.9. Applications of bioscrubbers to various emissions of industrial activities

Industrial activities	Volatile organic compounds	Waste gas flow rate ($m^3 h^{-1}$)
Car painting	Alcohols, glycols	12,000 (pilot unit)
Painting activities	Butyl acetate, butanol, xylene	54,000
Coating	Alcohols, glycols, ketones, glycol ethers, aromatics	–
Drinks can painting (steel)	Alcohols, glycols, toluene	57,000
Drinks can painting	Alcohols, glycols	48,000
(aluminum)	Alcohols, glycols, ethyl acetate	10,000
Pasteboard packaging for drinks	Ethanol, ethyl acetate, 1-methoxy propane-2-ol	150,000
Detergents and cosmetics	Alcohols, acrylate monomers, toluene	30,000
Conditioning paint residues	Ethyl acetate, acetone, methyl ethyl ketone, aromatics, alkanes	500
Aluminum foundry	Phenol	120,000
Chipboard production	Formaldehyde	500
Glass wool production	Phenol, formaldehyde	400
Printing	Alcohols, esters	60,000

from 400 to 150,000 $m^3 h^{-1}$. Different molecules (alcohols, ketones, aromatics, esters) are biodegraded alone or in complex mixtures. The solubility and the biodegradability of the molecules are important parameters to consider when adjusting the operating conditions of this treatment.

To illustrate these processes, an example is described here. At an aluminum can manufacturing facility, a bioscrubber was used to remove the gaseous emission from a painting workshop. The flow was about 31,000 $m^3 h^{-1}$, with a maximum close to 56,000 $m^3 h^{-1}$. In a first step, only 10,000 $m^3 h^{-1}$ was treated. The VOCs were a mixture of ethanol, ethyl acetate, acetone, methyl ethyl ketone, isopropanol and ethoxypropanol at concentrations of 0–10 g C m^{-3}. The effluent temperature was 50–60 °C. The system worked for 5,280 h $year^{-1}$. The performance of this process was found to be about 80% removal with an inlet concentration of 900 $mg m^{-3}$ and an outlet concentration of 150 $mg m^{-3}$. The level in the aqueous phase was 900 $mg L^{-1}$. There was some difficulty in dissolving the solvents, and a better transfer of pollutant into the liquid could improve the process. In the biodegradation tank, *Bacillus cereus* was mainly present in the activated sludge. The system produced 10 kg CO_2 h^{-1}. The energy balance gave a consumption of about 1,170 MWh $year^{-1}$. The investment was EUR 230,000, and the operating cost was EUR 68,000 $year^{-1}$.

13.4 By-Products Generated During Biological Treatments of VOCs

VOC biological treatments produce different wastes or by-products that have to be taken into consideration, either by recovery of the biosolids or by a definitive elimination of waste. This recycling could be followed by production of energy. It is the case for methane formation by methanogenesis, or incineration with cogeneration. Another use of biosolid recovery is in agriculture, by means of composting or land application. At the end of the chain, landfill disposal appears to be the ultimate solution.

13.4.1 Overview of Wastes and By-Products Generated

Carbon dioxide and water are the main gaseous compounds produced by aerobic biological treatment. CO_2 is well recognized as being responsible for the increase of the greenhouse effect. For this reason, they must be treated or recovered. Different solutions have been explored on a research scale: for example, sequestration in deep oceans or aquifers by hydrate gas formation, or use in photosynthetic processes to produce algae. In smaller quantities, other gaseous compounds like nitrogen and sulfur dioxide, bacterial aerosols, can be released by VOC biotreatment.

Among the liquid by-products, wastewater coming from biotrickling filters or leachate from biofilters can be transported to wastewater treatment plants (WWTPs), depending on their composition. Another solution is to resort to anaerobic digesters in order to produce a biogas. Activated sludge from bio-scrubbers can be incorporated into the classical schemes of sludge handling,

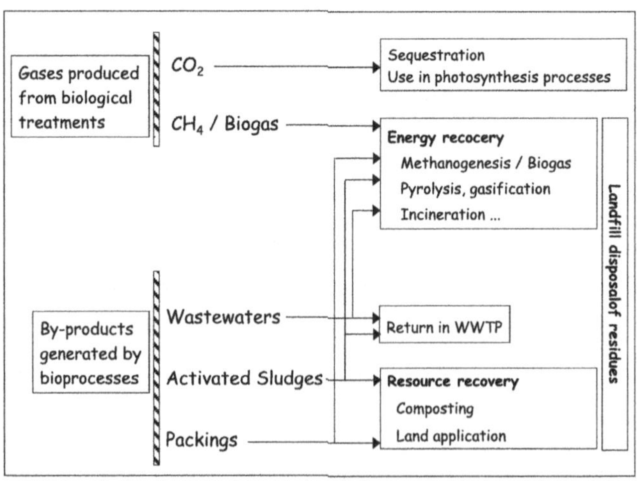

Fig. 13.7. By-products generated by biological treatments of VOCs and potential solutions for reuse or elimination

including their reuse in WWTPs if there is a nearby plant – anaerobic digestion, conditioning for heat treatments like incineration, pyrolysis, starved air combustion, and composting or land application.

Finally, biofilter packing materials represent the major proportion of solid end-products. They include soils, peat, wood bark, compost, dried sludge, and other organic materials that could be recovered or reused in ways similar to those for activated sludge. All of these by-products and their potential treatments are summarized in Fig. 13.7.

13.4.2 Energy Recovery

The recovery and reuse of organic wastes is an important element of waste management planning. Energy recovery and production offer a direct economic benefit, even if new treatments or disposals have to be investigated. In order to convert organic wastes into usable energy, an evaluation of the heating value of the waste is necessary. On this basis, if a waste has sufficient fuel potential, it could be economically recovered in energy production. In this chapter, the wastes considered could be activated sludge or some biofilter packing materials. The heating values of different wastes are reported in Table 13.10, and compared to fuel-gas. These technologies include mainly anaerobic digestion and methane production, pyrolysis, and gasification or incineration.

Table 13.10. Heat capacities of different organic wastes or fuels. (Adapted from Corbitt 1990)

Product	Energetic capacity ($kJ\,kg^{-1}$)
Activated sludge	19,800–23,300
Anaerobically digested primary sludge	12,800
Pine bark	22,000
Peat	9,000
Natural gas	53,000
Anthracite coal	29,500

13.4.2.1 Methanogenesis and Biogas Production

Methanogenesis is an anaerobic digestion of the waste organic part. The microbiology involved in this process includes facultative and anaerobic microorganisms, which convert organic materials into gaseous end-products and sludge in several steps. The biomass yield for an anaerobic process is much lower than that for an aerobic system, and thus less biomass is generated. The biogas produced is composed of methane (50–70%), which provides an economically valuable end-product, carbon dioxide (25–30%), and potentially hydrogen sulfide (0.5–5%). Depending on the technology and the operating conditions, the methane production ranges between 1 and 12 m^3 per m^3 of digester day^{-1} for an organic load ranging from 2 to 30 kg DOC m^{-3} day^{-1} (Scriban 1999). To ensure a suitable energy recovery or a combustion device, the biogas has to be treated in order to eliminate water, hydrogen sulfide, and potential traces of organohalides or heavy metals. Depending on the biogas quality, it is simply burned or could be introduced into the natural gas network. This process is particularly adapted to biological sludge treatment.

13.4.2.2
Pyrolysis and Gasification Processes

These technologies use high temperatures to alter organic materials and convert them into valuable alternative fuels. Pyrolysis is an old process that occurs when no oxygen is required, and no other reacting compounds are introduced into the reactor. Depending on the operating conditions, the by-products generated by pyrolysis are gas, including CH_4 (6–15%), H_2 (10–55%), CO (1–20%), CO_2 (5–25%), and potentially HCl, H_2S, and heavy metals, a liquid fraction containing oil, water and organic solvents, char, coke-derived compounds and ashes (Liu and Liptak 1997). Since the gaseous fraction is known to contain valuable fuels, these authors have shown that the other end-products of pyrolysis are sometimes difficult to recover due to their corrosiveness, moisture or difficulty in being pumped. On the contrary, gasification processes correspond to the addition of reacting gases in the reactor in order to increase the formation of new gaseous molecules (CO_2, CO, CH_4, H_2, N_2) and/or release heat. Gasification usually requires higher temperatures (850–1,400 °C) than those used for pyrolysis (750–900 °C) and, when the reaction is complete, all the solid is transformed into the gas phase. Some ashes are also produced with a quality close to incineration ash. Gasification seems then to be a simpler and more cost-effective technology.

13.4.2.3
Incineration

Due to the more stringent regulations applied to air pollution control and the increasing operating costs associated with residue management, incineration is more frequently used when other processes have failed. For these reasons, it appears to be the ultimate solution for the treatment of the waste organic part. The main advantages are the reduction of the total volume of waste by about 90%, and 75% by weight. Secondly, there is the production of hot gases and the recovery of energy in the form of steam, which may or may not be converted into electricity. As far as sludge is concerned, incineration requires a dewatering or a drying step in order to reach 35% of dry matter, for example, with a fluidized-bed incinerator (Liu and Liptak 1997). In all cases, different ashes are produced, accounting for about 10% of feed sludge volume – bottom ash or residue (85–90% of total ash), and fly ash (10–15% of total ash). Under European regulations, different classes of bottom ash have been defined, depending on their chemical composition. So, they can either be recovered in road techniques, for example, if their environmental impact is low, or be transformed by vitrification, for instance, in order to be disposed of in sanitary landfills. This latter solution is used for fly ash disposal.

13.4.2.4
Resource Recovery

Most of the solid organic wastes produced by the biological treatments of VOCs, dried sludge and biofilter packing materials (peat, soil, compost, wood bark), can be included in a resource recovery scheme in terms of land application or composting. Alternative solutions, like activated carbon production from activated sludge, could be an innovative option but have not yet been developed on a large scale (Rio et al. 2004). In France, about 60% of activated sludge produced in WWTPs is recovered in land application, and considered as fertilizers for agricultural soils. These disposals are severely controlled by regulations and laws, which define the conditions and the management of the application. Currently, composting is considered as a minor mode of treatment – only 6% of wastes are composted in France. This level is low in comparison to Austria (17%) or The Netherlands (20%). The composting process, however, has many positive points – it allows about 60% of the biodegradable organic part of waste to be reduced by aerobic degradation and, at the same time, the heat produced can kill the pathogens. Different technologies are available and have been tested; the simplest is the window method but more automated, closed systems exist like the rotating drum. The main disadvantage concerns the quality of the compost produced. Of course, this depends directly on the quality of the selective sorting of the waste in the early steps of the process. Incontestably, if this process is well conducted, the only end-product is economically valuable compost for use as an organic soil conditioner.

13.4.2.5
Sanitary Landfill Disposal

Since 2002, European law stipulates that only the ultimate residue of waste can be managed in sanitary landfills, except for radioactive, corrosive, explosive, and fuel wastes. It must be the last link in the chain of waste management. The landfill disposal takes place in natural impermeable sites (velocity in the soil less than 10^{-9} m s^{-1} over at least 5 m depth), ensuring a complete containment of waste. An impact study must include a geological and hydrogeological dimension. Leachate is collected by a network of drains, and a permanent laboratory carries out an analysis of wastes, leachate or trickling storm waters (Damien 2002). According to the biological treatment of VOCs, only some ashes produced from the incineration of biological sludge or residues and biofilter packing solids could be concerned by this kind of sanitary landfill disposal.

13.5 Conclusions

Due to continuing regulatory trends toward more stringent control of VOCs emissions, biological methods for contaminated air treatment have developed into environmental benign technologies over the last decade. However, further research in both microbiological and engineering areas is required to further optimize VOC treatment methods, and develop economically feasible and technologically advanced process.

References

Allen DG, Kong Z, Fulthorpe RR, Farhana L (2000) Thermophilic biofiltration of volatile organic compounds. In: Proc 93rd Annu Meet Exhibition Air and Waste Management Association, June 2000, Pittsburgh, Pap n° AE2e

Chou MS, Huang JJ (1997) Treatment of methyl ethyl ketone in air streams by biotrickling filters. J Environ Eng 123:569–576

Comiti J, Renaud M (1989) A new model for determining mean structure parameters of fixed beds from pressure drop measurements: application to beds packed with parallelepipedal particles. Chem Eng Sci 44:1539–1545

Corbitt RA (1990) Standard handbook of environmental engineering. McGraw-Hill, New York

Cotte F, Fanlo JL, Le Cloirec P, Escobar P (1995) Absorption of odorous molecules in aqueous solutions of polyethylene glycol. Environ Technol 16:127–136

Coulson JM, Richardson JF (1997) Chemical engineering, vol 2, 4th edn. Butterworth Heinemann, Oxford

Cox HHJ, Deshusses A (1999) Biomass control in waste air biotrickling filters by protozoan predation. Biotechnol Bioeng 62:216–224

Cox HHJ, Deshusses A (2001) Biotrickling filters in bioreactors for waste gas treatment. In: Kennes C, Veiga MC (eds) Bioreactors for waste gas treatment. Kluwer, Dordrecht, pp 99–131

Damien A (2002) Guide du traitement des déchets. Dunod, Paris

Devinny JS, Deshusses MA, Webster TS (1999) Biofiltration for air pollution control. Lewis, Boca Raton

Diks RMM, Ottengraf SPP (1991) Verification studies of a simplified model for the removal of dichloromethane from waste gases using a biological trickling filter, part II. Bioprocess Eng 6:131–140

Fass S, Vaudrey H, Vogel TM, Block JC (1999) Factors controlling the biodegradation of chemicals in soils. In: Baveye P, Block J-C, Goncharruk VV (eds) Bioavailability of organic xenobiotics in the environment: practical consequences for the environment. NATO ASI Series 2 Environment 64:93–117

Fortin NY, Deshusses MA (1999) Treatment of methyl-ter butyl ether vapors in biotrickling filters. 1. Reactor startup, steady state performance, and culture characteristics. Environ Sci Technol 33:2980–2986

Godish T (1997) Air quality. Lewis, Boca Raton

Hartman S, Tramper J (1991) Dichloromethane removal from waste gases with a biotrickling filter. Bioprocess Eng 6:83–92
Heinsohn RJ, Kabel RL (1999) Sources and control of air pollution. Prentice Hall, Upper Saddle River
Heinze U, Friedrich CG (1997) Respiratory activity of biofilms: measurement and its significance for the elimination of n-butanol from waste gas. Appl Microbiol Biotechnol 48:411–416
Kennes C, Thalasso F (1998) Waste gas biotreatment technology. J Chem Technol Biotechnol 72:303–319
Kennes C, Veiga MC (2001) Bioreactors for waste gas treatment. Kluwer, Dordrecht
Kirchner K, Wagner S, Rehm H-J (1992) Exhaust gas purification using biocatalyst (fixed bacteria monocultures) – the influence of biofilm diffusion rate (O_2) on the overall reaction rate. Appl Microbiol Biotechnol 37:277–279
Lebeault JM (1993) Traitements biologiques des déchets industriels. In: Scriban R (coordonnateur) Biotechnologie, 4th edn. Technique et documentation. Lavoisier, Paris, pp 884–901
Le Cloirec P (1998) Les composés organiques volatils (COV) dans l'environnement. Tec & Doc, Lavoisier, Paris
Le Cloirec P, Humeau P, Ramirez-Lopez EM (2001) Biotreatment of odours: control and performances of a biofilter and a bioscrubber. Water Sci Technol 44:219–226
Liu DHF, Liptak BG (1997) Environmental engineer's handbook, 2nd edn. Lewis, Boca Raton
Maier RM (2000) Microorganisms and organic pollutants. In: Maier RM, Pepper IL, Gerba CP (eds) Environmental microbiology. Academic Press, San Diego, pp 363–400
Morchadi J, Degorce-Dumas JR, Le Cloirec P (1994) Deodorization of waste gases containing VOC using suspended-growth bioscrubbers. Air Pollut II 1:487–496
Mpaniasc J, Baltziz BC (1998) An experimental and modeling study on the removal of monochlorobenzene vapor in biotrickling filters. Biotechnol Bioeng 59:328–343
Nester EW, Roberts CE, Nester MT (1995) Microbiology, a human perspective. Brown, Dubuque, Iowa
Oh YS, Bartha R (1994) Design and performance of a trickling air biofilter for chlorobenzene and o-dichlorobenzene vapors. Appl Environ Microbiol 60:2717–2722
Plaggemeier T, Lammerzahl O, Endgesser KH (1997) Purification of n-hexane polluted air by use of a biological trickling filter: monitoring the primary-degrading microflora. In: Prins WL, van Ham J (eds) Proc Int Symp Biological Waste Gas Cleaning. VDI, Düsseldorf, pp 257–269
Pol A, van Haren FJJ, Op den Camp HJM, van der Crift C (1998) Styrene removal from waste gas with a bacterial biotrickling filter. Biotechnol Lett 20:407–410

Rio S, Faur-Brasquet C, Le Coq L, Lecomte D, Le Cloirec P (2004) Preparation and characterisation of activated carbon from sewage sludge. Water Sci Technol 49:139–146

Roustan M (2003) Transferts gaz-liquide dans les procédés de traitement des eaux et des effluents gazeux. Tec & Doc, Lavoisier, Paris

Schindler I, Friedl A (1995) Degradation of toluene/heptane mixtures in a trickling-bed bioreactor. Appl Microbiol Biotechnol 44:230–233

Scriban R (1999) Biotechnologie, 5th edn. Tec & Doc, Lavoisier, Paris

VDI/VIN (1996) Handbuch Reinhaltung der Luft, vol 6. VDI, Düsseldorf

Zhu X, Rihn MJ, Suidan MT, Kim BJ, Kim BR (1996) The effect of nitrate on VOC removal in trickle bed biofilters. Water Sci Technol 34:573–586

Part IV

Biological Reactors – Case Studies

14 Odor Removal in Industrial Facilities

Vladimir Popov, Vitaliy Zhukov

14.1 Introduction

Air pollution, once the concern of only large urban settlements, is rapidly spreading worldwide and becoming one of the major challenges of modern civilization. Greenhouse gases, harmful and hazardous inorganic and volatile organic compounds (VOCs), nauseous odors – these are just some of the problems to be solved to ensure a healthy and comfortable environment for future generations. The principal sources of modern air pollution are vehicles, industrial sites, and municipal facilities dealing with water treatment, collection and disposal of waste, etc. (Table 14.1). There are several approaches to the problem. The preferred solution is, of course, the development of new "green" technologies that impose less burden on the environment, and the use of new, cleaner energy sources, including non-conventional ones. Nevertheless, efficient abatement technologies, i.e., end-of-pipe solutions, will in the near future remain an essential tool for reducing air pollution.

Whatever future advances we make in industrial production, one element of modern civilization will continue to aggravate air pollution. Mankind generates more and more waste, and its collection, disposal and recycling is a substantial factor in the sustainable development of our economies. All these activities generate various odors and smells that are detrimental to the environment and quality of life. These problems will persist indefinitely, and the only foreseeable remedy is the development of efficient and economical ways to treat these sources of air pollution.

Biological methods of air purification are indispensable when large and diluted volumes of VOCs and/or odor-laden air are to be treated (Cox and Deshusses 1998; Devinny et al. 1999; Cox and Deshusses 2000; Shareefdeen et al. 2003). Their main advantages over traditional methods (incineration, activated carbon adsorption, etc.) are well known and mainly reside in low operational costs. Moreover, they do not generate secondary waste streams and/or pollutants, and are readily accepted by regulating authorities and the general public. Biofiltration has been successfully used to neutralize odors of different origin (agricultural, municipal, industrial) containing sulfur and nitrogen compounds, and to remove an impressive number of various volatile

Shareefdeen/Singh (Eds.)
Biotechnology for Odor and Air Pollution Control

Table 14.1. Main sources of urban air pollution (major sources of pollutants are marked in *italics*)

Source	Pollutants	Remedies
Vehicles and public transport	*CO*, *NO_X*, SO_X, benz(a)pyrene, etc.	Alternative energy sources, "clean fuel", environmentally friendly engines, catalytic neutralizers
Industrial facilities	NO_X, SO_X, *VOCs*, odors	New technologies, re-profiling, closing or moving to a new location, end-of-pipe solutions
Municipal facilities (collection and disposal of waste, water treatment, etc.)	VOCs, *odors*	New technologies, end-of-pipe solutions

chemicals. These include solvents and diluents (BTEX, aliphatic hydrocarbons, esters, alcohols, chlorinated compounds, etc.), and components used in the industrial production of polymers, plastics and dyes (styrene, vinyl chloride, epichlorohydrine, formaldehyde, phthalates, cyclohexanone, etc.), including recalcitrant and poorly biodegradable ones (Devinny et al. 1999).

The difference between VOC and odor abatement is tenuous. It depends on individual circumstances and national legislation, and can be considered purely semantic. However, it is worth outlining some of the differences between the two principal biofiltration applications (Table 14.2).

14.2 Substrate Composition and Concentration

Standard VOC abatement application is usually confined to the removal from an airstream of a limited number of pollutants, often of close chemical structure and physicochemical properties. A common example is BTEX, aromatic poorly soluble compounds, while combinations of water-soluble alcohols and/or carbonyl compounds, usually not more than three to five substances altogether, are characteristic of flexographic printing applications (Popov et al. 2004). It is not uncommon that only one VOC has to be removed to comply with existing regulations. Thus, the use of specific monocultures adapted to the highly efficient destruction of a limited number of substrates is justified for VOC treatment, and may lead to advantageous results (Popov et al. 2000).

At the other extreme, even conventional malodorous airstreams may contain dozens of compounds that are highly diverse in structure and properties, which have a bearing on their elimination by microorganisms. As an example, Table 14.3 presents a composition of an air vent from a pet-food producing factory. More than 20 different chemicals were detected by

Table 14.2. Main differences between VOC and odor abatement biofiltration applications

Parameter	VOC abatement	Deodoration
Number and composition of compounds to be eliminated	Few substances, frequently close in chemical structure	Usually from 10 to 30 major components, very diverse in composition
Inlet concentration of volatile compounds to be abated	From several to several thousand $mg\,m^{-3}$	From several to several thousand $\mu g\,m^{-3}$
Outlet concentration of volatile compounds to be abated	Up to tens of $mg\,m^{-3}$. Depends on legislation	Should be minimized. Usually from parts to tens of $\mu g\,m^{-3}$
Rate of biomass accumulation	High or medium	Low or medium
Excess biomass control	Complicated. Could be a serious problem	Usually does not present a problem
Composition and formation of microbial consortium	Monocultures could be employed	A complicated consortium. Usually not less than 4–5 major strains
Success rating	Meets legislation	Meets public perception
Cost	Average and higher than average	Average and lower than average

GC/MS, of which 25% remained unidentified and could only be attributed to a certain class of compound. A vast variety of substrates present in a typical odor application requires a microbial consortium with diverse metabolic capability.

For a typical VOC control application, the relative composition of the emission is more or less stable, reflecting the intrinsic stability of a standard technological process. It does not mean that overall concentration cannot fluctuate substantially in terms of the absolute organic content – so-called spike emissions. These can be successfully dealt with introducing passive buffering systems (Popov et al. 2004). With odors, on the other hand, the composition of the emission is more labile. Minor alterations, e.g., in cooking temperature or the composition of the raw material, may result in considerable change in the relative concentration of the emitted substances, and hence the resulting odor.

14.3 Biomass Control

Relatively high (VOC applications) inlet concentrations of organic compounds enable fast biomass growth after initial inoculation of the system, leading to short acclimation periods. This advantage is reversed with time and results in

Table 14.3. GC/MS identification of major pollutants in the air exhaust of a pet-food factory (non-identified compounds are given in *italics*)

Retention time (min)	Compound	Retention time (min)	Compound
3.55	Hexaldehyde	7.24	1-Nonanol
4.30	Methyl propionate	9.10	*Hydrocarbon*
4.63	2-Methyl-1-propanol	11.30	2-*tert*-butyl-4-methoxyphenol
4.80	Valeric acid	11.60	2,6-di-*tert*-butyl-4-methoxyphenol[a]
5.05	Heptaldehyde	16.0	*Ester*
5.64	1-Heptanol	16.40	Dibutyl phthalate
5.81	Benzaldehyde	16.50	*Phthalate*
5.99	2-Pentyl furan	17.60	Methyl linoleate
6.71	1-Octanol	20.60	*Ester*
7.16	*Ester*	22.90	Di-(2-ethyl-hexyl) phthalate

[a] Major contaminant (> 80%)

serious problems with excess biomass growth, resulting in deterioration of the system performance.

In the case of odor control applications, low inlet concentrations preclude buildup of the biomass, resulting in longer startup periods but fewer problems with clogging. Moreover, in most odor control systems, autotrophic microorganisms, e.g., oxidizing hydrogen sulfide, are an essential part of the consortium. As a rule they are characterized by a lower growth rate compared to heterotrophic microflora. Taken altogether, this may result in problems with building up and preservation of the composition of the initial microbial consortium. Contrary to VOC applications where the main problems with biomass control are encountered at late stages due to the excess biomass buildup, in odor control cases the main problems are to be expected at initial startup phases, due to low biomass accumulation that can inhibit the initially optimized seeding consortium.

14.4 Compliance

Concentrations of typical VOCs both at the inlet and outlet of the biofiltration system can be easily monitored with conventional techniques. For example, GC equipped with FID usually provides sufficient sensitivity for a desired range of 1 – 1,000 mg m^{-3}, enabling continuous monitoring of the process, even remotely if required. On the other hand, identification of the substances responsible for bad odor is sometimes a challenge, due to their very low concentration, often below 1 ppm.

Compliance with local regulations for VOC control application is usually based on a predefined value for the end-of-pipe emission, quantified as total carbon, a value that can be easily monitored and instrumentally verified. On the other hand, odor control applications in the majority of countries rely not on instrumental values but on the specific perception of an odor by selected panelists, or even the local population. Hence, compliance limits sometimes are not specified but determined by the attitude of the local community. Even complete elimination of the majority of the pollutants does not guarantee success, as another odor that was previously hidden by some major contaminant may be de-masked. As it was noted recently, there could be no direct correlation between the concentrations of volatiles measured by GC/MS and sensed odor (Defoer et al. 2002). A specific issue with odors is individual perception and the synergistic and antagonistic effects on the human receptors of the variety of compounds present in the emission. Olfactometry, the method of choice for quantifying odors, is difficult to implement, and is labor-intensive and costly.

Removal of odors is considered technically to be somewhat easier than treating VOCs, especially poorly degradable ones. Either less complicated systems can be employed, e.g., open vs. closed biofilter systems, or smaller residence time can be used for the same type of device. This is reflected in the reduced capital costs and lower prices for the end user.

14.5 Modern Trends in Biofilter Development

Having evolved in the middle of the last century, biofiltration is now a fully mature technology with a wealth of commercial applications (Devinny et al. 1999). A number of construction principles have been implemented in the field. Three major trends in contemporary biofilter design can be outlined for the purpose of this chapter (bioscrubbers are not included). These are (1) open, often soil/compost biofilters, (2) enclosed container-type devices, often utilizing special proprietary media, and (3) trickling biofilter reactors based on the use of inert carriers. This is a very general classification that does not take into consideration several important features and characteristics. Many intermediary forms and types of biofilters have been proposed, reflecting the reality well known to practitioners that emissions are all different and require individual approach.

It cannot be determined a priori which type of biofiltration device will be the best solution for a particular application. In the long run, it is always a trade-off between efficiency, stability of operation, intrinsic limitations, such as size, weight, etc., which are inherent to the site, and the costs, both capital and operational. Both the advantages and disadvantages of the biofiltration devices outlined above are well known and have been thoroughly discussed

Table 14.4. Comparison of design complexity and advantages of biofilter types

Biofilter type/properties	Advantages
Open (soil/compost) biofilters	Simple, low capital cost
Enhanced and/or proprietary media	Resistant to compaction, better rheological properties, lower pressure drop, reduced weight
Enclosed, container type	Better process control, reliable assays, direction of the airflow can be reversed
Modular multilayer design	Smaller footprint
Specific, adapted microorganisms	Higher degradation rates
Trickling bioreactors	Full process control, enhanced efficiency per unit volume, small footprint

(Devinny et al. 1999). Some of the main features typical of various designs are summarized in Table 14.4.

Table 14.5 presents a selection of vendors of biofiltration equipment. It is not exhaustive and simply illustrates various trends in the field to date.

Classical examples of open soil biofilters are provided by Bohn Biofilter Corporation, USA (www.bohnbiofilter.com). They are characterized by long residence times, usually in the range of minutes, and hence large volumes of the media are required. They can vary in capacity from thousands to hundred of thousands of cubic meters per hour of treated air, and use relatively unsophisticated biofilter controls.

Enclosed container-type biofilters are the most abundant class of biofiltration devices existing at present, and are the standard in the industry for odor control applications. Designs range from relatively simple one-layer devices utilizing natural carriers (compost, wood chips, peat, etc.) to complicated machines with modular multilayer arrangements of filter layers (e.g., Biocube™ of Biocube, or Bioton™ of Monsanto Enviro-Chem Systems), sometimes with very sophisticated PLC controls, usually based on a specially processed and/or structured organic carrier with additions of inert materials that prevent compacting, channeling, and clogging (Bio-Reaction Industries, Bord na Móna, PPC, etc.). Media are often augmented with specific microflora (Bord na Móna, Bioton, Bowin Enviro engineering, BBK Bio Airclean). This class of biofilters is characterized by considerably shortened residence times in the range 20–60 s (with an average of about 30–40 s). To accommodate large airflows, container biofilters are arranged by many vendors in modules of 3,500–10,000 $m^3 h^{-1}$, which enables easy scaling-up and upgrade of the abatement system.

Trickling bioreactors based on synthetic supports and utilizing specialized cultures of microorganisms are not widespread at present, and are available

Table 14.5. Comparison of selected commercially available biofiltration plants

Vendor	Type	Media	Microorganisms	Capacity ($m^3 h^{-1}$)	Contact time (s)	Reported efficiency	Flow, hardware
(1) Open type							
Bohn Biofilter Corp. (USA)	Open-bed; soil, one bulk media layer	Adjusted local soil	Natural	3,000–185,000	50–60	"Without complaints"	Upflow. Blower, irrigation
Bowin Environeering Ltd. (Hong Kong)	Open-bed; one bulk medium layer; modular design from FRP	Proprietary Bikovent® aeration plates, made from polyethylene	Special active cultures	200–30,000	72–90	90–99%	Up-flow. Blower, humidifier, PLC-automatic
Matrix Environmental Technologies, Inc. (USA)	Open-bed, one bulk medium layer	Compost, activated carbon, bulking agents, buffering agents, and inorganic additives	Natural	~ 5,000 per open bed	Not reported	> 95%	Up-flow. Blower, top spray, humidifier
AMBIO Biofiltration Ltd. (Canada)	Open-bed; one bulk medium layer; small body units with bulk media	Wood chips	Natural	70–20,000	40–50	"Without complaints"	Up-flow. Blower, irrigation
(2) Container type							
Griffin Cardwell Ltd. (UK); Griffin & Company, Inc. (USA)	Open or close container (concrete, FRP, stainless steel); one medium layer	Specially prepared fibrous material	Not reported	Up to 25,000	Not reported	"Satisfactory"	Up-flow. Blower, preceding scrubber

Table 14.5. (continued)

Vendor	Type	Media	Microorganisms	Capacity ($m^3\,h^{-1}$)	Contact time (s)	Reported efficiency	Flow, hardware
BBK Bio Airclean A/S (Denmark)	Closed container from concrete or concrete elements; one bulk medium layer	BBK-Filter media, resistant to microorganisms; initially pasteurized	Specific selected	1,000–50,000	30–32	99–100%, < 20 ppm inlet H_2S	Down-flow. Blower, top spray
BIOTEG GmbH (Germany)	Double-walled container (plastic, steel); modular design	Wood chips	Natural	500–12,000	40–48	> 96%	Up-flow. Blower, preceding scrubber
Bio-Reaction Industries (USA)	Closed container, one bulk medium layer	Bio-Air SPHERES™					Down-flow. Blower, humidifier
Bord na Móna Environmental Ltd. (Ireland)	Closed container, one bulk medium layer; single units or modular design	MonaFil™ (peat base), MonaShell™ and MonaShell Ebf™ (seashells, synthetic)	Specially isolated and adapted cultures	3,500–15,000	36–40	97–99%	Up- or down-flow. Blower, top spray, humidifier
PPC, Inc. (USA)	Closed containers, one medium layer; modular design	Engineered media	Not reported	5,000–170,000	25–35	> 95%	Down-flow. Blower, top spray, humidifier, PLC-automatic

Table 14.5. (continued)

Vendor	Type	Media	Microorganisms	Capacity ($m^3 h^{-1}$)	Contact time (s)	Reported efficiency	Flow, hardware
Biocube, Inc. (USA)	Terete body with one or several (3–6) layers; modular design	Compost blend; patented Bio-Organic media (MSDS)	Not reported	340–8,450	From 26 (< 30 ppm H_2S) to 130 (> 200 ppm H_2S)	> 99%	Down-flow. Blower, humidifier with temperature control
Monsanto Enviro-Chem Systems, Inc. (USA)	Closed container; metal, plastic or concrete; modular design	Patented Bioton-Monsanto porous media	Special active cultures	1,600–80,000	16–28	> 95% odor removal	Blower, top spray; humidifier, PLC-automatic
(3) Trickling bioreactors							
BIOWAY (The Netherlands and USA)	Closed container; metal, plastic or concrete; modular design	Synthetic media PermaPac™ (includes nutrients)	Indigenous to the support medium	250–6,000	25–30	> 95%, 140–150 ppm inlet H_2S	Up-flow. Blower, top spray, PLC-automatic
Corain Impianti S.r.l. (Italy)	Container with several (2–7) trickling layers	Proprietary BioKey™ (optimal nutrients, pH, water content, etc.)	Not reported	350–15,000; up to 250,000	6–10 (HCOOH); 20–30 (odors)	> 95%	Up-flow. Blower, humidifier, trickling nutrients, pH correction

Table 14.5. (continued)

Vendor	Type	Media	Microorganisms	Capacity ($m^3\,h^{-1}$)	Contact time (s)	Reported efficiency	Flow, hardware
HanKi Industrial Ltd. (RK)	Stainless steel container with 2–6 trickling layers; modular design possible	Fibers or sponge with open pores	Specially isolated, selected and adapted bacterial strains	600–50,000	2.9–6.0	> 95% odor removal	Down-flow. Humidifier, trickling nutrients, pH correction
Sutcliffe Speakman PLC (UK); Innovation Biotechnologies (Russia)	Terete body with one or several (up to six) layers; modular design possible	Inert polymer or fibers	Specially isolated, selected and adapted bacterial strains	2,000–15,000	3–15	"Meets compliance limits"	Down-flow. Humidifier, trickling nutrients, pH correction

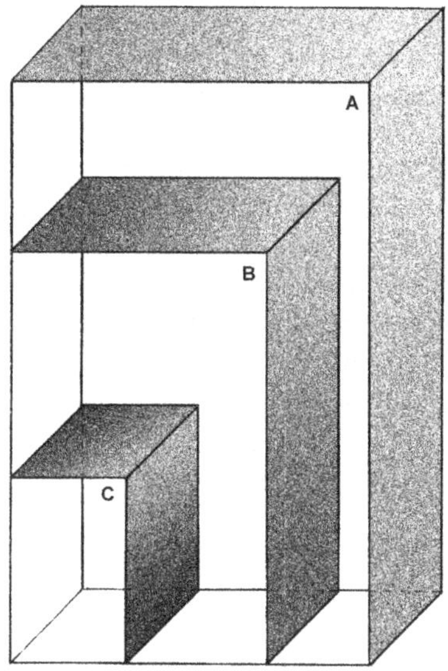

Fig. 14.1. Graphical representation of the comparison between an open soil/compost biofilter, operating at specific capacity of 50 h^{-1}, i.e., volume of air in m^3 passing through 1 m^3 of medium per hour (A), conventional enclosed container biofilter operating at 100 h^{-1} (B), and enhanced trickling bioreactor operating at 1,000 h^{-1} (C)

only from a few manufacturers, e.g., Bioway, Sutcliffe Speakman, Corain Impianti, HanKi Industrial. This type of biological filter has been mainly used for VOC control applications that are more demanding in terms of process control. However, recent examples provide illustrations of their successful application in odor treatment. Besides enhanced process control, they provide a number of other advantages including low contact times, sometimes as low as 10 s (Park et al. 2001; Popov et al. 2002; Gabriel and Deshusses 2003), resulting in an intense process that can be implemented in compact devices occupying much less space.

A graphic comparison between open soil/compost, conventional, enclosed container type and enhanced trickling filters in terms of residence time and volume of the carrier required to purify the same flow of laden air is presented in Fig. 14.1. The graph emphasizes the difference in intensity of the process of air purification between one type of device and the other.

The following section will provide several specific examples, with emphasis on biotrickling reactors that have not received as much attention as their more traditional counterparts. Case studies are presented, highlighting countries and examples (Republic of Korea and Russia) that are not often cited.

14.6 Case Studies

14.6.1 Odorous VOC: Formaldehyde Removal

Formaldehyde is widely used as a raw material in a number of industrial processes, in particular in the composite panel board industry and in the manufacturing of amino-molding powders and speciality resins. Formaldehyde has a disturbing smell and is classified as a putative human carcinogenic compound. Thus, not is only odor an issue but strict legislative limitations apply for emission into the environment. Formaldehyde is readily soluble, and subject to easy degradation by specific methylotrophic microorganisms capable of metabolizing C1 compounds.

Two European companies, Corain Impianti S.r.l. (Italy) and Sutcliffe Speakman PLC (UK), have developed a microbial process for industrial formaldehyde abatement based on the principles of biotrickling filtration (Table 14.6). Both processes rely on artificial inert carriers with a high specific area. In the case of Corain Impianti, it is a macroporous inorganic matrix coated with a thin layer of adsorbent activated material called BioKey™ (Cioci et al. 1997; Ferranti and Conca 2000; Ferranti 2001). Although the specific surface is not disclosed, it allows optimal micro-environmental conditions (pH, moisture, nutrients, etc.) for the growth and development of microorganisms. Sutcliffe Speakman

Table 14.6. Industrial formaldehyde abatement (n.a. data not available)

Parameter	BIP, UK	Sadepan Chimica, Italy
Vendor	Sutcliffe Speakman PLC	Corain Impianti S.r.l.
Biocatalyst	*Methylobacterium extorquens* on fibrous packing	Non-specified on BioKey™
Airflow (nominal; $m^3 h^{-1}$)	15,000	15,625 × 8 sections in four units
Dimensions	2.5 m in diameter ×4.5 m high	Four units, each 3.5 W × 12 L × 3.2 H
Operating weight	∼ 7 t	n.a.
Volume of the carrier (m^3)	9.6	26 × 8 sections
Bed arrangement	Six layers	One layer per section, 75 cm high
Contact time (s)	2.3	6–7.5
Bed temperature (°C)	30–35 °C	n.a.
Average formaldehyde load ($kg h^{-1}$)	1.6 ± 1.2	4.6 ± 2.3 total 0.58 ± 0.29 per section
Average efficiency (%)	96	91
Water consumption	∼ 1 $m^3 day^{-1}$	n.a.
Energy consumption	∼ 8 kWh	n.a.

(Popov et al. 2000, 2002) uses a specially structured fibrous carrier for cell immobilization with an extremely high specific surface of $\sim$ 4,000 $m^2\,m^{-3}$. Both plants are equipped with a system of pH correction and nutrient supply, and provide efficient removal of formaldehyde at low contact times, 3–7 s. The removal efficiency of 91–96% is achieved at 1.6–0.6 $kg\,h^{-1}$ loading of the pollutant per 15,000 $m^3\,h^{-1}$ module.

The Sutcliffe plant, installed at BIP Ltd. in Birmingham, is in the form of a cylinder that houses six shelves supporting fibrous carrier, each shelf equipped with an individual sprinkling system. The Italian plant at Sadepan Chimica, Viadana, is arranged as an eight-modular system each housing one bed of about 75-cm-deep BioKey™. Both biotrickling filters are preceded by wet scrubbers that remove particulates and condition the inlet airstream. The BIP plant was commissioned in 1997, and that of Sadepan Chimica in 1999. Since then, a number of other formaldehyde abatement plants similar in design and layout to that of Sadepan have been started in Italy by Corain Impianti: Chimica Pomponesco (120,000 $m^3\,h^{-1}$, December 1999), the SIA industrial plant in Viadana (240,000 $m^3\,h^{-1}$, January–October 2000), and the SIT plant at Mortara (120,000 $m^3\,h^{-1}$, 2001).

The applications cited present some of the most successful examples of efficient VOC/odor removal on an industrial scale. In August 2000, 3 years after the startup of the BIP plant, the customer advised that the system had never failed or required any servicing of the carrier. Besides high efficiency and consistent and stable operation, the technology presents large savings in operational costs. Some of the estimates performed by the owners of the equipment show that incinerators of the same capacity would be approx. ten times more expensive to run.

14.6.2 High-Performance/Enhanced Removal of Sulfur Compounds

Deodorizing of gaseous emissions is the most common area of application of biological filtration techniques. Quite often the source of the disturbing odor is sulfur-containing compounds such as hydrogen sulfide and mercaptans. Municipal wastewater treatment stations, waste collection, disposal and recycling, and rendering are just a few areas where sulfur-containing nauseous emissions require appropriate treatment.

All known types and classes of biofilters have been successfully used for this purpose, including open soil/compost and container biofilters. In general, conventional biofilters operate at 30–120 s contact time or < 100 volumes of air treated by one volume of the biofilter media per hour. Usually they provide 95–99% efficiency toward odor removal at H_2S concentration up to 20–100 ppm. In some cases, exceptionally high concentrations of H_2S – up to 5,000 ppm – have been successfully removed from the airstreams (www.bnm.ie/environmental/clean_air_solutions).

However, problems encountered during sulfur compound removal that adversely affect the long-term performance of biofilters are also well known.

The major one is acidification of the media due to sulfuric acid formation. To counteract the pH drop, special measures are usually taken. The general approach is to enhance the buffering capacity of the media by adding alkaline compounds, such as lime, or using a carrier base that itself has some alkaline properties. For example, the MonaShell™ carrier developed by Bord na Móna, specially for H_2S filtration, is based on seashells. Biotrickling gas-phase reactors equipped with a system of pH stabilization and nutrient control provide the most reliable way of ensuring the most advantageous environment for the efficient biodegradation activity of microorganisms. It is not surprising that some novel approaches have been introduced recently into the biofiltration of sulfur compounds.

Several research groups reported successful application of trickling bioreactors operating at very low contact times, around or below 3 s, for efficient abatement of H_2S and some other sulfur compounds from airstreams. A brief account covering the activities in Asia (Republic of Korea), Europe (Russia) and America (USA) is presented below.

14.6.2.1 USA

Recently, Deshusses (Gabriel et al. 2002; Gabriel and Deshusses 2003) have described a successful retrofitting of a chemical scrubber into a trickling biofilter for H_2S emission control. A former scrubber with the capacity of 16,300 $m^3\,h^{-1}$ was successfully converted into a biotrickling filter by minor alterations comprising a change of the packing material and replacement of the recycle pump. TriPack standard scrubber packing was replaced by open-pore polyurethane foam shaped in 4-cm cubes from Zander, Germany, with a surface area of 600 $m^2\,m^{-3}$ and density of 35 $kg\,m^{-3}$, allowing for very high mass transfer. Neither addition of any nutrients, nor additional pH control was required for efficient performance of the new plant, keeping the original scrubber contact times in a range of 1.6–2.3 s.

Long-term monitoring of the system showed > 97% efficiency at loadings of 15–95 $g\,H_2S\,m^{-3}\,h^{-1}$, meeting the compliance limit of 1 ppm_v as average outlet concentration over 24 consecutive hours for H_2S discharge. However, other malodorous components of the airstream contributing to the smell were not so efficiently degraded. Removal efficiency for CS, CS_2 and methyl mercaptan present in 67–193 ppb concentration was between 35 and 67%, resulting in only a 65 ± 21% reduction of smell measured by an odor panel. This required maintenance of the post-treatment of the exhaust gas.

A cost-benefit analysis performed by the end-user showed that annual savings in operating costs were around US$ 30,000, while the conversion procedure was estimated around US$ 40,000–60,000, showing an attractive economic benefit. The greater part of the expense was possibly due to the relatively high cost of the polyurethane foam used as the carrier (Deshusses and Webster 2000).

14.6.2.2 Republic of Korea

Close attention of the Korean government to environmental issues and changes in legislation have resulted in the rise of interest of commercial companies in biofiltration and odor control. More than 15 companies are active in the field, marketing various types of biofiltration equipment ranging from relatively simple to most sophisticated devices.

As an example, a range of biotrickling filters with capacities of 3,000–30,000 $m^3 h^{-1}$, developed by HanKi Industrial Ltd. in cooperation with scientists from the Bakh Institute of Biochemistry, Moscow, is presented here. These plants can be classified as container-type modular biotrickling gas-phase reactors utilizing a specially adapted consortia of microorganisms that can efficiently degrade a number of malodorous compounds, including sulfur-containing ones. The microorganisms are immobilized on a macroporous foam with a very high surface-to-volume ratio in the range 1,000–2,000 $m^2 m^{-3}$. The

Fig. 14.2. Trickling bioreactors manufactured by Han-Ki Industrial Ltd., RK. Model BioCat-50/250-I 12,000 $m^3 h^{-1}$ installed at water treatment plant in the town of Pusan. Recommended EBRT 2.0–3.0 s. **A** Front view; **B** rear view

plants are equipped with a preceding humidifier/heat exchanger that is incorporated into the biofilter frame, and a control and measurement system regulating various parameters, such as pH, nutrient supply, etc., and enabling continuous monitoring of the performance of the plant. The plants can be thermo-insulated, while the control system enables efficient biofilter performance under virtually any climatic conditions and operating regimes. It also enables long-term survival of the biofilter when it is not in use.

A stable and highly efficient consortium of microorganisms is an essential element of the technology. The strains utilized to build the consortium are very robust and comprise up to 60–90% of the total biomass, even after several years of biofilter operation. These microorganisms are highly stable, resistant to stress and invading microflora, and do not produce any secondary smell that may affect the biofilter performance.

Figure 14.2A, B presents the front and rear views of the trickling bioreactor rated for 12,000 $m^3 h^{-1}$ for deodoration of the exhausts of the water treatment plant in the town of Pusan. Figure 14.3 shows two 30,000 $m^3 h^{-1}$ machines installed at the Kayang sewage treatment plant, Seoul. The parameters of the two biotrickling filters are summarized in Table 14.7. Compositions of the emissions of Pusan and Kayang sites are very similar, and thus Table 14.8 presents data only for the Kayang plant that treated about three to ten times more concentrated emission. Table 14.9 provides an example from another plant for treating food waste in the town of Pusan, with a quite different composition of emitted volatiles. All the data presented show efficient (96–99%) removal of malodorous volatiles at contact times < 3 s.

At the Kayang plant, two bioreactors (30,000 m^3 each) with four driving fans (two work fans and two backup fans rated for 2,500 Pa) as well as all the other auxiliary equipment are located inside the interior of just one of the three concrete basins of the former compost biofilter. The decommissioned compost biofilter had a capacity of 48,000 $m^3 h^{-1}$ and comprised three basins (16,000 $m^3 h^{-1}$ per basin) as well as a separate building housing six high-

Fig. 14.3. Trickling bioreactors manufactured by Han-Ki Industrial Ltd., RK. Model BioCat-100/500-II 30,000 $m^3 h^{-1}$ at the Kayang sewage treatment plant, Seoul. Recommended EBRT 2.0–3.0 s

Table 14.7. Deodoration plants at waste-water treatment plants, HanKi Industrial Ltd., Kayang, Seoul and Pusan, RK

Parameter	Pusan BioCat-50/250-I	Kayang BioCat-100/500-II
Airflow (nominal; $m^3 h^{-1}$)	12,000	30,000
Dimensions (mm)	3,250 × 3,200 × 4,560	6,900 × 3,000 × 4,750
Footprint (mm)	4,180 × 3,650	8,100 × 3,600
Carrier	Foam packing	Foam packing
Bed arrangement	Three layers	Six layers
Operating weight (t)	7.6	12.3
Bed temperature (°C)	15–20	15–20
Pressure drop (Pa)	750	1,250
Superficial air velocity ($m s^{-1}$)	0.62	1.3
Contact time (s)	2.4	2.3
Start year	May 2001	December 2000
Average efficiency (%)	> 96–99	> 97–99

pressure fans, 6,500 Pa each. This illustrates the comparative intensity of the two similar processes – optimized trickling gas-phase reactor and conventional-type container biofiltration.

At present, HanKi Industrial markets a range of Bio-Cat plants for odor removal purposes with a throughput from several thousand up to 50,000 $m^3 h^{-1}$. Rough estimates of their operational parameters show that high-intensity biofiltration devices provide an attractive alternative to conventional biofiltration. The water consumption is around 0.008–0.014 $l m^{-3}$ purified air, and

Table 14.8. Average efficiency of waste air deodoration at the wastewater treatment plant, Kayang, Seoul, RK, using BioCat-100/500-II (30,000 $m^3 h^{-1}$, EBRT 2.3 s)

Component	Concentration ($\mu g m^{-3}$) Inlet	Outlet	Removal efficiency (%)
Ammonia (NH_3)	1,873	61	96.7
Trimethyl amine [$(CH_3)_3N$]	974	5	99.4
Acetaldehyde (C_2H_4O)	1,150	n.d.[a]	> 99.9
Styrene ($C_6H_5CH{=}CH_2$)	130	n.d.	> 99.9
Hydrogen sulfide (H_2S)	23,210	61	99.8
Methyl mercaptan (CH_3SH)	1,713	41	97.6
Dimethyl sulfide [$(CH_3)_2S$]	111	n.d.	> 99.9
Dimethyl disulfide [$(CH_3)_2S_2$]	n.d.	n.d.	–

[a] n.d., Not detected by employed assay technique

Table 14.9. Point assay of the efficiency of waste air deodoration by BioCat-50/250-I trickling bioreactor (12,000 $m^3 h^{-1}$, EBRT 2.4 s), food waste treatment plant, Pusan, RK

Pollutant	C_{in} (ppb)	C_{out} (ppb)	Removal (%)	MHS[a] (ppb)
Acetaldehyde[b]	115.6	1.3	98.9	1.5
Ethanol	418.0	3.3	99.2	94
Dimethyl sulfide(DMS)	10.6	0.0	100.0	3
n-Propanol	65.7	0.0	100.0	26,000
Isopropanol	58.2	0.0	100.0	38
n-Pentane	8.9	0.0	100.0	1,400
Dimethyl disulfide(DMDS)	9.1	0.0	100.0	3
2-Methyl pentane (MIBK)	2.9	0.0	100.0	7,000
Methyl ethyl ketone (MEK)	27.6	0.0	100.0	440
Hexane	9.3	0.0	100.0	1,500
Ethyl acetate	13.8	0.0	100.0	870
Methyl cyclopentane	1.1	0.0	100.0	1,700
2-Butanol	47.2	0.0	100.0	–
2-Methyl hexane	1.7	0.0	100.0	420
3-Methyl hexane	3.7	0.0	100.0	840
cis-1,3-dimethyl cyclopentane	23.3	4.7	79.9	–
trans-1,2-dimethyl cyclopentane	13.2	1.8	86.4	–
Trans-1,3-dimethyl cyclopentane	12.2	1.9	84.7	–
n-Heptane	7.0	0.0	100.0	670
Methyl cyclohexane	12.2	0.0	100.0	150
Ethyl cyclohexane	3.5	0.0	100.0	1,700
2,3-Dimethyl hexane	3.1	0.0	100.0	–
1,2,4-Trimethyl cyclopentane	0.8	0.0	100.0	110
Toluene	63.1	4.5	92.9	330
2-Methyl heptane	10.8	1.0	90.5	110
4-Methyl heptane	3.8	0.0	100.0	110
n-Octane	25.0	2.8	88.8	1,700
2,5-Dimethyl heptane	3.0	0.0	100.0	–
2,6-Dimethyl heptane	4.6	0.0	100.0	–
Ethyl cyclohexane	36.7	4.1	88.7	1,700
Ethyl benzene	23.0	0.0	100.0	170
m,p-Xylene	26.3	0.5	97.9	41
o-Xylene	18.6	0.0	100.0	380
n-Nonane	9.2	0.0	100.0	2,200
n-Decane	6.5	0.0	100.0	870
D,L-Limonene	729.5	0.0	100.0	38
Total/average	1,977	85.3	95.7	–

[a] Minimal level of human sensitivity

[b] Pollutants that exceed MHS are indicated in *italic*

electricity (without fan) is from 0.2 (no heating) to 0.5 W m^{-3} (with heating). At high H_2S concentration (> 10–12 ppm), 5–6 mg Na_2CO_3 m^{-3} purified air is required. The plants can operate without servicing for months. The approximate cost-breakdown for the plant rated for 15,000–30,000 m^3 h^{-1} is as follows: main unit and humidifier, 50–60%; carrier, ~ 25%; control unit and sensors, 5–7%; pumps and fans, 12–15%; biomass, < 1%. The overall cost to the end-user is estimated around US$ 6 m^{-3} h^{-1} of installed capacity.

14.6.2.3 *Russia*

Russian legislation in the field of air emissions is as strict as the European one, and in some cases even stricter. Unfortunately, the poorly structured economic reforms of the 1990s as well as the financial crisis of 1998 delayed wide introduction of biofiltration techniques into the industrial field. Russia was one of the pioneers of the high-intensive biofiltration. Several groups of scientists and engineers have been active in the field since the 1980s. Plants with contact times in a range < 10 s to control VOCs and treat odors have been erected in Russia since the early 1990s (Popov et al. 2002; Vinarov et al. 2002). Industrial plants to control styrene emission and ethyl acetate-containing emissions have been described in detail elsewhere (Popov et al. 2000). We are reporting here on the first Russian low-contact time gas-phase bioreactor to control malodorous emissions of a wastewater treatment plant.

Kur'anovsk aeration station is the biggest Moscow wastewater treatment plant. With a capacity of 400,000 m^3 day^{-1}, it occupies 500 hectares and is located close to densely populated urban areas. Notwithstanding the absence of significant complaints from the local population, the administration of the station has nevertheless considered a range of preventive measures to reduce potential malodorous air emissions. A number of options have been tested in past years, including conventional biological methods for air treatment. However, the large volumes of air to be treated prompted a search for a high-intensity solution. The Bakh Institute of Biochemistry, Moscow, in co-operation with Mosvodokanal (the sole Moscow water supply and wastewater treatment organization), started a long-term project aimed at providing an economic solution to the problem. As a starting point, a version of the deodoration plant successfully implemented by Bakh Institute in Korea and for the MARS pet-food facility (Zhukov et al. 1998) was adopted.

Pilot tests were carried out during the third and fourth quarters of 2003 at the residue sedimenting and dewatering site at the Kur'anovsk station, characterized by the highest local level of H_2S emission. A voluminous fibrous carrier with high specific surface area was used as a support for the microbial consortium. The results of the tests verified earlier observations concerning the possibility of removal of high H_2S concentrations at relatively low contact times (< 3 s; Fig. 14.4). In about a month, the pilot plant reached its optimal performance and operated stably until the end of the year. The average H_2S concentration was 12.6 mg m^{-3}, while average removal was 95%.

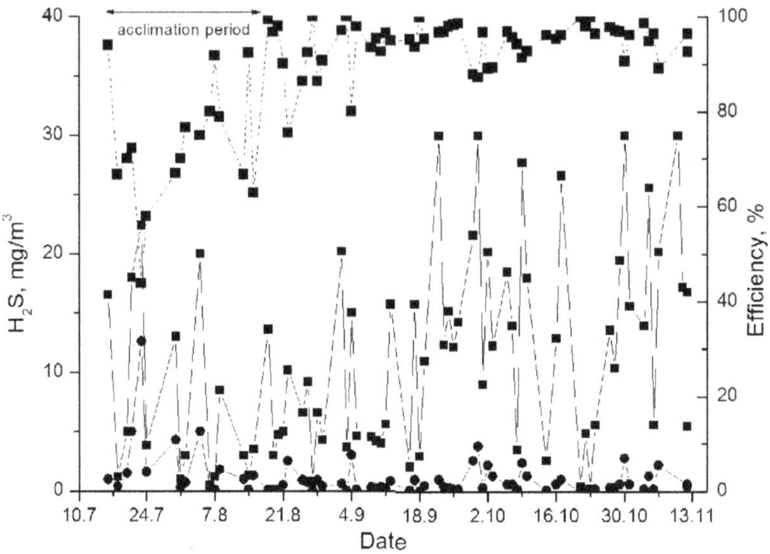

Fig. 14.4. Startup and performance of the trickling bioreactor for odor control operating at Kur'anovsk water treatment plant, Moscow. EBRT 3.0 s

Following successful pilot tests, a full-scale prototype system rated for 10,000–12,000 $m^3\,h^{-1}$ is under construction and will be put into operation in 2004. It will also serve as an experimental site for further improvement in the technology.

14.7 Conclusions

A number of general criteria can be formulated for biofiltration plants. They should (1) comply with local safety and environment legislation, (2) operate with minimal manual interference, (3) allow easy and infrequent maintenance, (4) have long service life, should be non-susceptible to clogging and develop low backpressures, (5) comprise stable and robust microbial consortia, and (6) be cost-competitive against existing alternative technologies. Plants dedicated to odor removal should satisfy some additional requirements. The most important is that they should employ microorganisms and materials that themselves do not generate additional annoying odors.

Recent advances in the field of biofiltration show that high-performance abatement of odors originating both from municipal facilities as well as various industries and businesses at contact times $< 3-6$ s is feasible. Introduction of new media, enabling fast mass transfer as well as efficient consortia of microorganisms, facilitate both reduction of contact times and overall dimensions and footprints of industrial abatement plants.

References

Cioci F, Lavecchia R, Ferranti MM (1997) High-performance microbial removal of ethanol from contaminated air. Biotechnol Tech 11:893–898

Cox HJ, Deshusses MA (1998) Biological waste air treatment in biotrickling filters. Curr Opin Biotechnol 9:256–262

Cox HJ, Deshusses MA (2000) Biotrickling filters. In: Kennes C, Veiga MC (eds) Bioreactors for waste gas treatment. Kluwer, Dordrecht, pp 99–131

Defoer N, Bo D, Langenhove V, Dewulf J, Elst V (2002) Gas chromatography-mass spectrometry as a tool for estimating odor concentrations of biofilter effluents at aerobic composting and rendering plants. J Chromatogr A 970:259–273

Deshusses MA, Webster TS (2000) Construction and economics of a pilot/full-scale biotrickling filter reactor for the removal of VOCs from polluted air. J Air Waste Manage Assoc 50:1947–1956

Devinny JS, Deshusses MA, Webster TS (1999) Biofiltration for air pollution control. CRC Lewis, Boca Raton

Ferranti MM (2001) Formaldehyde biological removal from exhaust air in the composite panel board industry from pilot tests to industrial plant. In: Proc 35th Int Particleboard Composite Materials Symp, 2–5 April 2001, Pullmann, WA

Ferranti MM, Conca F (2000) Formaldehyde biological removal from exhaust air in the composite panel board industry from pilot tests to industrial plant. In: Proc Air and Waste Management Association 93th Annu Conf, 18–22 June 2000, Salt Lake City, Utah

Gabriel D, Deshusses MA (2003) Retrofitting existing chemical scrubbers to biotrickling filters for H_2S emission control. Proc Natl Acad Sci USA 100:6308–6312

Gabriel D, Cox HHJ, Brown J, Torres E, Deshusses MA (2002) Biotrickling filters for odor treatment at POTWs: full-scale experience with a converted scrubber. In: Proc Air and Waste Management Association 95th Annu Conf, 23–27 June 2002, Baltimore, Pap no 43558

Park SJ, Nam SL, Choi ES (2001) Removal of odor emitted from composting facilities using a porous ceramic biofilter. Water Sci Technol 44:301–308

Popov VO, Bezborodov AM, Cross P, Murphy A (2000) Industrial trickling bed biofilters for abatement of VOCs from air emissions. In: Wise D (ed) Remediation of hazardous waste contaminated soils, 2nd edn. Decker, New York, pp 449–473

Popov VO, Bezborodov AM, Cross P, Jackson W (2002) Design, construction and long-term performance of novel type of industrial biotrickling filters for VOC control. In: Proc Air and Waste Management Association 95th Annu Conf, 23–27 June 2002, Baltimore, Pap no 42602

Popov VO, Bezborodov AM, Cavanagh M, Cross P (2004) Evaluation of industrial biotrickling filter at the flexographic printing facility. Environ Prog (in press)

Shareefdeen Z, Herner B, Webb D, Wilson S (2003) Biofiltration eliminates nuisance chemical odors from industrial air streams. J Ind Microbiol Biotechnol 30:168–174

Vinarov AI, Rodysheva ZN, Smirnov VN, Sokolov DP (2002) Stability of microbial association in the process of industrial biofiltration cleaning of gaseous discharges. Appl Biochem Microbiol (Russia) 38:523–528

Zhukov VG, Rogojin IS, Ushakova NA, Zagustina NA, Popov VO, Bezborodov AM (1998) Development of the technology of air deodoration and its field testing with the use of a pilot plant. Appl Biochem Microbiol (Russia) 34:370–376

15 Odor Removal in Municipal Wastewater Treatment Plants – Case Studies

Todd S. Webster

15.1 Introduction

This chapter presents eight different case studies from both designers and end users of biological odor control technologies at municipal wastewater treatment plants. These case studies represent typical applications in the field. However, since odors are generated from the various stages of the wastewater treatment process, as well as at collection and pumping stations, the system designs and airstreams treated are individually unique. The studies are provided based on published information and personal communication with the system designers and end users. No endorsement for any of the systems is implied. For each case study, details regarding system design, operation, and performance are provided. Also, a summary table for each application is included. The reader is encouraged to seek out more information with the individual designer and/or end user of each system.

15.2 An Odor Control Biofilter Located Within a Sewer Manhole Cover

In this case study, odor complaints occurred at an apartment complex within the City of San Diego. A gravity sewer line runs down a canyon and vents odorous sewer air from one of its manholes located on the parking lot of the complex, approximately 9 m from the balconies of the nearest apartments.

The City of San Diego had tried several methods to resolve the odor problems at this site. An activated carbon drum was installed at the parking lot, but the residents were not happy with the sight of it and complained about the blower noise. The City then started chemical treatment of the line. Since the reaction of the chemical was flow-sensitive, the City had to treat the whole line upstream of the manhole to eliminate hydrogen sulfide (H_2S) emissions at the site. In addition to being quite costly, the chemical treatment was not very effective for the treatment of organic sewer odors, and the City continued receiving complaints. The last resort seemed to be the sealing of the manhole, but this caused odors to back up into the vent pipes of homes. Finally, the City decided to install a Bioteg Manhole Biofilter in this manhole in 2001, which allowed them to stop their chemical treatment and reopen the manhole, allowing it to vent.

Shareefdeen/Singh (Eds.)
Biotechnology for Odor and Air Pollution Control

15.2.1
Design

Bioteg Manhole Biofilters have been used successfully throughout Europe for the last 8 years. They were originally developed in cooperation with collection system supervisors to provide a cost-effective and environmentally friendly solution to spot odor problems along sewer lines. The original design was then adjusted to fit the American manhole design and dimensions.

The Bioteg Manhole Biofilter is a passive manhole insert that is suspended on a stainless-steel supporting ring underneath the manhole cover (Fig. 15.1). A flexible rubber gasket seals the gaps between the manhole frame and the filter. The odorous air from the sewer system enters the filter through its perforated bottom and is released through outlet vents in the top. The filter is covered to prevent the media from being contaminated by dirt, sand or oil from the street. The filter features a drainpipe with an odor flap that drains water and dirt runoff from the street back into the sewer system. Odors are biologically degraded as the sewer air passes through the filter material.

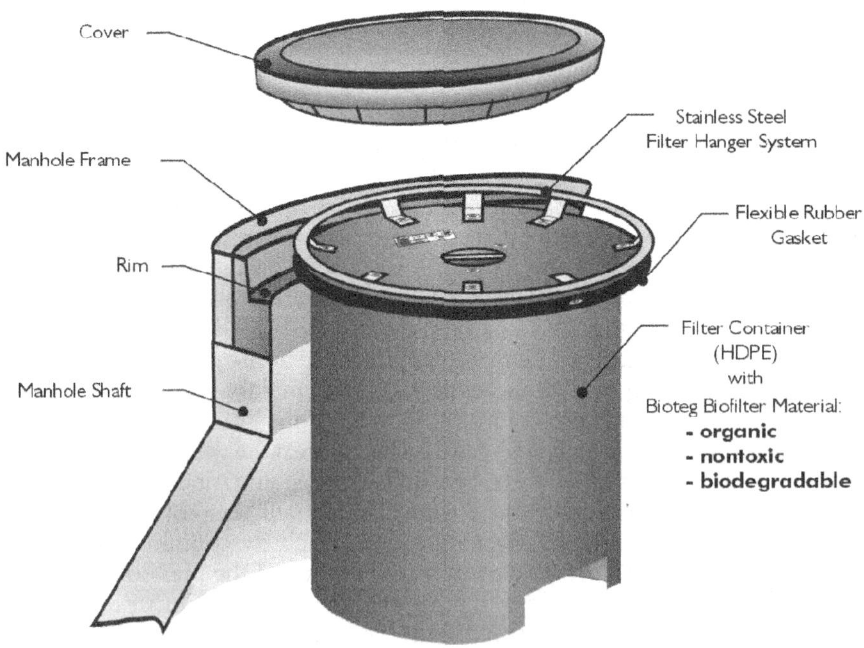

Fig. 15.1. Bioteg Manhole Biofilter schematic. (Drawing courtesy of Bioteg Biofilter Systems; San Diego, CA)

15.2.2 Operation

This passive biofilter operates automatically, with no or very little maintenance. The odorous air from the sewer system enters the filter through its perforated bottom. The very low pressure drop of the biofilter material allows the air to gradually rise up and through the filter. The odors are absorbed into the biofilm surrounding the media and degraded by microorganisms. The clean air then exits the filter through vent outlets in the top and is released through the manhole cover into the atmosphere. The biofilter system does not require power or a water hookup. However, the medium does need to be kept moist, since the odor-degrading microbes only thrive in a humid environment. Under normal conditions and in moderate climates, the medium maintains its moisture from the incoming humidity in the sewer system, which is usually near saturation. Some manholes, however, "breathe" fresh air in at certain time periods. This can be critical in dry climates, since it will eventually dry out the filter medium from top to bottom. This phenomenon was observed during the summer at the site in San Diego. The City re-moisturized the filter, and established a maintenance plan to go out to the site every 4–6 weeks to manually pour 4–7.5 l water over the filter. The filter performance fully recovered (I. Mello 2003, pers. comm., Bioteg Biofilter Systems, LLC, San Diego, CA).

The biofilter material is expected to last up to seven years, depending on the environmental conditions and use. After this time, the filter material can be easily replaced. The used filter material is non-hazardous and can be readily composted, with no need for further treatment or disposal.

15.2.3 Performance

The system provides an economical and effective solution to unpleasant street-level sewer odors (Table 15.1). They are, however, limited by their size or the size of the manhole. After application of the unit, odor complaints have ceased. The biofilter reactor effectively eliminates sewer odors from sites with low to medium inlet concentrations (> 95% removal). At sites with high inlet concentrations, the filters will effectively reduce concentrations to very low levels, which in most cases are sufficient to eradicate odor problems. The system is also capable of treating organic sewer odors that are often the main odor source from manholes. Unlike activated carbon, the system continues to perform efficiently even after treating extremely high spikes in concentrations, without the need for a medium change out.

15.3 Multiple Biofilter Application Treating Odors from a Headworks Operation

Pima County Wastewater Management (Arizona) has constructed a new 2×10^6 gallon per day addition to the wastewater treatment plant in Green

Table 15.1. Characteristics of the Bioteg Manhole Biofilter

Reactor type	Bioteg MBF-22 Manhole Biofilter
Owner and location	City of San Diego, CA, Mission Center Road
Manufacturer	Bioteg GmbH Germany
Year of installation	2001
Type of airstream	Sewer off-gas
Reactor dimensions and construction type	0.56-m diameter HDPE vessel, 0.6 m in height
Medium type and volume of medium	Bioteg organic bpc BT-50 biofilter material, approx. 0.11 m^3
Height and number of layers of medium	Approx. 0.51-m medium depth
Airflow rate	Approx. 8.5 $m^3 h^{-1}$
Empty bed residence time	48 s
Pressure drop	Less than 0.5-cm water gauge through complete unit
Average bed temperature	N/A
Contaminants treated	H_2S, organic sewer odors
System controls	None
Design and acceptance criterion	Design based on manhole dimensions, airflow rate not to exceed 15 $m^3 h^{-1}$, best results at concentrations $<$ 50 ppmv
Capital costs	\$ 1,000
Approximate operating costs/treatment costs per 1,000 m^3 off-gas treated	N/A
Typical performance	$>$ 95% removal efficiency of H_2S from concentrations below 30 ppmv 85–95% removal efficiency of H_2S from concentrations above 30 ppmv Simultaneous removal of organic sewer odors Media life is not affected by high spikes in concentrations

Valley, Arizona. The close proximity to residential neighborhoods was an impetus for Pima County to construct a form of plant odor control. For this reason, in October, 2003, Shaw Environmental provided a dual P-600 biofilter to treat the off-gases generated from the headworks of the treatment plant.

15.3.1 Design

For this application, the Shaw Environmental dual P-600 Biofilters were designed to treat 3,570 $m^3 h^{-1}$ (1,785 $m^3 h^{-1}$ each) of air at hydrogen sulfide concentrations up to 10 ppmv (Fig. 15.2). The reactors are bottom loaded and contain approximately 17 m^3 of media (each). The lower media level consists of a 0.1-m layer of porous, inorganic, granular medium (HUMFIL) that is placed on top of a mesh mat. Above this inorganic layer, an organic, biologically active layer of medium (VAMFIL) is added to a height up to 1.2 m. VAMFIL medium is a proprietary mixture of select tree barks and compost that have been treated to provide a maximum level of porosity and biological activity, low pressure drop, and predictable long life. The upper medium layer is also amended with chemical fertilizer(s) that is washed into the lower sections of the bed with the water from the upper spray irrigation system during startup. Irrigation systems are placed within and above the media, thereby assuring that moisture and nutrients are directed to the specific locations within the biofilter where they will provide optimum benefit. The dual P-600 reactors are covered with a pre-manufactured FRP cover section. This minimizes the effects of weather on the operation of the system.

Fig. 15.2. Shaw Environmental dual P-600 biofilter assembly treating headworks off-gases. (Photograph courtesy of Shaw Environmental, Inc., Baton Rouge, LA)

15.3.2 Operation and Performance

The system began operation in October 2003 (Table 15.2). Because the plant was newly constructed and minimal wastewater had been treated, the biofilters were subject to hydrogen sulfide concentrations of less than 1 ppmv from the headworks operation. For this reason, it was necessary to spike the system with compressed hydrogen sulfide gas to properly acclimate the system. The spiking system provided 10–20 ppmv during the 3-week acclimation period. During this time, daily inlet and outlet hydrogen sulfide concentration measurements were obtained using a Jerome Hydrogen Sulfide Analyzer (Tempe, AZ). Initial testing results demonstrated approximately 50% removal of the generated spike. After 10 days of operation, the system performance was greater than 99% removal of the hydrogen sulfide odor.

An interesting point to note about this site was that the dual P600 system was delivered late in 2002. Due to plant construction issues unrelated to the odor control devices, the biofilters were not operated for at least 10 months after delivery. During this time, the operators at the plant would occasionally check the bed moisture and, if necessary, spray the media with a potable water source. The robustness of the biofilter technology is clearly demonstrated based on the long period of no operation, followed by the startup and rapid acclimation of the system to treat the artificially generated odors.

15.4 Multiple Biofilter Application (High Flow) at a Wastewater Pumping Station

The City of Birmingham (Alabama) retained BIOREM Technologies Inc. (Guelph, ON, Canada) to provide a biological odor control system for a wastewater pumping station. After a continuous (2-year) laboratory pilot study, concentration ranges and empty bed residence time relationships were established for the complete removal of hydrogen sulfide from waste airstreams. From such testing, a total of six biofilter units were installed at this site for a total treatment of 51,000 $m^3\ h^{-1}$ of waste air containing primarily odorous hydrogen sulfide.

15.4.1 Design and Operation

This biofilter system is designed to operate with a 20-s empty bed residence time, using permanent BIOSORBENS™ media. The inlet hydrogen sulfide concentration varies in the range 5–25 ppmv. Each reactor contains a single 1.5-m layer of medium, and a total medium volume of 45 m^3 (Fig. 15.3).

15.4.2 Performance

This biofilter system effectively removes more than 99% of the hydrogen sulfide (Table 15.3). Regular monitoring of the system is an essential component of

Table 15.2. Characteristics of the Shaw Environmental multiple biofilter application

Reactor type	Biofilter
Owner and location	Pima County, Green Valley Wastewater Treatment Plant, Green Valley, Arizona
Manufacturer	Shaw Environmental
Year of installation	2003
Type of airstream	Air from the headworks of a wastewater treatment plant
Reactor dimensions and construction type	Two P-600 reactors in parallel, FRP body with steel plenum
Medium type and volume of medium	Wood bark-based medium (VAMFIL). Each unit has approx. 17 m^3 of medium
Height and number of layers of medium	Media height is 1.2 m (0.1-m inorganic layer of media and a 1.1-m organic layer of media)
Airflow rate	Total airflow is 3,570 $m^3\,h^{-1}$
Empty bed residence time	$\sim$ 34 s per reactor
Pressure drop	Pressure drop is less than 1-cm water gauge per reactor. Over time, this value will increase
Average bed temperature	16–35 °C
Contaminants treated	Hydrogen sulfide at a concentration less than 10 ppmv
System controls	Water irrigation timers and nutrient delivery system (pumpless). Low- and high-pressure switches to shut down the system in case of low flow conditions
Design and acceptance criterion	Designed to remove greater than 99% of hydrogen sulfide
Capital costs	Typical dual P-600 units are \$ 110,000. An additional \$ 140,000 was added for extensive performance monitoring, a spiking system, and specialized parts to meet strict specifications of the contract
Approximate operating costs/treatment costs per 1,000 m^3 off-gas treated	Labor costs are minimal and plant water is used to maintain appropriate soil moisture. Factoring in electricity, water, nutrients, and manpower, operating costs are approximately \$ 0.06 per 1,000 m^3 off-gas treated
Typical performance	Greater than 99% removal of hydrogen sulfide

Fig. 15.3. BIOREM multiple biofilter treatment system. (Photograph courtesy of Biorem Technologies, Inc., Guelph, Ontario, Canada)

preventative maintenance, reducing potentially costly and time-consuming corrective actions. Parameters such as odor or contaminant removal efficiency, process air moisture, relative humidity, process air temperature, airflow, airflow distribution, pressure differential across the biofilter media, and media bed temperature are routinely monitored.

In order to maintain high removal and smooth operation of the biofilter, a monitoring service package is also provided for this system by the vendor. At quarterly time intervals, the vendor supplies a medium sampling kit with complete sampling instructions. Representative medium samples obtained by the reactor owner are analyzed at the vendor's laboratory for parameters such as bacterial enumeration, medium moisture content, pH of the biofilter material, nutrient characterization, mineral deposit quantification, and particle-size analysis.

15.5 A Single Biofilter Application (Low Flow) at a Wastewater Pumping Station

A large pump station is located in a remote, high-profile, expensive home area of Chapel Hill, North Carolina. The wastewater treatment plant was receiving numerous complaints from nearby residents. City officials discussed treatment options, including building their own in-ground biofilter. A prepackaged biofilter system, designed by Bio Reaction Industries, was the final choice to expedite odor reduction at the pumping station.

Table 15.3. Characteristics of the BIOREM multiple biofilter application

Reactor type	Biofilter
Owner and location	Jefferson County, Shades Valley, AL
Manufacturer	Biorem Technologies Inc., Guelph, ON, Canada
Year of installation	2002
Type of airstream	Municipal air
Reactor dimensions and construction type	Six BASYS™ units; stainless-steel closed vessels in parallel; dimensions $12 \times 2.4 \times 2.4$ m ($L \times W \times H$)
Medium type and volume of medium	Inorganic permanent BIOSORBENS™ media, 45 m^3 $unit^{-1}$
Height and number of layers of medium	1.5 m, single layer
Airflow rate	8,500 $m^3 h^{-1}$ per unit ×6 (51,000 $m^3 h^{-1}$)
Empty bed residence time	20 s
Pressure drop	N/A
Average bed temperature	0–43 °C
Contaminants treated	5 ppmv H_2S, average (25 ppmv H_2S, maximum)
System controls	N/A
Design and acceptance criterion	99% removal of H_2S
Capital costs	$ 1.3 million (Canadian)
Typical performance	Greater than 99% removal of hydrogen sulfide

15.5.1 Design and Operation

The owner requested a small pilot system to evaluate the performance of the system. An Apollo S50 system was delivered in January 2002 for a 60-day trial. During this period of operation, the neighbors' complaints stopped, despite the fact that the unit was undersized and only removing a portion (51 $m^3 h^{-1}$) of the entire airstream through the biofilter.

As a result of the success of the pilot system, a larger airflow design (340 $m^3 h^{-1}$) was agreed upon to meet the needs of the oversized wet well after calculating sufficient air changes (12×). The Apollo S200 was delivered and installed in March 2002, where it continues in present operation (Fig. 15.4). No added chemicals or nutrients are required for this system. Long-term operating issues have been primarily prevented by an annual maintenance contract provided by the vendor.

Fig. 15.4. BioReaction Industries single Apollo S200 biofilter system. (Photograph courtesy of BioReaction Industries, Tualatin, OR)

15.5.2 Performance

Performance criteria for the design were complete odor elimination at the fence-line of the property (Table 15.4). This was achieved within the first 48 h of startup. System operation was tested in August 2002, with an Arizona Instruments Jerome analyzer. Over a 2-day period, 19 data points were obtained. The results demonstrate greater than 99% removal of the inlet hydrogen sulfide concentrations (Fig. 15.5).

Table 15.4. Characteristics of BioReaction Industries biofilter

Reactor type	Biofilter
Owner and location	OWASA, Chapel Hill, NC
Manufacturer	BioReaction Industries, Tualatin, OR
Year of installation	2002
Type of airstream	Odors from two wet wells at lift station
Reactor dimensions and construction type	First system tested was 1.2 m in diameter by 1.5 m tall, of FRP construction; second system installed was 1.8 m in diameter by 2.1 m tall, of FRP construction
Medium type and volume of medium	3.2 m^3 of Bio-AIRSPHERES, an organic medium manufactured by Bio-Reaction
Height and number of layers of medium	One 1.2-m layer Bio-AIRSPHERES
Airflow rate	340 $m^3 h^{-1}$
Empty bed residence time	34 s
Pressure drop	25.4-cm water gauge through water sump, 12.7-cm water gauge through Bio-AIRSPHERES media (note: these pressure drops valid for Apollo S systems only; Apollo A systems on airflows > 340 $m^3 h^{-1}$ operate at approx. 20.3-cm water gauge)
Average bed temperature	18 °C
Contaminants treated	H_2S and other organics
System controls	Continuous monitoring of airflow, air and bed temperature, pressure drop and automated watering/blowdown
Design and acceptance criterion	Design was based on successful pilot study and wet well sizing calculations
Capital costs	$ 16,470 (incl. add-on's at customer request – gel coating, power requirements/upgraded control panel)
Approximate operating costs/treatment costs per 1,000 m^3 off-gas treated	$ 1,530 annually for electrical costs, or $ 0.36 per 1,000 m^3 off-gas treated. This accounts for full-time blower operation, with intermittent irrigation pump usage. No addition of chemicals or nutrients necessary. Maintenance costs are minimal, averaging < 1 h $week^{-1}$ of operation
Typical performance	Greater than 99% removal of hydrogen sulfide Other organics reduced to non-detection levels

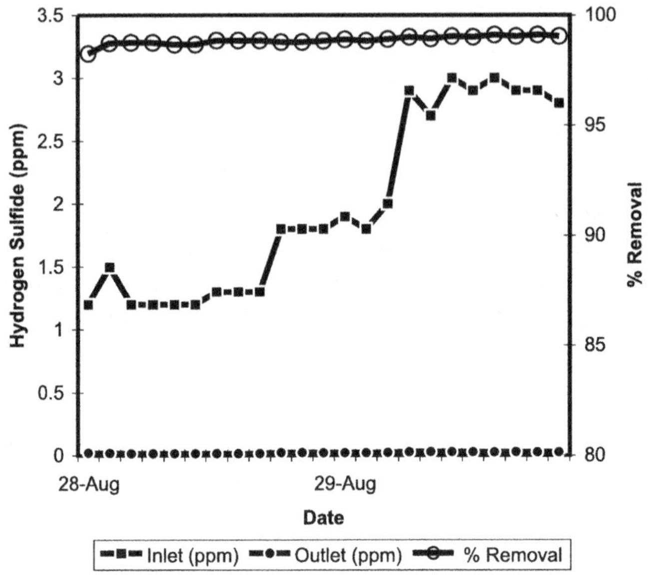

Fig. 15.5. Hydrogen sulfide removal versus time for the Apollo S200 biofilter system after 5 months of operation. (Graph courtesy of BioReaction Industries, Tualatin, OR)

15.6 Single Biofilter at a Wastewater Pumping Station Operated Under Varying Air Temperatures

The Long Quarter Pumping Station in Baltimore County, Maryland, has received persistent odor complaints from residents in the neighborhood. The pumping station is located in a low spot and receives wastewater from the surrounding area, as well as wastewater from other pumping stations. Without ventilation, static hydrogen sulfide concentrations in the wet well have been observed to exceed 200 ppmv during the summer months.

To ensure that there would be no further odor complaints from the residents, the County decided to employ an active odor control system that ventilates and treats the airborne odors at the pumping station. Considering the air volume of the wet well and an air exchange rate of four air changes per hour, it was determined that a 1,360–1,700 $m^3\,h^{-1}$ CFM system would be required.

15.6.1 Design and Operation

Baltimore County decided to test a Biocube Model BMF1000114 Biofilter (Fig. 15.6). Other alternatives such as chemical scrubbing and activated carbon adsorption were considered. Biofiltration was chosen because of its low operating cost and its safe, chemical-free nature. The vendor provided the system

Fig. 15.6. Biocube biofilter system treating low flow waste gases. (Photograph courtesy of Biocube, Inc., Victor, NY)

on a rental basis, with a purchase option after a 4-month trial period. During this 4-month period, the vendor agreed to perform tests to demonstrate the system's effectiveness.

Installation of the biofilter began in late August 2002, and the system was made fully operational by September 2002 (Table 15.5). Unfortunately, installation and instrumentation problems hindered the testing protocol, so reliable test data were not provided until late November. Using Odolog Hydrogen Sulfide Meters provided by Detection Instruments, Inc. (Phoenix, AZ), inlet and outlet H_2S levels were analyzed. A 0–1,000 ppmv meter was used on the inlet side of the system, and a 0–50 ppmv meter was used on the outlet side.

15.6.2 Performance

At normal November (winter time) concentration levels, over a 48-h period, the average inlet H_2S concentration was only 4.06 ppmv, with a maximum of 26 ppmv. During the same time period, the average outlet H_2S concentration was 0.02 ppmv, with a peak of 0.1 ppmv. Thus, the average removal was 99.5%, and the removal at peak load was 99.6%. The inlet temperature of the wet well gas varied in the range 15–19 °C, while the biofilter gas outlet temperature varied from 26 to 28 °C (Heuckeroth 2002). The outlet temperature correlates very well with the biofilter core temperature. The temperature rise is mostly due to heat transfer from the humidifier water heater.

Although these data demonstrated exceptionally good cold-weather operation, Baltimore County required data that demonstrated performance at summer conditions in order to approve the system. Using a low-pressure regulator,

Table 15.5. Characteristics of the Biocube biofilter

Reactor type	Compost-based biofilter with temperature and humidity controls
Owner and location	Baltimore County, Long Quarter Sewerage Pumping Station
Manufacturer	Biocube, Inc., Victor, NY
Year of installation	2002
Type of airstream	Sewerage gas (primarily hydrogen sulfide)
Reactor dimensions and construction type	3.7-m OD/3.45-m ID ×2.7 m tall, insulated dual wall vinyl ester fiberglass with internal media supports molded from 35,000 kPa tensile strength GE Noryl PPX resin
Medium type and volume of medium	Proprietary compost-based mixture
Height and number of layers of medium	Four layers, 0.3-m media depth per layer, 2.8 m^3 of medium per layer
Airflow rate	1,700 $m^3\,h^{-1}$
Empty bed residence time	24 s
Pressure drop	10–15 cm water gauge
Average bed temperature	29 °C summer, 21 °C winter
Contaminants treated	0–200 ppmv H_2S
System controls	Temperature and humidity controls
Design and acceptance criterion	Acceptance criteria: > 99% H_2S removal at 200 ppmv, 1,700 $m^3\,h^{-1}$ Actual performance at acceptance test: 99.5% at 213 ppmv, 1,700 $m^3\,h^{-1}$
Capital costs	$ 89,000
Approximate operating costs/treatment costs per 1,000 m^3 off-gas treated	Costs for media were estimated at about $ 1,200 per year, i.e., $ 0.08 per 1,000 m^3 off-gas treated. Electricity and maintenance costs were not available but were estimated to be similar to those of a carbon adsorption system that would otherwise be used if the biofilter was not available
Typical performance	> 99.5% at 0–200 ppmv, 1,700 $m^3\,h^{-1}$

compressed H_2S gas was metered into the inlet of the system to simulate worst-case inlet concentrations (near 100 ppmv). It is noteworthy that the spikes above 200 ppmv H_2S measured by Baltimore County in the past were under non-ventilated conditions. Actual concentrations under ventilated conditions have been projected to be less than 100 ppmv H_2S. Hence, the compressed hydrogen sulfide was supplied near 100 ppmv.

On the first day of testing, the biofilter operated at about 23 °C and achieved approximately 98% removal. Adjustments were made to the system to increase the operational temperature of the process. The temperature increased from 23 to 29 °C during a 6-h period, with inlet H_2S concentrations remaining relatively stable during this period, in the range 94–148 ppmv with an average of 122 ppmv. Concurrently, the system performance increased from 98.4 to 99.7% removal. These data clearly show the importance of temperature on the biological process. From these data, it is demonstrated that to produce greater than 99% removal during the summer months, an operational temperature above 27 °C would be required. This is not problematic, as the air temperatures in the sewer are often above 27 °C during the summer months (Heuckeroth 2002).

An additional test was performed to evaluate the biofilter performance under spike loading conditions. Previous testing showed that the concentrations in the wet well under normal conditions can have spikes in concentration that are 2–5 times the average. It was important for the County to know how well the biofilter could handle spike loads. The system was operated in the range 40–50 ppmv, and then subjected to a 200-ppmv spike. During the administration of the inlet spike, a handheld instrument was used for adjustment. The concentrations on this instrument actually went above 300 ppmv for a short period of time (< 1 min). The outlet showed 97.5% removal at the peak of the spike, with a background removal exceeding 99%. During this test, the ambient temperature that day had fallen to below 0 °C, and the system heater was only able to maintain a 25 °C operational temperature in the biofilter. Although there was a slight breakthrough during the spike load, at higher operational temperatures it is estimated that there would be no breakthrough. Indeed, tests previously performed at other sites of the vendor have demonstrated 99% removal with spikes above 500 ppmv H_2S.

Finally, to supplement the cold-weather performance data collected in late 2002, a warm-weather test was performed during the summer of 2003 to determine the performance capabilities of the biofilter under optimal temperature conditions (the ambient air temperature was 27 °C). The results from this test confirmed that the system meets the worst-case design criteria of 1,700 $m^3 h^{-1}$, 200 ppmv H_2S with $> 99\%$ removal. This test also demonstrated that operation above 1,700 $m^3 h^{-1}$ is not advisable, even under low concentration conditions, as performance was below 96% removal (R. Heuckeroth 2003, pers. Comm., Heuckeroth Associates, Saugerties, NY).

15.7 Biofiltration of Odors at a Biosolids Handling Facility

The Los Angeles County Sanitation District (LACSD) operates one of the largest biosolids dewatering facilities in the United States at its Joint Water Pollution Control Plant (JWPCP), producing over 1.5×10^6 kg day^{-1} of anaerobically digested biosolids. At the biosolids handling facility, during dewatering, conveying, storage and loading to offsite transportation vehicles, significant amounts of odor are produced and released to the atmosphere.

The biosolids processing facility is particularly close to residential neighborhoods, commercial retail outlets, and major thoroughfares. The close proximity to sensitive receptors was the impetus for the LACSD to initiate an odor assessment study that would lead to methods for controlling biosolids odors at the JWPCP. The odor study included modeling that indicated the possibility of producing offsite odors – thus, that odor controls would be necessary. Various odor control technologies were compared, and biofiltration was selected as the optimal control method. Full-scale modular biofilters capable of treating over 34,000 $m^3\,h^{-1}$ of air were installed to control odors. Testing was conducted for 1 year to ascertain the performance of three different media to effectively remove odor and air contaminants. The most efficient odor-removing media, determined during the testing of the modular biofilters, would then be used in large biofilter systems designed to treat 340,000 $m^3\,h^{-1}$ of air from the biosolids handling facility.

15.7.1 Design and Operation

The LACSD fabricated the modular biofilters out of used, $12 \times 2.4 \times 2.4$ m cargo containers (Fig. 15.7). The biofilters are bottom loaded, and the tops of the cargo containers have been removed to minimize restrictions to airflow. A water sprinkler system is placed on the top for water addition. The biofilter system utilizes two 35-Hp blowers to push the air through the biofilters. Two humidification tanks, each 19,000 l, are located upstream to the biofilters to

Fig. 15.7. Los Angeles County Sanitation District multiple biofilter arrangement to treat off-gases from a biosolids handling facility. (Photograph courtesy of the Los Angeles County Sanitation Districts, Los Angeles, CA)

saturate the incoming air and remove large particulates before entering the biofilter.

Ten cargo containers were converted to biofilters, providing an empty bed residence time of 60 s for 34,000 $m^3 h^{-1}$ of air (Table 15.6). Each cargo container has a surface area of 30 m^2, with an air loading rate for each biofilter of 110 $m^3 h^{-1} m^{-2}$ medium. The medium depth of the biofilters is 6 ft. Three different types of media were tested to determine their effectiveness in the removal of odors, eliminating odor-causing compounds, removing volatile

Table 15.6. Characteristics of the LACSD biofilter reactors

Reactor type	Biofilter
Owner and location	Los Angeles County Sanitation Districts; Joint Water Pollution Control Plant (JWPCP) Carson, CA
Manufacturer	Los Angeles County Sanitation Districts
Year of installation	2002
Type of airstream	Air from a biosolids handling facility
Reactor dimensions and construction type	Ten reactors were constructed from $12.2 \times 2.4 \times 2.4$ m cargo containers. Tops of the containers were removed to minimize backpressure on airflow
Medium type and volume of medium	Three types of media were used: lava rock and composted biosolids, perlite and composted biosolids, and wood shavings, nugget bark and composted biosolids. Each unit had approx. 54 m^3 of medium
Height and number of layers of medium	Media height is 1.8 m, with only one bed layer
Airflow rate	Total air flow is 34,000 $m^3 h^{-1}$ or approx. 3,400 $m^3 h^{-1}$ per container unit
Empty bed residence time	$\sim$ 60 s
Pressure drop	Pressure drop is 23 cm water gauge for the lava-rock medium, and about 5–7.6 cm water gauge for the perlite and wood media
Average bed temperature	16–29 °C
Contaminants treated	Organic sulfur compounds, 1–3 mg sulfur m^{-3}: methyl mercaptan, dimethyl sulfide, and dimethyl disulfide. Other compounds present include ammonia and very low levels ($<$ 20 ppbv) of aromatics

Table 15.6. (continued)

Reactor type	Biofilter
System controls	Inlet temperature and outlet temperature and humidity are data logged, and two units media are monitored using soil moisture probes
Design and acceptance criterion	Design based on pilot plant tests at site. Designed to remove 90% of inlet odors
Capital costs	$ 100,000 for all ten units
Approximate operating costs/treatment costs per 1,000 m^3 off-gas treated	Labor costs are minimal, and plant water is used to maintain appropriate soil moisture. Electrical cost is approx. $ 30,000 for the two 30-Hp blowers
Typical performance	Greater than 90% removal of methyl mercaptan and ammonia Greater than 80% removal of dimethyl sulfide and dimethyl disulfide Greater than 70% of very low concentrations of aromatics

organic compounds (VOCs), and to ascertain changes in operational parameters over time. The types of biofilter media tested and their composition are lava rock (84%) and composted biosolids (16%), perlite (70%) and composted biosolids (30%), and wood shavings (65%), composted biosolids (20%) and nugget bark (15%). The first several biofilters were put into operation at the end of March 2002, and several biofilters were put into service each succeeding week until all were in operation by the mid-April 2002.

15.7.2 Performance

Odors associated with biosolids include ammonia, hydrogen sulfide, reduced sulfur compounds, amines, volatile fatty acid compounds, and VOCs. Inlet ammonia concentrations to the biofilters were low, averaging slightly below 2 ppmv. The perlite and wood media achieved better than 95% ammonia removal, while the lava-rock media removed greater than 90% of the ammonia. Hydrogen sulfide concentrations were very low in the inlet airstream, below 100 ppbv.

Gas chromatograph analyses for reduced sulfur compounds (using South Coast Air Quality Control Districts' Method 307-91) identified methyl mercaptan, dimethyl sulfide and dimethyl disulfide present in the air coming from the biosolids handling facility. All the media achieved effective removal of methyl mercaptan ($>$ 87%). Dimethyl sulfide removal was less effective, with the wood media achieving the best removals (80%). All three of the biofilter media appeared to be less effective in the removal of dimethyl disulfide (65% for the perlite), but the concentrations were very low (< 36 ppbv).

Odor sampling analysis was performed for the different types of biofilter media on a regular basis. The wood medium provided the best removals, with elimination of 90% of the inlet odor that averaged 1,470 D/T (dilution-to-threshold). The lava-rock and perlite media required longer startup periods (> 8 weeks) before good odor removal was achieved. The perlite and wood media were more consistent in odor removal than the rock media.

During the 1-year study of the biofilters, two areas of operational management became integral to maintaining dependable odor removal. The first was removing weeds from the top of the biofilters. If plants are allowed to grow on the top of the biofilter media, the roots allow air to channel through the biofilter media, and treatment does not occur. To eliminate the small plants before they become too large, weeding should be performed on a weekly basis. The second maintenance concern is the tendency of the biofilter media to shrink away from the sides of the cargo containers, allowing untreated air to escape along the walls. It is necessary, approximately once a month, to ensure that the media maintain contact at the location where they meet the walls of the cargo container.

From the testing results, two new, large biofilters are being constructed to treat the air from the JWPCP biosolids handling facility. They will be put into service in the middle of 2004. These biofilters will handle approximately 340,000 $m^3\,h^{-1}$ of air, with empty bed residence times of over 1 min. The air loading rates to these units will be lower than the modular biofilters at 55–73 $m^3\,h^{-1}\,m^{-2}$ medium. Wood-based media were selected for use in the biofilters, because they are effective in removing odors, and are less expensive and easier to obtain than the perlite-type media. However, the wood media will degrade faster than the other media, so they will require more frequent replacement (Morton et al. 2003).

15.8 An Intermittent Water Addition Biotrickling Filter Reactor

The neighborhood around the South Walton, Florida wastewater treatment plant is extremely sensitive to bad odors. This wastewater treatment plant is located in the middle of a residential housing area and close to a hotel and the beach (< 0.8 km). Numerous tourists visit South Walton during summer, when the generation of odors from the wastewater treatment plant is the greatest.

15.8.1 Design

A vendor, Bioway, was contracted to install a Zerochem™ odor control system (Fig. 15.8). The system treats an airflow of 1,700 $m^3\,h^{-1}$ containing numerous

Fig. 15.8. Bioway intermittent trickling filter odor control system. (Photograph courtesy of Bioway America, Inc., Marlton, NJ)

odorous compounds (Table 15.7). The average hydrogen sulfide concentration is 60 ppmv (max. 120 ppmv), and the odor concentration is about 4,000 D/T. Acceptable performance criteria for the odor control system are 99% removal of hydrogen sulfide (or lower than 0.5 ppmv), 90% removal of total organic reduced sulfur compounds, and 90% odor removal. Additional important criteria are the operational costs, the capability of the bioreactor to perform while subjected to acid conditions, maintenance requirements, and minimal offsite odor impact.

15.8.2 Operation

During the startup of this system, a 300-l vessel containing an acid-resistant pump and a water-level control was used. The vessel was first filled with water, nutrients, and microorganisms (combination of prepared inoculum and activated sludge). The pump recirculated discontinuously the water over the media in the bioreactor in order to have the microorganisms grow on the inert media. The water level of the 300-l vessel automatically filled back up when the water level reduced due to water evaporation from the bioreactor. The relative humidity of the outgoing air from the bioreactor was saturated with water (relative humidity is 100%), while the incoming air was normally not saturated (relative humidity is approximately 50–80%). During the first week of startup, the airflow was set at approximately 50% of the design airflow to minimize untreated odors from leaving the bioreactor. After the first week, the airflow was set at 100% of the design airflow (1,700 $m^3\,h^{-1}$). After a period of 3 weeks, the 300-l vessel (with the recirculation pump) was disconnected and removed, and the system control was changed over from startup operation to normal operation.

During normal operation, water (effluent water from the wastewater treatment plant) is discontinuously added to the bioreactor every 10–15 min. During the first 2 months of operation, nutrients were added to the influent water because the plant effluent was relatively clean. In subsequent months, the nutrient feed was stopped and only plant effluent water was used. The pH of the drain water from the bioreactor has been maintained at a pH of 2. Minimal maintenance is performed on the bioreactor, other than checking the blower operation and the strainers (removing small amounts of residual suspended

Table 15.7. Characteristics of the Bioway intermittent biotrickling filter reactor

Reactor type	Zerochem™
Owner and location	Municipality of South Walton
Manufacturer	Bioway bv, Ede, The Netherlands
Year of installation	2002
Type of airstream	Waste gas from the inlet channel, grit chamber, screening room, grit classifier and auxiliary pump station
Reactor dimensions and construction type	Zerochem™ type ZC3000 with control panel; housing material FRP, with PVC lining inside for the protection against sulfuric acid
Medium type	Permapac™ (structured, open, inert media, which is resistant to low pH conditions)
Height and number of layers of medium	Two layers
Airflow rate	Operated at 1,700 $m^3 h^{-1}$ and designed up to 2,000 $m^3 h^{-1}$
Pressure drop	< 75 Pa (< 0.76 cm water gauge)
Average bed temperature	20 °C
Contaminants treated	Odorous compounds, including hydrogen sulfide and reduced sulfur compounds
System controls	Automatic water and nutrient feed by PLC. Alarm generation for different unwanted situations, for example, low or no water flow or low nutrient level
Design and acceptance criterion	–
Capital costs	$ 66,400 – includes fan, installation on site, freezing protection, and startup

Table 15.7. (continued)

Reactor type	Zerochem™	
Approximate operating costs/treatment costs per 1,000 m^3 off-gas treated	Depreciation over ten years with an interest rate 5%:	\$ 8,600
	Energy (fan: 2,000 $m^3 h^{-1}$ at 0.3-in. water and control panel: 0.4 kW at \$ 0.04 kWh^{-1}):	\$ 251
	Water (none; effluent water is used):	\$ 0
	Nutrients (none; effluent water is used):	\$ 0
	Chemicals (none):	\$ 0
	Operating costs (1/3 h per week at \$ 50 h^{-1}):	\$ 833
	Media change replacement:	\$ 0
	Maintenance (1.5% per year of the investments cost):	\$ 996
	Total:	\$ 10,680
	→ \$ 0. 61 per 1,000 m^3 *off-gas treated* (at 2,000 $m^3 h^{-1}$)	
	> 99% hydrogen sulfide removal	
	> 90% reduced sulfur compound (other than H_2S) removal	
	> 95% *overall odor removal*	

solids from the plant effluent water). Once a week, the bioreactor is checked by the plant operator by measuring the inlet and outlet hydrogen sulfide concentrations, the pH of the drain-off water, and checking the smell of the air coming from the bioreactor.

15.8.3 Performance

During startup, the pH of the bioreactor drain water and the inlet and outlet hydrogen sulfide concentrations were monitored. The pH in the bioreactor started to drop after the first day of operation, demonstrating that the removal of reduced sulfur compounds had begun. After 16 days of operation, the hydrogen sulfide removal was greater than 99% (data not shown).

After the startup period, the bioreactor performance was measured by analyzing:

- Online inlet and outlet hydrogen sulfide concentrations (using Odalogs, Phoenix, AZ; type with the range 0–200 ppmv for the inlet air, and the type with the range 0–50 ppmv for the outlet air).

- Odor concentrations of the inlet and outlet air. Samples were collected in a Tedlar bag and sent to a laboratory (St-Croix Sensory Inc., Lake Elmo, MN) for determination of the odor concentration, using the ASTM 679-99 standardized method at an airflow of $20\,\mathrm{l\,min^{-1}}$.
- Reduced sulfur compounds other than hydrogen sulfide. Samples were collected and measured by Air Toxics Ltd. (Folsom, CA), using standardized method ASTM D 5504. This method uses GC-MS and measures the concentrations of 19 different organic reduced sulfur compounds.

The system performance results for 2 and 4 months after startup are provided in Table 15.8. These results demonstrate reactor performance of greater than 99% removal of hydrogen sulfide, and 96% removal of odors.

In early 2004, after approximately 9 months of operation, the airflow to the reactor was increased beyond design to $2{,}590\,\mathrm{m^3\,h^{-1}}$. After 3 weeks of acclimation at the new airflow rate, with a cooler air temperature of 12 °C,

Table 15.8. Bioreactor removal efficiencies of hydrogen sulfide, reduced sulfur compounds and odors after 2 and 4 months after startup

Pollutant	Unit	Inlet concentration	Outlet concentration	Removal (%)
Hydrogen sulfide (48 h before taking bag samples)	ppbv	10,000–87,000	100–500	99
Hydrogen sulfide (in bag samples)	ppbv	17,000	320	98
Organic reduced sulfur compounds				
Methyl mercaptan	ppbv	430	16	
Carbonyl sulfide	ppbv	0	16	
Isopropyl mercaptan	ppbv	0	13	
Total reduced sulfur	ppbv	430	45	90
Odor concentration (8 weeks)[a]	D/T	5,500	690	87
Odor concentration (4 months)	D/T	7,800	310	96
Performance at higher airflow	D/T	7,800	310	96
Hydrogen sulfide	ppmv	9.4	0.013	99.9
Odor concentration	D/T	5,700	540	90.5
Outlet air temperature	F		54	

[a] At 20% more airflow (2,040 instead of $1{,}700\,\mathrm{m^3\,h^{-1}}$)

the Zerochem™system removed more than 99.9% of the H_2S (< 0.1 ppmv H_2S in outlet air) from the inlet airstream (at 9.4 ppmv H_2S). The outlet odor concentration was 540 D/T, which resulted in more than 90% odor removal. The system performed well, considering the cooler weather operating conditions (N. Kraakman 2004, personal communication, Bioway America, Inc., Marlton, NJ).

15.9 Long-Term Operation of a Biotrickling Filter Reactor

The Los Angeles County Sanitation District (LACSD) has typical odor generation issues at its largest wastewater treatment facility, the Joint Water Pollution Control Plant (JWPCP). Space limitations are an issue, so LACSD decided to construct a full-scale biotrickling filter reactor to treat the off-gas emissions from the headworks operation.

Pilot study data indicated that a lava rock-based biotrickling filter provided the best hydrogen sulfide and odor removal performance over a long duration. The pilot lava-rock biotrickling filter reactor was operated for four years at empty bed residence times ranging from 12 to 15 s and with inlet H_2S concentrations of 4–380 ppmv. The average inlet concentration during that time was 70 ppmv, and the reactor achieved over 99% removal of H_2S. Even at H_2S loadings over 50 g H_2S m^{-3} h^{-1}, the lava-rock reactor was demonstrated to consistently remove over 99% H_2S. The average inlet odor intensity was at a D/T value of approximately 20,000. The reactor reduced the inlet odors by 99%.

15.9.1 Design

A new unit was designed so that it would be large enough to replace a full-size caustic scrubber unit at the JWPCP, which was treating 42.5 m^3 min^{-1} of air (Fig. 15.9). However, the unit was also designed to be a research unit, having features that normally would not be used in a plant odor scrubber. Earlier pilot plant testing clearly showed that a lava-rock medium was the most effective packing medium, and this was chosen to be used in the new unit. Consequently, the new full-size unit had to be designed to hold rock having a total weight of approximately 9,000 kg. Because the unit is designed to be operated in Southern California, a seismically active area, the unit and its base pad had to meet seismic zone 4 requirements, along with being able to withstand 160 km min^{-1} wind gusts. The unit was placed at a location near the JWPCP's headworks that historically has H_2S concentrations in ducted off-gases ranging between 80 and 120 ppmv. The unit was designed with a conservative empty bed detention time of 14 s. The unit was also designed to have two separate beds, so that different types of media could be simultaneously studied and/or one of the beds could be removed and the biotrickling filter operated with one-half the normal detention time.

Fig. 15.9. Los Angeles County Sanitation District biotrickling filter reactor for odor removal. (Photograph courtesy of the Los Angeles County Sanitation Districts, Los Angeles, CA)

15.9.2 Operation and Performance

During start up, the unit was filled with 430 gallons of secondary effluent. An additional 20 gallons of primary effluent was added to the unit to seed the unit with *Thiobacilli* bacteria, and the pH of the water in the unit was lowered to approximately 3.0 (to optimize *Thiobacillus* sp. growth) by the addition of sulfuric acid. The secondary effluent, primary effluent, and sulfuric acid were allowed to mix for several hours before odorous air was introduced into the unit. The unit was operated with an empty bed detention time of 14 s. The first few measurements indicated that the system was able to remove 30% of the H_2S present in the influent air. After 24, 48, and 72 h of operation, the unit was achieving H_2S removals over 66, 98, and 100%, respectively.

Table 15.9. Characteristics of the LACSD biotrickling filter reactor

Reactor type	Biotrickling filter
Owner and location	Los Angeles County Sanitation Districts; Joint Water Pollution Control Plant (JWPCP) Carson, CA
Manufacturer	Los Angeles County Sanitation Districts
Year of installation	1997
Type of airstream	Air from the headspace of an aerated channel feeding wastewater to a battery of primary sedimentation tanks
Reactor dimensions and construction type	One reactor that is 1.8 m in diameter and 7.6 m high. The unit was constructed out of FRP
Medium type and volume of medium	Lava-rock medium use, and the volume was 10 m^3
Height and number of layers of medium	Media height is 1.8 m in the first layer, and 2 m in the second layer
Airflow rate	Total airflow is 2,550 $m^3\,h^{-1}$
Empty bed residence time	14 s
Pressure drop	Pressure drop is 5–7.6 cm water gauge
Average bed temperature	16–29 °C
Contaminants treated	Unit designed to treat H_2S at concentrations of 10–300 mg sulfur m^{-3}. Can remove some aromatics at 1–10 ppmv
System controls	Daily measurements of airflow, water flow, pH, H_2S, and system pressure
Design and acceptance criterion	Design based on pilot plant tests at site. Designed to remove 99% of inlet H_2S
Capital costs	$ 40,000
Approximate operating costs/treatment costs per 1,000 m^3 off-gas treated	Labor costs are minimal, and plant water is used to maintain appropriate soil moisture. Electrical cost is approx. $ 3,500 for recirculation pump and blower
Typical performance	Greater than 99% removal of hydrogen sulfide Greater than 20% removal aromatics

Since the short startup phase, the full-scale biotrickling filter has maintained near 100% removal of H_2S (Table 15.9). The average H_2S loading into the unit is fairly low, at approximately 10 g H_2S $m^{-3}\,h^{-1}$ (2.8 g H_2S $ft.^{-3}\,h^{-1}$).

Measurements along the reactor length demonstrate that at a point about 66% along the length of the unit (providing a 9.5-s contact time), the average removal is about 98%. This contact time is within range (a few seconds) of the contact times of many of the caustic scrubbers used at the JWPCP. This suggests that biotrickling filters can achieve the same H_2S removals as chemical scrubbers, in units of roughly the same size (Morton and Caballero 1998).

Removals of aromatic compounds in the biotrickling filter are as high as 40–50% of the inlet concentrations. Such removal extends the life of the downstream activated carbon units from the typical 20 days before change-out to over 30 days before change-out.

Results from the long-term operation of the unit (> 5 years) have demonstrated no decline in treatment performance, minimal bed pressure increase, and no/minimal deterioration of the lava-rock media (R. Morton 2003, personal communication, Los Angeles County Sanitation Districts, Los Angeles, CA). Such a system convincingly demonstrates the long-term effectiveness and robustness of biotrickling filter reactors at wastewater treatment plants.

15.10 Conclusions

There are numerous applications of biological air-phase treatment to remove odors from municipal wastewater treatment plants. The eight case studies provided in this chapter demonstrate the diversity of options available to wastewater treatment plant managers, engineers, and operators. In all of the applications, the ease of operation, high level of performance, and robustness of the technology are clearly demonstrated. Costs per system type vary depending on site-specific conditions and the necessary performance requirements. The reader is encouraged to contact the specific vendor and/or end user for more detailed information.

References

Heuckeroth R (2002) Odor control performance study of the biocube BMF biofilter at the Baltimore County Long Quarter Pumping Station. Prepared for Biocube, Inc.,Victor, NY

Morton R, Caballero R (1998) Using full scale biotrickling filters for the removal of hydrogen sulfide and odor from wastewater treatment facilities' air streams. In: Proc USC-TRG Conf Biofiltration, 22–23 October 1998, Los Angeles, CA, pp 107–114

Morton R, Ackman P, Acosta G, Tang C, Horvath R, Stahl J (2003) Full scale testing for odor control at the Los Angeles County Sanitation Districts' Joint Water Pollution Control Plant Biosolids Handling Facilities. In: Proc WEF Tech Exhib Conf, 11–15 October 2003, Los Angeles, CA. Water Environment Federation

16 Biotrickling and Bioscrubber Applications to Control Odor and Air Pollutants: Developments, Implementation Issues and Case Studies

Bart Kraakman

16.1 Introduction

Air pollution has become more and more of an environmental and medical concern. The effects of air pollution on all life forms have become better understood, and therefore legislation controlling the emissions of air pollutants (VOCs, toxics and odor) has proliferated.

In recent years, biological techniques have been applied more frequently for industrial applications, because they eliminate many of the drawbacks of classical physical-chemical techniques (van Groenenstijn and Hesselink 1993; Devinny et al. 1999; Kennes and Veiga 2001). The disadvantages of the traditionally used air-treatment techniques are high energy costs (incinerators), the use of chemicals (chemical scrubbers), and the formation of waste products (incinerators, scrubbers, activated carbon).

Conventional organic medium biofilters have been often applied to treat very different types of airstreams. Many of these installations have been successful, but there also have been a number of setbacks. Conventional biofilters have serious limitations though in treating polluted airstreams that contain compounds that are difficult to degrade biologically, or which result in a build-up of intermediates or acid degradation products from the biological degradation process.

Bioscrubber and biotrickling technologies have made it possible to extend the application of biological waste gas treatment technology. Better process control, higher degradation capacities, lower operational costs, and smaller footprints are some of the advantages of biotrickling and bioscrubber installations. In this chapter, the advantages and limitations are discussed. Recent developments and issues related to implementations of biotrickling and bioscrubber reactors are discussed and examples of full-scale applications are given.

Shareefdeen/Singh (Eds.)
Biotechnology for Odor and Air Pollution Control

16.2
Definitions, Advantages and Limitations

16.2.1
Definitions

Bioscrubbers for waste gas treatment can be defined as a two-reactor unit and water recirculating between the two reactor units. In the first unit, the pollutants are absorbed in the water phase; mass transfer is from the gas phase to the liquid phase. In the second unit, the pollutants are degraded by microorganisms. The microorganisms in the second unit are suspended in the water or immobilized on a packing (see also Chap. 8).

Biotrickling reactors for waste gas treatment can be described as a one-reactor unit with packing on which water is usually recirculated. The absorption of pollutants from the gas phase to the liquid phase, and degradation of the pollutants take place in one and the same unit. The microorganisms are immobilized on a packing, and suspended in the recirculated water (see also Chap. 7).

In biotrickling and bioscrubber reactors, gas contaminants are first absorbed in a free liquid phase before they are biodegraded. In this free liquid phase, it is possible to measure and control continuously the important process conditions like pH, nutrient concentrations, biomass concentration, salt content, and removal of possible toxic degradation products. The control of these important process parameters is thereformuch easier than in biofilters.

16.2.2
Advantages of Biotrickling Filters and Bioscrubbers versus Biofilters

The liquid phase in a biotrickling or a bioscrubber reactor is very easily accessible. Therefore, important process conditions like pH, salt content, nutrients, toxic intermediates or end products of microbial degradation are much easier to analyze and to control. Also, the liquid phase itself (the water content) can be better controlled to obtain the optimal water content through the bioreactor system, and to minimize drying out of the packing material or the biofilm on the packing material. Heslinga (1994) mentioned that probably 50–75% of the problems with conventional biofilters are related to a poor control of the water content in the biofilter media. Biotrickling filters and bioscrubbers do not require an upfront humidifier, like biofilters do, to increase the inlet air humidity up to a preferable 100%.

The media in biofilters needs to be replaced frequently, due to the deterioration of the (usually organic) media, or the increasingly worse process conditions like pH decrease, nutrient depletion, or extensive increase in biomass, salt content or pressure drop. In biotrickling and bioscrubber reactors, no medium change-out is required, since inert packing material is used and process conditions can be better controlled. Operational costs can be saved with biotrickling or bioscrubber reactors, since up to 40% of the operational cost of a biofilter is typically related to the medium change-out.

Biofilters usually have a media layer of 1 or 2 m to prevent excessive air velocities through the media, which easily results in high pressure drop or airflow preferences. The footprint of biotrickling and bioscrubber reactors is normally much smaller, since they usually contain a more open packing that can be more than 2 m high.

The investment costs of biofilters, biotrickling and bioscrubbers reactors are different for each application, and therefore difficult to compare. Both investment and operating costs of biological air-treatment systems are generally considerably less than classical physical-chemical techniques but differ greatly depending on the type of airstreams. Examples of different cost comparsions have been given by Schroeder (2002), van Durme et al. (2002) and Diehl et al. (2003).

16.2.3 Disadvantage

An important disadvantage of biotrickling and bioscrubber reactors (Table 16.1) is the higher complexity to construct and to operate. The startup of biotrickling and bioscrubber reactors is also more complicated, since the inert medium does not contain microorganisms at the start. As soon as the microorganisms are present, they can be washed out by the required drainage of the process water; a problem that is encountered with full-scale operations of bioscrubbers. Pressure-drop control is more complex especially with biotrickling reactors. Compared to a biofilter, higher air velocities through the media result more easily in a higher pressure drop. When biotrickling or bioscrubber reactors are applied for the treatment of pollutants like hydrogen sulfide, ammonia or chlorinated compounds, the degradation produces acid end products in the drain water, which needs further processing. Biotrickling and bioscrubber reactors make a better process control possible, which require, on the other hand, measurement or control instrumentation.

16.3 Recent Developments

In recent years many biotrickling and bioscrubber installations have been developed and applied to clean different waste gases. Table 16.2 shows examples of full-scale applications of biotrickling and bioscrubber installations developed and installed over the last 10 years. Developments have been on different aspects, and some of these developments are discussed below.

- Full-scale applications operating at extremely low pH are developed to reduce water, nutrient chemical requirements (Kraakman 2001; Gabriel and Deshusses 2002; Kraakman 2003; Gabriel et al. 2003). Operating a biotrickling reactor at low pH requires no chemicals for neutralization. Therefore, these biological air-treatment applications become cost-effective as compared to classical physicochemical techniques.

Table 16.1. Advantages and disadvantages of biotrickling and bioscrubber reactors versus biofilters

	Advantages	Disadvantages
Biofilters	Often less complex to build and to operate	Poor process control Not applicable for the treatment of pollutants that are degraded to acid end products Require medium change-out (increase inoperational costs) Require a relatively large footprint
Biotrickling reactors	Good process control possible Applicable for pollutants that are relatively poorly water-soluble or degraded to acid end products A relatively small footprint required Lower operational costs in general No pre-humidifier required	More complex to build and to operate
Bioscrubber reactors	Good process control possible Applicable for pollutants that are degraded to acid end products Relatively small footprint required Lower operational costs No pre-humidifier required	More complex to build and to operate Less applicable for compounds that are poorly water-soluble

A full-scale bioscrubber application operating at rather unusually high temperature conditions was presented by Sperka and Dussing (2003). Cox and Deshusses (2000b) showed that thermophilic conditions can have advantages such as lower biomass accumulation. Biomass accumulation becomes important at high loadings, and could be minimized operating a bioreactor at thermophilic temperature, since the yield of the bacteria used was lower at high temperature.

Efforts are made to improve the solubility of pollutants that are poorly water-soluble. For example, Hammervold et al. (1995) showed that the addition of powdered activated carbon increased removal efficiency of a bench-scale bioscrubber system, and could help to buffer shock loads of pollutants in the bioscrubber. Two-liquid phases using non-degradable water-immiscible silicon oil are proposed by van Groenenstijn and Lake (1999) in biotrickling reactors and bioscrubbers (Schippert 1989). The oil improves the solubility of poorly water-soluble pollutants, but full-scale applications do not seem to be feasible as the silicon oil is very expensive. Barton et al. (2002) showed that the Henry-coefficient is greatly influenced by the occurrence of biomass, either dead or alive. The Henry's law constants in a gas–water/biofilm phase differ from the Henry's constants in a gas–water phase. They showed that the

Table 16.2. Examples of full-scale applications and onsite pilot studies of biotrickling (*BT*) and bioscrubber (*BS*) installations developed and installed over the last 10 years

Industry	Pollutants	Gas flow ($m^3 h^{-1}$)	BT, BS	Reference
Full-scale applications				
Chemical industry (amino molding powder production)	Formaldehyde	15,000	BS	Popov et al. (2000)
Printing industry	Alcohols/acetates	14,000	BS	Popov et al. (2000)
Municipal WWTP	H_2S, mercaptans and odor	Various	BT	Kraakman et al. (1997, 1998); Kraakman (2001)
Municipal WWTP	H_2S and odor	6,000	BS	Hansen (1997); Hansen and Rindel (2001)
Municipal WWTP	H_2S	800	BT	Van Durme et al. (2002)
Municipal WWTP	H_2S	500	BT	Perkins et al. (2002)
Municipal WWTP	H_2S, mercaptans and odor	16,300	BT	Gabriel et al. (2002); Gabriel and Deshusses (2003)
Industrial WWTP	H_2S, mercaptans	700	BT	Kraakman et al. (1998); Kraakman (2001, 2003)
Viscose industry (fiber and sponge manufacturing)	CS_2/H_2S	51,000	BT	Kraakman (2001, 2003)
Rock wool manufacturer	Phenol and ammonia	12,000	BT	Rydin et al. (1994)
Painting industry	Solvent mixture (acetone, acetates, alcohols, and aromatic compounds)	1,200–3,000	BT	Wittorf at al. (1997)

Table 16.2. (continued)

Industry	Pollutants	Gas flow ($m^3\,h^{-1}$)	BT, BS	Reference
Industrial WWTP	H_2S from biogas	450	BS	Janssen et al. (1997)
Fish feed production	Odor	40,000	BS	Rasmussen et al. (1994)
Chemical industry (monomer production)	Odor/organic monomers	62,500	BS	Diehl et al. (2003)
Cigarette production	Odor/NH_3	41,000	BS	Frohlich (1994)
Pulp and paper industry	Formaldehyde/organic acids	2,000	BS	Sperka and Dussing (2003)
Semiconductor industry	Solvent mixture	10,000	BT	Bogenrieder (2003)
Chemical industry	Styrene	6,000	BS	Poppe and Schippert (1992)
Chemical industry (cellophane production)	CS_2/H_2S	44,200	BT	Cox and Deshusses (2000a)
Cigarette production	Odor	160,000	BT	Cox and Deshusses (2000a)
Field pilot or optimization studies				
Plastic processing (bath manufacturing)	Styrene	340	BT	Webster et al. (1999)
Chemical industry (rayon fabrication)	CS_2/H_2S	400	BT	Melse and Kraakman (1998)
Resin-laden factory	Phenol/formaldehyde and ammonia	1,000	BT	Arnold et al. (1997)
Municipal WWTP	Odor/H_2S	500	BT	Kraakman et al. (1997)
WWTP	Odor	Various	BT	STOWA (2000)

Table 16.2. (continued)

Industry	Pollutants	Gas flow ($m^3\,h^{-1}$)	BT, BS	Reference
Petrochemical industry	H_2S and SO_2	160	BS	Janssen et al. (1997, 2000)
Chicken manure processing plant	Odor	300	BT	Kraakman and Koers (1996)
Composting plant	Odor	200–1,300	BT	Fischer (1994)
Painting industry	Solvent mixture (ketones, acetates, alcohols, aromatic compounds)	1,000	BT	Oosting et al. (1994)
Chemical industry (fiber production)	N,N,-dimethylacetamide	2,800	BT	Waalewijn et al. (1994)
Plastic processing (bath manufacturing)	Styrene, toluene	360	BT	Choi et al. (1998)
Chemical industry (sponge manufacturing)	CS_2/H_2S	3,400	BT	Hugler and Acosta (1997)
Municipal WWTP	Odor	50	BT	Loo et al. (2003)
Municipal WWTP	Odor	100	BT	Morton and Caballero (1996)
Municipal WWTP	H_2S and traces of VOCs	3,000	BT	Cox and Deshusses (2000a)
Chemical industry	Dichlorethane, isopropanol, epichlorhydrine, styrene	1,000	BT	Buser (2002)
Spray painting	Toluene, xylenes, methyl ethyl ketone, n-buthyl acetate	40–150	BT	Webster et al. (1998)

occurrence of a biofilm influence the mass transfer of many pollutants from the waste gas, which is very important to know for design and operation.

Different groups have worked on modeling biotrickling and bioscrubber systems (Ramesh and Devinney 2004). Better models, especially on the biofilms, are made and include, besides degradation of pollutants, processes like growth, death, and predation.

Characterization and identification of actively degrading microorganisms in a biological waste gas treatment system can be nowadays better performed using gene probes or phospholipid fatty acid (PLFA) analysis in combination with isotope-labeled substrates (Lipski and Alterndorf 2002; Steele et al. 2004).

These research and development activities resulted in improvements of design, construction and operations of biotrickling and bioscrubber reactors, of which the possibilities seem without limits.

16.4 Robustness

When biological air-treatment techniques are applied in industries where air pollution control is very strictly regulated, a good control of the biological systems is a prerequisite. A good understanding and a good control of the important process conditions and it's risk in a bioreactor system is necessary. One way to limit the risks of biological air treatment is to determine its robustness.

Robustness can be defined to reflect the ability of the biological system to deal with fluctuations, and the ability to recover after operational failures. Biological systems are influenced by parameters such as pH, water content, temperature, and concentrations of the carbon sources, energy sources, and nutrients. Quantification of the robustness of a biological air purification system would be helpful to designers and operators of these systems.

Robustness may be quantified by determining the risk of negative effects on the biological system for each possible upset, multiplying by the frequency of the upset, and summing over all possible upsets.

The risk of all negative effects on a biological system (R) can be defined as:

$$R = S(p \times E) ,$$

where p is the probability of occurrence of a fluctuation, and E the negative effect of the fluctuation.

The probability of occurrence of a fluctuation (p) can be expressed as the expected number of occurrences per year (number year^{-1}) or as the percentage of operating time during which it is likely to occur. The negative effect of the fluctuation (E) can be expressed as the loss of removal efficiency (%), the loss of total removal (kg day^{-1} or kg year^{-1}), or the impact on the people living near the installation (e.g., the number of occurrences during which the concentration of the emitted airstream exceeds the odor threshold in the neighborhood).

The microbiological community in a bioreactor will face fluctuations related to the production process (changes of air inlet concentrations and air conditions such as flow and temperature). These fluctuations result from continuous or discontinuous processes, irregular unplanned shutdowns (production process failures), planned shutdowns (holiday/maintenance), and diurnal fluctuations. There may also be fluctuations related to the operation of the air purification system, for example, associated with loss of control of the nutrient supply, water recirculation flow, or the pH of the support medium. Quantification of robustness of a full-scale biotrickling reactor operating at an industrial plant is illustrated by Kraakman (2003).

Microbial responses to stress conditions are being investigated in various laboratories worldwide. Biotrickling and bioscrubber reactors using mixed cultures as inoculum are self-optimizing, with species becoming dominant that are most competitive under the environmental conditions existing in a full-scale biological air-treatment system. Unfortunately, this self-optimizing adaptation process seems to be relatively slow, and is likely to take months.

A biological air-treatment system seems to be resilient to changes in inlet concentrations due to increased mass transfer rates with corresponding degradation rates (Kraakman 2001; Schroeder 2002). Little impact on performance is observed when the airflow containing pollutants is stopped for periods up to 12 h (Schroeder 2002). Choi et al. (1998) showed that the effect of starvation is minimal if the duration of starvation is 48 h. Cox and Deshusses (2002) showed that recovery is less than 24 hours after starvation periods from 2 to 9 days. Kim et al. (2004) studies the impact of interchanging of different VOCs on the performance of a tricklingfilter. Nutrient depletion results in decreased removal within days, but recovery is rapid when nutrients are added (Schroeder 2002; see also Sect. 16.6.2).

When robustness is quantified and the risks are understood, extra measures can be taken to reduce the risks and to smoothen the implementation at full-scale (Kraakman 2003).

16.5 Missing Gaps for Future Developments

Biological air purification technology has been demonstrated and accepted by regulators as being most cost-effective and reliable for the treatment of low-concentration waste gases from mostly organic (waste or food) processing industries (e.g., wastewater treatment plants, composting plants, rendering plants, animal housing, pulp and paper industry, various food processing industries). These applications treat airstreams with relatively low concentrations of mostly odorous pollutants. Biotrickling and bioscrubber reactors have extended this field of biofiltration over the last years. New markets for biological air-treatment technology can be explored by improving the existing technology. Development should be focused on issues like:

Robustness (flexible to process fluctuations/failures)

Large pollutant loadings

High temperatures

Halogenated compounds

Poorly water-soluble compounds

Higher degradation capacities can result in an increase of pressure drop over the packing, preference airflow through the packing, and reduction of removal efficiencies due to biomass accumulation. Different solutions have been tried to prevent biomass accumulation, like chemical washing (Cox et al. 1999), periodic switching of flow direction (Song and Kinney 2000), use of higher organisms (Cox et al. 1999; Woertz et al. 2002), altered temperatures (Cox and Deshusses 2000c) or pulses of high air pressure (Heits et al. 1997). Unfortunately, application of these improvements to full-scale bioreactor systems is difficult.

To increase performance, the rate-limiting step (the biological degradation rate or the mass transfer rate) should be better determined for different waste gas streams – when mass transfer is limiting, it should be made easier to find out what exactly is limiting (oxygen, pollutant, or accumulated degradation intermediates or products), and where exactly limitation occurs (water film, biofilm, air distribution, etc.).

Flow characterization through the media of a biotrickling or bioscrubber reactor is important, since gas flow, liquid flow and gas velocity have an important impact on process parameters like mean gas residence time, gas dispersion in the reactor, and pressure drop over the system (Humeau et al. 2000). These parameters are important to scale up and to operate a reactor at optimum conditions.

Models sometimes fail to predict short-term operation stability and long-term performance. Further research is required to transform the technology of biofiltration from an empirical practice to a more theoretical and scientifically based technology.

16.6 Case Studies

16.6.1 Odor Removal from Waste Gas Emissions at an Anaerobic Wastewater Treatment Plant with a Purspring Bioreactor

16.6.1.1 Introduction of the Site and the System

In an anaerobic wastewater treatment plant, septic conditions cause the emission of malodorous gases. These odor emissions caused problems in the neighborhood of this industrial plant. At the plant, beer is made and the wastewater is treated onsite at the plant. Increasing production made the emissions to the neighborhood worse, and treatment of these gases was proposed.

16.6.1.2
Design Comments

The following design criteria were followed: an airflow of 1,200 $m^3 h^{-1}$ (phase 1) + another 6,200 $m^3 h^{-1}$ (phase 2) containing hydrogen sulfide concentrations up to 200 ppm. Acceptance criteria were the following:

- Criterion 1: 99% H_2S removal at all time (500 ppm at phase 1, and 82 ppm at phase 2);
- Criterion 2: outlet odor concentration less than 170 MOU h^{-1} (at less than 16,200 MOU h^{-1} on the inlet).

16.6.1.3
Operation of the System

A startup vessel contained an acid-resistant pump and a water-level control. The vessel was first filled with water, nutrients, and microbial culture (the culture used was a combination of prepared inoculum and activated sludge). After a period of 4 weeks, the startup kit (with the recirculation pump) is disconnected and removed, and the system control is changed over from startup operation to normal operation. During normal operation, water with nutrients is discontinuously added to the bioreactor (see Fig. 16.1), operating in up-flow. The water used for this biotrickling reactor is effluent water from the aerobic wastewater treatment, which treats the effluent water from the anaerobic

Fig. 16.1. Bioreactor treating odorous air from an anaerobic wastewater treatment plant

wastewater treatment system. This water contained enough nutrients to operate the biotrickling reactor without a nutrient dosing system. During normal operation, no special attention has to be paid to the bioreactor, other than checking the blower operation, the screen filtering the effluent water, and the alarm signals present. The pH of the discharged water is continuously monitored, and the water amount is automatically adjusted by the control system.

16.6.1.4 Performance

Only phase 1 is implemented and, during normal operation, the plant operator checks the bioreactor weekly, by visual inspection and alarm signals. The bioreactor performance is measured by analyzing:

- online inlet and outlet H_2S concentrations (using Odalogs; the 0–1,000 ppm range type for the inlet air, and the 0–50 ppm range type for the outlet air);
- odor concentrations of the inlet and outlet air (during approximately half an hour, air samples were collected in a Tedlar bag and sent to a laboratory (TNO, Apeldoorn, The Netherlands) for determination of odor concentration, using the EN13725 standardized method);
- reduced sulfur compounds other than hydrogen sulfide (measured by Nijmegen University, using gas chromatography). The results are presented in Tables 16.3 and 16.4.

Table 16.3. Removal efficiencies of reduced sulfur compounds after 6 weeks from the startup date (n.m. not measured)

Pollutant	Unit	Inlet concentration	Outlet concentration	Removal
Hydrogen sulfide	ppm	800	1.7	> 99%
Organic reduced sulfur compounds				
Methyl mercaptan	ppb	2,260	145	94%
Carbonyl sulfide	ppb	n.m.	< 4	
Dimethylsulfide	ppb	452	220	51%
Dimethyldisulfide	ppb	68	20	71%
Dimethyltrisulfide	ppb	n.m.	10	
Total reduced sulfur	ppb	> 2,780	< 399	86%
Odor removal efficiencies of the bioreactor after approximately 10 weeks and 25 weeks from the startup				
Odor concentration	MOU h^{-1}	7,000	200	97%
Odor concentration	MOU h^{-1}	5,000	30	99%

n.m. = not measured

Table 16.4. Case study of odor removal at an anaerobic wastewater treatment plant (*WC* water column)

Reactor type	Purspring bioreactors	
Owner and location	Brewery in The Netherlands	
Manufacturer	Bioway bv (Ede, The Netherlands)	
Year	2002	
Type of airstream	Waste gas from an anaerobic wastewater treatment plant	
Reactor type	Purspring	
Medium type	Permapac type BrO	
Airflow rate	1,200 $m^3 h^{-1}$ (phase 1) +6,200 $m^3 h^{-1}$ (phase 2)	
Pressure drop	< 100 Pa (< 0.4 in. WC) at 1,200 $m^3 h^{-1}$	
Average bed temperature	15–25 °C	
Contaminants treated	Hydrogen sulfide	
System controls	Automatic water feed by PLC. Alarm generation for different unwanted situations like temperature, water pressure, water flow, low pH and air pressure	
Design criteria	99% removal at 500 ppm hydrogen sulfide less than 170 MOU h^{-1}	
Capital costs	$ 385,000, including fan, ductwork, installation onsite and startup	
Approximate operating costs per year	Depreciation (over 20 years):	$19,250
	Interest rate (5%):	$11,643
	Energy (fan: 7,400 $m^3 h^{-1}$ at 1.6 in. WC and control panel: 0.5 kW at $0.04 kWh^{-1}):	$874
	Water (none; groundwater is used):	$0
	Nutrients:	$0
	Chemicals (none):	$0
	Operating costs (0.5 h per week at $50 h^{-1}):	$1,300
	Media change replacement:	$0
	Maintenance (1% per year of the investments cost):	$ 3,850
	Total:	$36,917
	Costs per 1,000 m^3 off-gas treated:	$0.57
Typical performance	> 99% hydrogen sulfide removal 97% odor removal	

16.6.2 H_2S Removal from Stripped Groundwater with a Purspring Bioreactor

16.6.2.1 Introduction of the Site and the System

The neighborhood around this brewery plant is very sensitive to bad odors. The plant is located close to a residential housing area and next to a busy road. The brewery installed equipment to pump and treat groundwater for the production of the beer. The groundwater contained a relative high concentration of hydrogen sulfide, since swamps are present in the area. Two strippers strip the groundwater, and each stripper was connected to a bioreactor treating the odorous air from the stripper.

16.6.2.2 Design Comments

The following design criteria were followed: an airflow of 34,000 $m^3\,h^{-1}$ containing hydrogen sulfide concentrations of 31 ppm (max. 60 ppm). Acceptance criteria were based on the following:

- $>$ 80% removal of hydrogen sulfide at 60 ppm

16.6.2.3 Operation of the System

A startup kit was used in the beginning. This vessel contains an acid-resistant pump and a water-level control. The vessel was first filled with water, nutrients, and culture consisting of prepared inoculum and activated sludge. The pump recirculates the water over the media in the bioreactor in order to have the microorganisms grow on the inert media inside the bioreactor. After a couple of weeks, the startup kit (with the recirculation pump) was disconnected and removed, and the system control is changed over from startup operation to normal operation. During normal operation, water with nutrients is discontinuously added to the bioreactor (Fig. 16.2) operating in up-flow. The water used in the bioreactor was the stripped groundwater after membrane filtration, which is also used as water source for the beer production. Addition of nutrient is therefore very important, since the filtrated water contains nothing but water ions. During normal operation, no special attention has to be paid to the bioreactor, other than checking the blower operation, the water valves, and the nutrient dosing system. The pH of the discharged water is continuously monitored, and the fresh water with nutrients is automatically adjusted by the control system.

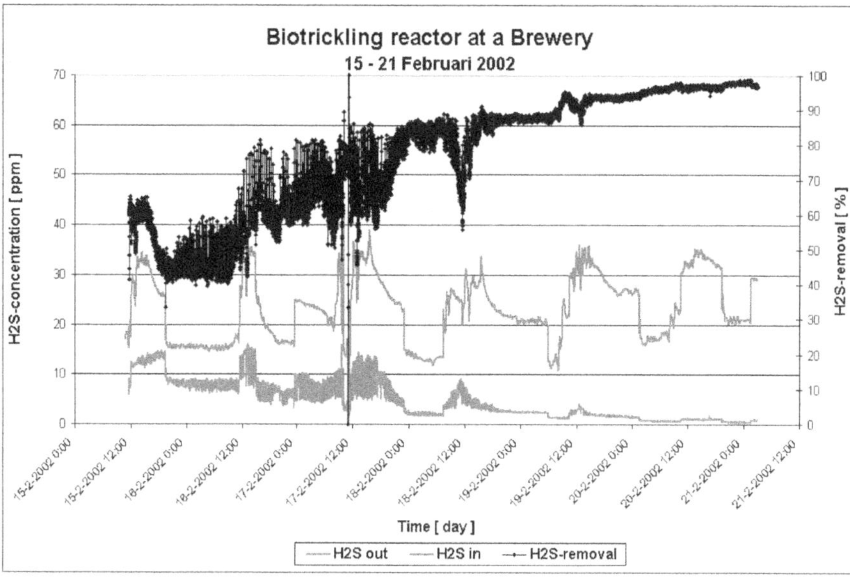

Fig. 16.2. Bioreactor treating H_2S from stripping towers stripping groundwater recovering after a failure of the nutrient dosing system

16.6.2.4 Performance

During normal operation, the plant operator checks the bioreactor weekly, by visual inspection and the alarm signals present. Figure 16.3 shows an example

Fig. 16.3. Bioreactors (*right*) treating odorous air from the strippers (*left*) at a brewery

Table 16.5. Case study of H_2S removal stripped groundwater

Reactor type	Two Purspring bioreactors	
Owner and location	Brewery in Florida	
Manufacturer	Bioway America Inc. (New Jersey, USA)	
Year of installation	Spring 2000	
Type of airstream	Waste gas from two strippers stripping groundwater	
Reactor type	Purspring	
Medium type	Permapac type BrS	
Airflow rate	Approx. 34,000 $m^3 h^{-1}$	
Pressure drop	< 400 Pa (< 1.6 in. WC)	
Average bed temperature	Ambient air (5–35 °C)	
Contaminants treated	Hydrogen sulfide	
System controls	Automatic water and nutrient feed by PLC. Alarm generation for different unwanted situations like temperature, water pressure, water flow, nutrient flow, low pH and air pressure	
Design criteria	> 80% removal at 60 ppm	
Capital costs	$640,000, including fan, ductwork, installation on site and startup	
Approximate operating costs per year	Depreciation (over 20 years):	$32,000
	Interest rate (5%):	$19,355
	Energy (fan: 34,000 $m^3 h^{-1}$ at 1.6 in. WC and control panel: 0.5 kW at $0.04 kWh^{-1}):	$2,446
	Water (none; groundwater is used):	$0
	Nutrients:	$2,500
	Chemicals (none):	$0
	Operating costs (0.5 h per week at $50 h^{-1}):	$1,300
	Media change replacement:	$0
	Maintenance (1% per year of the investments):	$6,400
	Total:	$64,001
	Costs per 1,000 m^3 off-gas treated:	$0.21
Typical performance	> 95% hydrogen sulfide removal	

of the performance of the bioreactor, which is measured by analyzing online inlet and outlet H_2S concentrations (using an H_2S-meter of Industrial Scientific, type STX70 calibrated at 25 ppm). The figure shows the recovery after a failure of the nutrient dosing system. The removal efficiency of hydrogen sulfide was low (40–60%) at the start of the online monitoring. After performing maintenance to the nutrient dosing system (16 February around 12:00 h), the bioreactor started to recover until it reached its original removal efficiency of $> 95\%$. The results are presented in Table 16.5.

16.6.3 V-Spring Bioreactor System Treating CS_2 Emissions at a Fungicide Manufacturing Plant

16.6.3.1 Introduction of the Site and the System

This fungicide manufacturing plant had to be in compliance with regulations of the authorities, which implies a reduction of the CS_2 emissions. Different waste gas abatement techniques were considered, including active carbon, incineration, and biofiltration. Biotrickling was proposed, because this eliminate many of the drawbacks of classical physical-chemical techniques, especially in investment and operating costs.

16.6.3.2 Design Comments

The following design criteria were followed: an airflow of 8,500 $m^3\,h^{-1}$ containing carbon disulfide concentrations of 100–300 ppm. The acceptance criteria are based on 97% removal of carbon disulfide, with an average outlet concentration of less than 10 ppm. The waste gas from this fungicide production plant contains very small amounts of fungicide dust particles, and also compounds other than CS_2. Therefore, different laboratory studies and bench-scale studies are performed to first find out whether biological treatment of this waste gas is possible. The restrictions and important process parameters had to be determined. The results show that dust has to be removed from the airstream, and that pretreatment of the waste gas is necessary to remove some of the traces of VOC pollutants present in the waste gas. A range of compounds other than CS_2 is present at concentrations lower than 1 ppm, among these a couple of chlorinated compounds that are, however, necessary to remove from the waste gas as much as possible.

A biotrickling bioreactor system (V-spring reactor) was designed and built. The system consisted of two bioreactors, and an upfront scrubber and a control container (Fig. 16.4). Microorganisms capable of degrading CS_2 at extreme low pH are cultivated and used as inoculum for startup of the bioreactor system. A CS_2 dosing system is installed to make it possible to cope with

Fig. 16.4. Bioreactors treating CS_2 from waste gas at a fungicide manufacturing plant

irregular productions. During production shutdowns, CS_2 can be dosed in small amounts to keep the microorganisms active, and to shorten startup times at restarts of production.

16.6.3.3 Operation of the System

The bioreactor system was started up using the CS_2 dosing system, which was actually installed to deal with the irregular production. Ambient air with CS_2 was led through the bioreactor system, resulting in the growth of microorganisms. As soon as the bioreactors were fully grown, the ambient air is shut off, and process waste gas from different process units was led through the bioreactors.

During normal operation, caustic is dosed in the upfront scrubber to remove unwanted compounds. Water and nutrients are dosed discontinuously to the bioreactors, and the continuous measurement of the pH and acid content of the process water in the biotowers controls the system. The production schedule can be discontinuous, which gives very discontinuous loadings of CS_2 to the bioreactor. To prevent reduced CS_2 removal efficiencies, the CS_2 dosing system was used to keep the microorganisms active during production stops.

16.6.3.4 Performance

The removal of CS_2 was at first slow, and problems were encountered to obtain a stable operating system. Additional site research was combined with laboratory studies to improve the system stability. It was found that the stability problems were related to dust accumulation in the scrubber, waste gas sucking from an unwanted source, and suboptimal pretreatment in the upfront scrubber. This resulted in unwanted compounds in the waste gas (most likely chlorinated compounds), which caused the unstable performance.

After optimization, the system operated according to the specification (see Tables 16.6 and 16.7). The total CS_2 emission was below 10 ppm at the design inlet CS_2 loadings. It is seen that 12 h upfront resuming of the production, the CS_2 dosing system should be started in order to be able to eliminate removal efficiency due to discontinuous production. This will be further optimized to minimize CS_2 dosing during production interruptions. The biotrickling system showed to be robust for situations like the power blackout of 8 h (power blackout USA and Canada, 14 August 2002), and fresh-water interruptions of at least 2 days. The operational costs are for this case relatively high due to the pre-treatment of the waste gas and the extreme low required outlet concentrations.

Table 16.6. CS_2 removal in the biotrickling reactors measured on different days by gas chromatography

Measurement	Inlet CS_2 conc. (ppmv)	Outlet CS_2 conc. (ppmv)	Removal (%)
1	273.0	2.6	99.1
2	315.4	2.0	99.4
3	403.7	9.4	97.7
4	456.7	7.2	98.4
5	456.2	13.3	97.1

16.7 Conclusions

Due to the better process control, biotrickling and bioscrubber reactors extended the application field of biological air-treatment technology. Nowa-

Table 16.7. Case study of biotrickling reactors treating waste gas at a fungicide production plant

Reactor type	V-spring Bioreactor system	
Owner and location	Fungicide manufacturer	
Manufacturer	BIOWAY bv (Ede, The Netherlands)	
Year of installation	2002	
Type of airstream	Waste gas from the process reactors, drying screens and vacuum pumps	
Reactor type	Two V-spring bioreactors with upfront scrubber and control container	
Medium type	Permapac type V	
Airflow rate	8,500 $m^3 h^{-1}$	
Pressure drop	< 100 Pa (< 0.4 in. WC)	
Average bed temperature	20–25 °C	
Contaminants treated	Carbon disulfide (200 ppm) and traces of other VOCs (< 5 ppm)	
System controls	Automatic water and nutrient feed by PLC. Alarm generation for different unwanted situations: temperature, water pressure, water flow, air pressure, salt content, liquid levels in recirculation tank and scrubber	
Design and acceptance criterion	97% removal of carbon disulfide with an average outlet concentration of less than 10 ppm	
Capital costs	$1,190,000, including upfront scrubber/deduster, process control container, installation onsite and startup	
Approximate operating costs per year	Depreciation (over 20 years):	$59,500
	Interest rate (5%):	$35,989
	Energy (fan: 8,500 $m^3 h^{-1}$ at 0.5 in. WC and control panel: 2.5 kW at $0.04 kWh^{-1}):	$1,398
	Water (10 $m^3 day^{-1}$ at $2 m^{-3}):	$7,300
	Nutrients:	$5,423
	Chemicals (NaOH):	$12,500
	Operating costs (2 h per week at $50 h^{-1}):	$5,200
	Medium change replacement:	$0
	Maintenance (1.5% per year of the investments cost):	$17,850
	Total:	$145,160
	Costs per 1,000 m^3 off-gas treated:	$1.95
Typical performance	> 97% carbon disulfide removal	

days, different biotrickling and bioscrubber reactors are applied in different fields to treat waste gases. New applications have been seen in recent years, and the development of these biological air-treatment technologies is still underway. An important aspect for implementation is robustness. Robustness can be determined, and measures can be taken to reduce the risks. Case studies showed successful applications under different industrial situations.

References

Arnold M, Lehtomaki J, Roine J (1997) Biological treatment of resin-laden air. In: Prins WL, van Ham J (eds) Proc Int Symp Biological Waste Gas Cleaning, 28–29 April 1997, Maastricht. VDI, Düsseldorf, pp 75–82

Barton JW, Jones SA, Vodraska CD, Davison BH (2002) In: Proc USC-TRG Conf Biofiltration, 31 October–1 November 2002, Los Angeles, pp 69–76

Bogenrieder C (2003) Reinigung losemittelbeladener Abluft in der Halbleiterindustrie mit dem Biotropfkörper-Verfahren. Koll Biologische Ablufttreinigung, Leipzig, VDI Ber 1777, pp 257–265

Buser M (2002) Biotricklingfilter zur Reinigung von VOC- und CKW-haltiger Abluft. WLB Wasser Luft Boden 5:93–95

Choi D-S, Devinney JS, Deshusses MA (1998) Performance of a pilot-scale biotrickling filter under non-steady state conditions. In: Proc USC-TRG Conf Biofiltration, 22–23 October 1998, Los Angeles, pp 159–168

Cox HHJ, Deshusses MA (2000a) Biotrickling filters for air pollution control. In: Bitton G (ed) The encyclopedia of environmental microbiology. Wiley, New York

Cox HHJ, Deshusses MA (2000b) Thermophilic biotrickling filtration of ethanol vapors. In: Proc USC-TRG Conf Biofiltration, 19–20 October 2000, Los Angeles, pp 159–166

Cox HHJ, Deshusses MA (2000c) Chemical removal of biomass from waste air biotrickling filters: screening of potential chemicals. Water Res 33:2383–2391

Cox HHJ, Deshusses MA (2002) Effect of starvation on the performance and re-acclimation of biotrickling filters for air pollution control. Environ. Sci. Technol. 36:3069–3073

Cox HHJ, Nguyen TT, Deshusses MA (1999) Predation of bacteria by protozoa *Tetrahymena pyriformis* in toluene-degrading cultures. Biotechnol Lett 21:235–239

Devinney JS, Deshusses MA, Webster TS (1999) Biofiltration for air pollution control. Lewis, Boca Raton

Diehl B, Koster D, Ostermann E (2003) Biologische Abluftreinigung in der chemischen Industrie – Einsatzmöglichkeiten, Betriebserfahrungen und Perspektiven. Koll Biologische Abluftreinigung, Leipzig, VDI Ber 1777, pp 83–91

Fischer K (1994) Bioreaktor zur Abluftreinigung einer Kompostierungsanlage im mesophilic und thermophilic Temperaturbereich. In: Proc Biological

Waste Gas Cleaning Conf, 9–11 March 1994, Heidelberg. VDI Ber 1104, pp 181–191

Frohlich S (1994) Entscheidungshilfen und Kriterien für den Einsatz von Biowäschern zur Abluftreinigung und Geruchsminderung. In: Proc Biological Waste Gas Cleaning Conf, 9–11 March 1994, Heidelberg. VDI Ber 1104, pp 81–93

Gabriel D, Deshusses MA (2003) Performance of a full-scale biotrickling filter treating H_2S at a gas contact time of 1.6 to 2.2 seconds. Environ Prog 22:111–118

Gabriel D, Cox HHJ, Brown J, Torres T, Deshusses MA (2002) Biotrickling filters for POTWs Air treatment: full-scale experience with a converted scrubber. In: Proc Odor and Air Toxic Emissions 2002 Conf, 28 April–1 May 2002, Albuquerque, New Mexico

Hammervold RE, Overcamp TJ, Grady CPL, Smets BF (1995) A sorptive slurry bioscrubber for the control of acetone. Air Waste Manage 50:954–960

Hansen NG (1997) Bioscrubber and biofilter for air purification at wastewater treatment plants. In: In: Prins WL, van Ham J (eds) Proc Int Symp Biological Waste Gas Cleaning, 28–29 April 1997, Maastricht. VDI, Düsseldorf, pp 397–412

Hansen GN, Rindel K (2001) Bioscrubber for treating waste gases from waste water treatments plants. In: Kennes C, Veiga MC (eds) Bioreactors for waste gas treatment. Kluwer, Dordrecht, pp 285–298

Heits H, Laurenzis A, Werner U (1997) Biologische Abgasreinigung mit kontrolliertem Biomassenaustrag im periodisch ruckgespulten Rieselbettreaktor. Gefahrstoffe-Reinhaltung Luft 5:153–158

Heslinga DC (1994) Biofiltration technology. In: Proc Biological Waste Gas Cleaning Conf, 9–11 March 1994, Heidelberg. VDI Ber 1104, pp 13–18

Hugler W, Acosta C (1997) Biological removal of carbon disulfide waste air streams. Environ Prog 18:173–177

Humeau P, Baleo JN, Bourcier J, Le Cloirec P (2000) Flow characterization in a gas-liquid column: application to bioscrubber for deodorization of waste gas. Water Sci Technol 41:191–198

Janssen AJH, de Hoop K, Buisman CJN (1997) The removal of H_2S from air at a petrochemical plant. In: In: Prins WL, van Ham J (eds) Proc Int Symp Biological Waste Gas Cleaning, 28–29 April 1997, Maastricht. VDI, Düsseldorf, pp 359–364

Janssen AJH, Dijkman H, Janssen G (2000) Novel biological processes for the removal of H_2S and SO_2 from gas streams. In: Lens PNL, Hulshoff Pol L (eds) Environmental technologies to treat sulfur pollution. IWA, London

Kellner C, Vitzthum R (1997) Biologische Abluftreinigung mit bei dem Biotropfkörperverfahren Umsetzung vom Pilotversuch bis zur Grosanlage mit toluolbeladener Abluft bei einem Druckfarbenhersteller. In: Prins WL, van Ham J (eds) Proc Int Symp Biological Waste Gas Cleaning, 28–29 April 1997, Maastricht. VDI, Düsseldorf, pp 61–66

Kennes C, Veiga MC (2001) Bioreactors for waste gas treatment. Kluwer, Dordrecht

Kim D, Cai Z, Sorial SA (2004) Impact of interchanging VOCs on the performance of trickling-bed air biofilter. In: Proc. USC-CSC-TRG Conf. Biofiltration, 20–22 October 2004, Redondo Beach, pp 35–41

Kraakman NJR (2001) New bioreactor system for treating sulphur- or nitrogen-compounds. In: Kennes C, Veiga MC (eds) Bioreactors for waste gas treatment. Kluwer, Dordrecht, pp 269–284

Kraakman NJR (2003) Full-scale biological treatment of industrial CS_2-emissions at extreme conditions. The robustness of a biological system and its risks to the waste gas purification. Environ Eng 22:79–86

Kraakman NJR, Koers B (1996) Biological odor removal with a biotrickling-biofilter combination. In: Proc Int Symp AWMA, 23–28 June 1996, Nashville

Kraakman NJR, Koers B, Oosting R, Tessel P (1997) H_2S removal using a new type of biotrickling filter. In: Prins WL, van Ham J (eds) Proc Int Symp Biological Waste Gas Cleaning, 28–29 April 1997, Maastricht. VDI, Düsseldorf, pp 209–214

Kraakman NJR, Melse RW, Koers BB, van Dijk J (1998) Biological treatment of waste gases containing H_2S in combination with either odor or CS_2. In: Proc USC-TRG Conf Biofiltration, 22–23 October 1998, Los Angeles, pp 91–98

Lipski A, Altendorf J (2002) Identification of the degrading micro-organisms in biofilters by in-situ isotope labeling. In: Proc USC-TRG Conf Biofiltration, 31 October–1 November 2002, Los Angeles, pp 61–67

Loo YY, Wu Y, Chai JW, Koe CC (2003) Optimizing biotrickling filters for sewage odor treatment – a physical design perspective. J Inst Eng Singapore 43:6–9

Melse RW, Kraakman NJR (1998) Biological treatment of waste gases containing H_2S and CS_2 combined with the production of concentrated NaOH and H_2SO_4. In: Proc Symp Forum of Applied Biotechnology, September 1998, Gent. Meded Fac Landbouwwetenschappen 63:1841–1847

Morton R, Caballero R (1996) The biotrickling story. Water Environ Technol, June, pp 39–45

Oosting R, Urlings LGCM, van Riel PH, van Driel C, Maas A (1994) Entfernung von Losemitteln aus der Abluft von Spritzlackieranlage mit einem Rieselbettreaktor. In: Proc Biological Waste Gas Cleaning Conf, 9–11 March 1994, Heidelberg. VDI Ber 1104, pp 193–201

Perkins MA, Woolsey MS, McMillen JS (2002) Bioscrubbers achieve performance goals at TRG at central wastewater treatment plant. In: Proc Odor and Air Toxic Emissions 2002 Conf, 28 April–1 May 2002, Albuquerque, New Mexico

Popov V, Bezborodov A, Murphy A, Cross P, Jackson W (2000) Industrial trickling bed biofilters for abatement of VOCs from air emissions. In: Proc USC-TRG Conf Biofiltration, 19–20 October 2000, Los Angeles, pp 75–82

Poppe W, Schippert E (1992) Das KCH-Biosolv-Verfahren in Kombination mit einem Biowascher herkömmlicher Art – eine Verfahrensentwicklung zur Abluftreinigung für wasserlösliche und schwer wasserlösliche Schadstoffe. In: Dragt AJ, van Ham J (eds) Biotechniques for air pollution abatement and odor control policies. Elsevier, Amsterdam, pp 71–76

Ramesh J, Devinney JS (2004) A review of biofilter models. In: Proc. USC-CSC-TRG Conf. Biofiltration, 20–22 October 2004, Redondo Beach, pp 245–258
Rasmussen HH, Hansen NG, Rindel K (1994) Treatment of odorous nitrogen compounds in a bioscrubber comprising simultaneous nitrification and denitrification. In: Proc. Biological Waste Gas Cleaning Conf, 9–11 March 1994, Heidelberg. VDI Ber 1104, pp 491–497
Rydin S, Dalberg P, Bodker J (1994) Biological waste gas treatment of air containing phenol and ammonia using a new type of trickling filter. In: Proc Biological Waste Gas Cleaning Conf, 9–11 March 1994, Heidelberg. VDI Ber 1104, pp 231–237
Schippert E (1989) Das Biosolv-Verfahren von Keramchemie zur Absorption von Schwer wasserlöslichen Lösemitteln. VDI Ber 735:161–177
Schroeder ED (2002) Trends in application of gas-phase bioreactors. Re/Views Environ Sci Bio/Technol 1:65–74
Song J, Kinney KA (2000) Effect of vapor-phase bioreactor operation on biomass accumulation, distribution, and activity: linking biofilm properties to bioreactor performance. Biotechnol Bioeng 68:508–516
Sperka G, Dussing G (2003) Genehmigung und Überwachung einer Abluftreinigungsanlage mit Biowäscher bei einer Produktionsanlage für mitteldichte Faserplatten (MDF). Koll Biologische Abluftreinigung, Leipzig, VDI Ber 1777, pp 183–189
Steele J, Ozis F, Fuhrman JA, Devinney JS (2004) Structure of microbial communities in ethanol biofilters. In: Proc. USC-CSC-TRG Conf. Biofiltration, 20–22 October 2004, Redondo Beach, pp 245–258
STOWA (2000) Biologische luchtzuiveringssystemen op rwzi's. STOWA, Utrecht
Van Durme GP, Gilley AD, Groff CD (2002) Biotrickling filter treats high H_2S in a collection system in Jacksonville, Florida. In: Proc Odor and Air Toxic Emissions 2002 Conf, 28 April–1 May 2002, Albuquerque, New Mexico
Van Groenestijn JW, Hesselink PGM (1993) Biotechniques for air pollution control. Biodegradation 4:283–301
Van Groenestijn JW, Lake ME (1999) Elimination of alkanes from off-gasses using biotrickling filters containing two liquid phases. Environ Prog 18:151–155
Waalewijn E, Meijer CD, Weber FJ, Duursma UC, van Rijs P (1994) Practice runs of a biotrickling filter in the exhaust gases from a fibre factory. In: Proc Biological Waste Gas Cleaning Conf, 9–11 March 1994, Heidelberg. VDI Ber 1104, pp 545–550
Webster TS, Togna AP, Guarini WJ, McKnight L (1998) Treatment of VOC emissions from a spray paint booth application using biotrickling filtration.In: Proc USC-TRG Conf Biofiltration, 22–23 October 1998, Los Angeles, pp 41–50
Webster TS, Cox HH, Deshusses MA (1999) Resolving operational and performance problems encountered in the use of a pilot/full-scale biotrickling filter reactor. Environ Prog 18:162–172

Wittorf F, Knauf S, Windberg HE (1997) Biotrickling-reactor: a new design for the efficient purification of waste gases. In: In: Prins WL, van Ham J (eds) Proc Int Symp Biological Waste Gas Cleaning, 28–29 April 1997, Maastricht. VDI, Düsseldorf, pp 329–335
Woertz JR, van heiningen WNM, van Eekert MHA, Kraakman NJR, Kinney KA (2002) Dynamic bioreactor operation: effect of packing and mite predation on toluene removal from off-gas. Appl Microbiol Biotechnol 58:690–694

Part V

Future of Biotechnology

17 Future Prospects of Biotechnology for Odor Control

Fethiye Ozis, Arash Bina, Joseph S. Devinny

17.1 The Growing Need for Odor Control

The world's populations are simultaneously growing and migrating to the cities. This produces rapid expansion of urban areas into the surrounding countryside where they encroach on land dedicated to waste treatment, industry, and agriculture. Farmers are seeking increased efficiency through measures that grow animals in dense concentrations nearer the point of consumption. Rapid development of new materials and new industries creates a host of exotic chemical discharges. Odors form the burgeoning fast food industry are a special problem – coffee and fried chicken may smell great when you are ready to eat, but the continuous smell of roasting beans or stale grease becomes offensive. All of these trends mean that populations are more often living close to odorous facilities, and it means that odor control is an ever more pressing air quality problem. Cries from the farmers and treatment plant operators of "We were here first!" will have little effect.

With respect to odors, ugliness is entirely in the nose of the beholder. We have evolved to recognize some compounds as associated with danger: the smell of human waste is offensive because those individuals who avoided contact with it were less likely to die of disease, and the smell of spoiled food prompts us to avoid eating meals that could threaten food poisoning. The perception that an odor is foul is the body's way of telling the brain that it is best to go elsewhere.

In modern society, we frequently encounter odors that no longer represent a health risk – it is not dangerous to be downwind of a dairy. But while the nose may not provide accurate warning of the presence or absence of a health risk in modern society, the evidence suggests that unpleasant odors can cause adverse physiological and neurogenic responses. The study by Schiffman et al. (2000) found that the most frequently reported health complaints related to odors include eye, nose, and throat irritation, headache, nausea, diarrhea, hoarseness, sore throat, cough, chest tightness, nasal congestion, palpitations, shortness of breath, stress, drowsiness, and alterations in mood. Certainly, unpleasant odors can cause human nuisance and physical discomfort.

Noxious and foul odors also have economic consequences. Foul odors can reduce property values in affected neighborhoods from 15 to as much as 90% (Kleemeier et al. 2002; Weida and Hatz 2002; Anstine 2003). The offensive odors

Shareefdeen/Singh (Eds.)
Biotechnology for Odor and Air Pollution Control

affect the quality of life, so that fewer people and businesses may be attracted to the region, potentially lowering property values.

Even where health and economic consequences cannot be demonstrated, the social and political consequences remain. Businesses that release odors may find their futures threatened by an angry local community.

Thus, growing population, the trend towards urbanization, urban sprawl, new factories, factory farms and fast food have combined to make odor a leading issue in environmental protection. The public demand for control has never been greater, and the need to protect the public from nuisance and perceived health and safety threats is rapidly growing. It is unquestionable that the general market for odor control is expanding, and probable that the expansion will be rapid in coming years.

17.2 Biotechnology is an Important Alternative

While it seems certain that the odor control market will expand, we must also ask what portion of it will be served by biotechnology. Customers can choose from many technologies, and assessing the prospects of biological systems requires that we determine what portion of the odor market can be taken from the competition. Historically, combustion, adsorption (activated carbon), and absorption (chemical wet scrubbing) were the primary technologies used to remove odorous compounds from air discharges. While the first biofilters were tried at least as long ago as the 1950s, it was not until the mid-1990s that biotechnology took its place in the US market. Each odor control technology has strengths and weaknesses, and the advantages and disadvantages (Ozis 2002) of competing odor control technologies can be summarized (Table 17.1).

As has been demonstrated in the previous chapters of this book, biofilters and biotrickling filters can be effective and economical in many applications. Biological treatment is appropriate for large flows because the challenge of managing a biological system is much the same for a large or small system, and management costs per unit flow become relatively high for small flows, although there are exceptions, such as the systems designed for rag drying by Bio-Reaction Industries (Stewart et al. 2000). It is more likely to be competitive for low concentrations of contaminant because high VOC concentrations may make incineration possible without additional fuel. The contaminant, of course, must be biodegradable and water-soluble, but these thresholds are not very high: benzene is commonly considered a low-solubility compound, but it is readily removed in biological treatment systems. Biological systems generally have a low specific activity – rates of contaminant destruction per unit volume of reactor – and the reactors must correspondingly be larger than competing devices for a given load. Thus, biological systems are not competitive where space limitations are severe. Biofilters have not yet been applied to waste streams including inert particulates like soot or dust. These non-biodegradable materials will be captured, but will accumulate in the biofilter until it clogs, so that packing replacement or cleaning will be necessary. Biotrickling filters

Table 17.1. Comparison of odor control technologies

Advantages	Disadvantages
Combustion: odorous compounds are burned	
Small, simple system	High capital/fuel/operating costs
Can handle very high concentrations and shock loads	Secondary waste stream (possible NO_x or SO_x production)
Consistent performance	Poor public image
99% Removal efficiency	Low inlet concentrations require fuel supplementation at high cost
Easy monitoring	High costs per m^3 of air
Adsorption: activated carbon traps odorous compounds	
Short handling time and small unit footprint	High operating and energy costs
Can handle very high concentrations and shock loads	Moderate capital costs
Consistent performance	Modifications required for high relative humidity waste streams
Can achieve high removal efficiencies	High carbon replacement and regeneration costs
	Disposal of used carbon as hazardous waste may be required
Wet scrubbers: chemical reactions oxidize odorous compounds	
Smaller and cheaper installation than biofilter (except for H_2S)	Chemicals must be continuously supplied to the system at significant cost
Less than 90% removal efficiency	Scrubber chemicals may be hazardous
	Liquid effluent at a very high or low pH may be produced
Biofilters and biotrickling filters: fixed microorganisms on porous packing degrade odorous compounds	
Creates no toxic end products, with no potential disposal problems (except degraded packing)	Relatively large space requirement
Low energy costs and moderate capital costs	Less suitable for high loads
Very low operating costs	Less operating and regulatory experience
Wide applicability for various odors and VOCs	Lower removal efficiencies, may not be best-available control technology
Public view of "natural" process is favorable	Requirement for operator sophistication, familiarity with biological systems

may be more workable in such situations because the moving water phase can collect the particles and carry them to a wastewater discharge.

There is a new aspect of waste air treatment that may give biological systems a further edge. The public wants technologies that combat odor in a safe, natural and sustainable way. The biological treatment systems work at room temperature without toxic chemicals. They are safer than incineration or chemical scrubbing; they use less energy, do not produce hazardous end products, and can be characterized as a "natural" process. This is a mix of well-defined technical advantages and vaguely defined, perceived advantages, but all together, the fit of biological treatment with the aims of sustainability may constitute an important competitive advantage. Odor-causing industries face growing scrutiny for their environmental stewardship. Well-maintained facilities can be managed not only to benefit environmental and long-term economic productivity but also to improve public perception of that specific industry. The growing desire for sustainability will favor biological systems as an alternative for odor control.

17.3 Possible Obstacles

Three critical factors will be of importance in determining the ultimate success of biological air treatment processes. First, it is necessary that sufficient experimental data and field experience be accumulated, analyzed, and published so that biological methods become a low-risk choice. There have been some failures of biofilter installations, especially when they have been designed and installed by consultants without specific experience, and it takes only a few to make the biological approach seem too risky to consultants and their customers. Many buyers, particularly small businesses, are facing make-or-break decisions when they buy air pollution control equipment– a costly failure may easily put them out of business. Biological treatment system manufacturers must be able to design devices appropriate to needs of the client, and to have a high degree of certainty that the installation will meet regulatory requirements.

Secondly, a workable regulatory environment must be developed. In general, regulators share the enthusiasm for biological approaches: they like nothing more than to tell the regulated community that there is a cheap and dependable means of complying with air quality standards. But biofilters and biotrickling filters depend on the complexities of a microbial ecosystem that is, in turn, governed in some poorly understood way by environmental conditions. What rules can be established to ensure that systems have been properly designed and installed? What criteria are appropriate to determine whether they are being properly operated? How should output sampling be done on an open biofilter, where the air escapes in an irregular way over a large area? Can biological systems be effectively monitored for regulation without raising costs to the point where other, more easily assessed technology becomes more economical?

Some biological treatment systems have failed because of inadequate maintenance. The third challenge for the industry is to determine appropriate methods for monitoring and maintaining biological systems. They will fail, for example, if they are given either too much water or too little. Is it possible to choose the appropriate amount for each application, and to maintain it in the face of varying environmental and operating conditions? How can they determine how the amount of water needed changes as the reactor packing ages? How often should the packing be replaced?

Moore (1991) has discussed the general challenges of marketing new technologies, dividing potential buyers with similar interests and demands into market segments. "Technology enthusiasts" are always interested in new technologies. "Visionaries" see the new technology as a beneficial tool for their purposes. "Pragmatists" have an interest in technology that can work for them, and they share one third of the market. "Conservatives" are the group that has no intention of using a new technology until it is regulated and presented as standard. "Skeptics" will only use the new technology when there is no other alternative. The challenge for biological treatment is to develop experience, establish regulatory methods and rules, and create operating standards so that at least the "enthusiasts", "visionaries" and "pragmatists" are satisfied.

17.4 Current Successes

Biotechnology for odor control has been used in different markets such as animal feed houses, baking, fish processing, food production, pulp and paper, wastewater treatment, foundries, tobacco product manufacturing, swine production, and dairy farming. Nicolai et al. (2000) have compiled useful information for biofilter design criteria. They have listed airflow rates from different industries. There is also some information about common reactor packing materials and their properties, such as porosity, moisture capacity, nutrient capacity, and useful life. Empty bed retention times (EBRTs) are suggested for different facilities, and the relationship between EBRT and pressure drop is investigated. They have provided a step-by-step design approach for reactor sizing, fan selection, moisture control, and other system characteristics.

A well-maintained biological system can reduce odors by 95% or more (Boyette and Bergstedt 1998; Schiffman et al. 2000). In some cases, biofilters have resulted in the elimination of odor complaints (Alix 1998) and in other cases, improved operation and biofilter renovation resulted in a reduction of odor complaints (Haines and Brandt 1999).

Heating of animal tissue during the process of rendering generates odorous compounds. The performance of biofiltration in removing these odors was investigated by Luo (2001) using pilot-scale biofilters. Efficiencies were between 29.7% and 99.9%. Different packing materials such as sand, finely and coarsely crushed wood bark, and a bark/soil mixture were used for the biofilters. The bark and sand biofilters substantially reduced odors for about 3 years at an air loading rate of 0.148 m^3 air per m^3 of packing per min.

17.4.1 Wastewater Treatment Plant Odor Control

The most important odor produced by wastewater treatment plants (WWTPs) is H_2S. Biotreatment is particularly effective for removing this gas. The sulfide is oxidized to sulfuric acid, which can be washed out of the packing. Oxidation is performed by chemoautotrophs that do not release the exopolysaccharides that create thick biofilms, so the layer of active organisms remains thin and clogging does not occur, even at high treatment rates. Results show that even in retention times near 5 s, biotrickling filters could remove 95% of the inlet H_2S. Wu performed three experiments (Wu et al. 2001). In this study, a pilot-scale biotrickling filter was used at a local WWTP. In the first experiment, the author used H_2S concentrations of about 2–3 ppm. At a retention time of 15 s, the removal efficiency was about 95–99%. The second experiment investigated high inlet H_2S concentrations. As the concentration was increased from 20 to 100 ppm, the removal efficiency was reduced from 90 to 70% at 5-s retention time but it stayed about 98% for retention times higher than 10 s. In a third experiment, the filter was tested for resistance to shock loading. At 5-s retention time, the concentration was increased from 80 to 150 ppm for 10 min, and then it was reduced to 80 and 50 ppm. The efficiency was reduced from 73 to 60% but then it recovered. Even further improvement in biotrickling filtration of H_2S was reported by Gabriel and Deshusses (case study 1).

There are other odorous emissions from WWTPs. Biosolids are abundant sources of food for microorganisms in the form of proteins, carbohydrates, and amino acids. Microorganisms degrade the biosolids under anaerobic conditions, forming odorous compounds such as mercaptans, ammonia, organic amines, and organic fatty acids. These are mixed with the odorous compounds already present in the settled sewage. Because large amounts of biosolids are handled, odorous releases are inevitable.

Increasing population density will put ever greater loads on wastewater treatment plants, even as it means more residences and businesses will be built close to facility fences. Biological methods will be a safe and cost-effective way of dealing with this problem.

17.4.1.1 Case Study No. 1

Gabriel and Deshusses (2003) converted one of the scrubbers in Orange County Sanitation District to a biotrickling filter. The scrubber was one of two at a facility that receives 80 MGD of wastewater. The packing material was 4-cm cubes made of polyurethane foam with 10–15 pores per linear inch, with a surface area of 600 m^2/m^3 and a porosity of 0.97. This filter treated 16,000 m^3/h of air with an empty bed contact time of 1.6–2.2 s. H_2S removal efficiency for inlet concentrations of about 30 ppm was higher than 95%, while that for concentrations around 60 ppm was about 90%. This biotrickling filter was also able to remove significant amounts of organic sulfur compounds, ammonia and VOCs present in the air, and this capability suggests other practical uses for this biotrickling filter. The conversion of a chemical scrubber to a biotrickling filter can overcome many safety issues of chemical compounds, and also would result in significant operating cost savings. Because there are many wastewater treatment plants that currently have chemical scrubbers, their conversion to biological systems represents a significant potential market.

17.4.1.2
Case Study No. 2

Hillsborough County, Florida, has 19 biofilters in its system, most of them at pumping stations. Serving the Tampa area, the Hillsborough County wastewater system includes ten treatment plants and 500 pumping stations. In 1992, the county began experimenting with biofiltration. Based on the success of the pilot units, Hillsborough County installed a biofiltration system at the Sun City wastewater treatment plant in 1998. The plant receives 4.5 MGD of wastewater. During a plant expansion, the county installed a biofilter that treats 3,400 cfm (cubic foot per minute) of fouled air from the headworks, anaerobic selector basins, and a pumping station that is part of the solids processing operation. Hillsborough County has installed another system (130 cfm) at its Van Dyke plant, which handles 500,000 MGD. Unlike the plants in Somerset and Hillsborough County, the South Cobb facility is located in an industrial area. The plant, which handles 23 MGD, had never had a dedicated odor control unit. The online biofilter, built in 1996, handles 4,000 cfm and costs US$ 75,000 to design and build. Another biofilter will be added to the system to treat 16,000 cfm, and will cost the county US$ 300,000.

17.4.1.3
Case Study No. 3

Georgia composting facility composts over 400 t/day (t, ton) of municipal solid waste and biosolids within a few hundred feet of neighboring homes and businesses. The facility's operating permit requires it to meet an odor performance standard of 10 dilutions to threshold units at the site boundary. The facility is designed with a ventilation system to contain, direct, and clean the air. Odor emissions in the building are captured at critical points, and nearly all the air entering the building leaves through the facility's biofilter. This system covers a total area of about one acre, and it consists of six humidification towers and seven biofilter beds. Each biofilter bed is 40 in. deep, 20 ft. wide and 280 ft. long. Two fans, together running 80,000 cfm, feed three of the beds, and five fans operating at 150,000 cfm serve the other four beds. The biofilter packing is wood chips with a broad particle-size distribution, mixed with 10% compost. The packing is replaced in the biofilters approximately every 18 months when the pressure drop exceeds 8–10 in. of water.

17.4.2 Swine Industry

Odors from swine production facilities arise basically from manure decomposition. Odor from fresh manure is generally less offensive than odor released when manure undergoes anaerobic decomposition. Rotting feed materials may also contribute significantly to the odor problem. Some food processing by-products fed to livestock are also particularly offensive. Dead animals and pesticide sprays may also contribute to odors.

Biofilters can reduce on average 95% of hydrogen sulfide and 65% of ammonia emissions from swine industry facilities. Knowing that the biofilters are efficient and cost-effective for this matter, designing them for different conditions and odor concentrations still remains a challenge. Air ventilation for summer rates from swine facilities ranges from 120 to 500 cfm per animal space (8 $ft.^2$ per animal).

17.4.2.1 Case Study No. 1

Two biofilters were designed and constructed in Minnesota in a 700-sow, deep pit facility in November 1997. Construction costs are reported to be US$ 0.22 per pig, presuming a 3-year life for thebiofilter packing. Odor, hydrogen sulfide, and ammonia were reduced by 93, 92, and 57%, respectively. The pressure drop ranges between 0.02 and 0.25 in. of water from minimum ventilation to maximum ventilation. Another biofilter was installed in a dairy unit in 1999 to treat air from a manure collection pit. The manure comes from a 1,200-cow barn and is collected in a pit. The efficiencies for odor, hydrogen sulfide, and ammonia were 94, 88, and 76%, respectively.

17.4.2.2 Case Study No. 2

A study was carried out by Mann et al. (2002) regarding treatment of odor from swine barns during subzero ambient temperatures. Four biofilters were constructed near a swine barn. Two of them had wood chips and compost packing in a 1 : 1 ratio, and the other two contained the same packing in a 3 : 1 ratio. EBRTs for the biofilters ranged from 5.2 to 6.7 s. Maximum and minimum ambient temperatures ranged from 9.2 to −34.2 °C. The temperature of the barn had been 20 °C, and the temperature within the biofilter had been measured to be 16.3 °C with a standard deviation of 1.8 °C. The odor reductions were observed to range between 56 and 94%. Authors suggest that biofiltration may be an effective way of odor reduction in subzero conditions. The inconsistency in the reductions may be because of the bed temperature or inadequate residence time.

17.5 Technology Developments

The generally increasing importance of odor problems and the advantages of biological treatment indicate that its future market share will be substantial. However, how will the characteristics of biological treatment systems and practices in the industry change?

17.5.1 Rational Design

Modeling of the various phenomena governing biological treatment systems has been advancing rapidly. However, the design of the bioreactors is still done largely by "rule of thumb" (this does not apply to appurtenances like ducting and blowers, which can be designed according to principles that have been well developed for other applications). Choice of the reactor packing is frequently governed by the personal tastes of the designer, or by the availability of some cheap local material. We are not yet able to say, for example, whether treatment of benzene and toluene vapors is better done on compost or wood chips. Choice of grain size in the packing is largely haphazard. Prediction of the life of the packing, or even whether it will fail by deterioration leading to compaction, or by biomass growth leading to clogging is seldom done. Designers seek reports of apparently successful previous installations and copy them, perhaps making small changes based on a recent research result.

In the future, improved "rational design" techniques will gradually develop. After determining the composition of the waste air, the designer will choose an optimum combination of mean pore size, packing composition, and operating characteristics for each application. The knowledge needed to do this will be developed by further research on topics such as the mechanisms of clogging (Nukunya et al. 2002; Ozis et al. 2003), the kinetics of biofilm growth (Song and Kinney 2002), and fundamental microbial ecology. It is a big step from such fundamental theory to design of an application, but understanding of biological treatment microbial ecology is growing.

17.5.2 Reliability

Better design will contribute to the most needed improvement in biological air treatment – improved reliability. Certainly, the industry is accumulating a track record, and as more consultants gain experience and more systems are proven, we will approach the time when virtually every biofilter and biotrickling filter works as planned.

Kraakman (2003) has suggested a list of steps that should be taken to improve the reliability of biological treatment systems.

17.5.2.0.1
Alarms.
Critical processes should be controlled by two levels of alarms. The first should notify the operator of a problem, and the second should automatically put the system into the least risky mode of operation. Increasing temperatures in the input air, for example, may first sound the alarm, then automatically shunt the air to the atmosphere to avoid killing the microorganisms.

17.5.2.0.2
Inoculum.
Several modes of failure, particularly high temperatures or low water content, can either kill the microorganisms or put them into a dormant state. Kraakman (2003) suggests that an active culture of appropriate organisms should be maintained at all times, so that a system in which the microorganisms have been inadvertently killed could be immediately reinoculated through the irrigation system. Maintenance of this inoculum in a chemostat separate from the working biological treatment system would be a modest effort.

17.5.2.0.3
Feed Augmentation During Downtime.
Industrial processes are commonly episodic. Many involve batch processing that releases high concentrations of contaminants in bursts. Even continuous processes are frequently shut down at night, on weekends, and on holidays. In many cases, the microbial ecosystems in biological air treatment have proven stable with respect to short interruptions, returning rapidly to activity when the contaminant reappears. For longer breaks, however, and for some particular pollutants, periods of a few hours during which treatment is poor have been seen at the resumption of operation. This can result in a significant degradation of overall treatment efficiency.

During interruptions, microorganisms using the contaminant for food may stop growing, reduce the concentrations of critical enzymes, enter dormancy, and even die. Problems will be particularly severe if airflow is stopped and the system goes anaerobic. Thus, at least a moderate flow of air should be maintained during downtimes, and it may also be appropriate to artificially provide a modest concentration of contaminant.

17.5.2.0.4
Backup Equipment.
Certain equipment, such as air fans and irrigation pumps, are critical for system operation. If they fail, treatment stops and the microbial ecosystem begins to decline. Where possible, these should be designed in multiple parallel units so that when one fails, the others can take the load. Procedures should be established for dealing with such failures. In one example, site personnel observed that the sprinkle irrigation system for a biofilter had failed. They called a sprinkler repair firm, which was not accustomed to the need for

rapid response and so did not come to the site for several days, and there was substantial damage to the performance of the biofilter. Procedures should be established so that both the site personnel and the responding company understand what needs to be done and how quickly a repair should be made.

17.5.2.0.5
Failure Modes.
Equipment should be designed so that failures put the system into the least damaging mode of operation by default.

17.5.2.0.6
Specify Standard Operating Procedures.
Proper procedures for operation and maintenance contribute to reliability. Every system installed should come with a manual. This means the biosystem vendors need to learn exactly what is required to keep their systems running, and carefully inform the owners. The owners must follow through by performing the maintenance and ensuring proper training of the personnel responsible for the systems. Ideally, vendors should continue to be involved in the operation of the system for a period after it is installed. Part of this lies in a commitment by the builder to fully inform the customer. The customer would like to hear, and so the builder would like to say, that biological treatment systems require little or no maintenance. However, an honest description of needed maintenance, and a clear recognition by the customer that maintenance must be done are a necessary first step for system reliability.

17.5.3
Inert Packing

A key element in bioreactor design is the material that is used to support the biofilm. In most cases, these are granular solids that are packed in the bioreactor with a random orientation of the grains. In some cases, more sophisticated materials, such as polyurethane foam, have been used to provide the surface and porosity that are basic to biofilter or biotrickling filter operation. The randomly packed granular materials may further be either "natural" organic matter such as wood chips, composted yard waste or peat, or may be inert (non-biodegradable) materials such as plastic shapes or fine gravel. Many biofilter problems are associated with failures of the organic packing materials. Some of these degrade rapidly in an environment, which, after all, is maintained in a condition intended to be ideal for biodegradation. Wood chips have been used in biofilters intended to control hydrogen sulfide, even as we know that sulfide causes rapid deterioration of wood–treatment with sulfide is the first step in converting wood to wood pulp. As organic packing materials loose their structural strength, compaction of the bed may close the pores. Fine particles may be released and wash downward to form a pore-clogging sludge.

As we have recognized these problems with organic packing, we have become less appreciative of their advantages. Organic packing materials, especially compost, may come with vigorous inoculum of microorganisms adept at biodegradation of complex organic compounds. But it has been repeatedly demonstrated that even totally inert packing materials can grow an appropriate microbial culture if a mixed microbial inoculum and the proper conditions are provided. The slow biodegradation of organic materials is also considered valuable because it releases nutrients: nitrogen compounds, phosphate, sulfate, trace minerals, and possibly complex compounds such as vitamins. However, once again, this has not proven critical. Nutrients can be added to biological systems with the irrigation water, and experience with inert packing material has shown that vigorous treatment is possible without an organic nutrient source– nutrients can be supplied with the irrigation water. There may be unknown compounds crucial to the success of individual microbial species, but we are clearly able to support effective treatment consortia with the nutrients we do know.

Organic packing materials are also difficult to specify. There are little data on the particle-size or pore-size distributions for the commonly used materials, and they are produced using a variety of methods. Composted yard waste, for example, varies according to the character of the original waste, the time of composting, water content and temperature during composting, and sieving. Each producer uses different methods, and even for a single producer, batches will differ in character.

Inert materials, such as plastic shapes or ground lava rock, can be essentially permanent, and their pore-size distribution will not change as treatment proceeds (except through the accumulation of biomass). They can be assembled to produce any desired mean pore size, and there may be considerable control over the pore-size distribution. Many of the packing materials are lighter or structurally much stronger than compost, so that deeper beds can be constructed without fear of compacting the lower layers. This in turn allows construction of biological systems with a smaller footprint, addressing one of the primary disadvantages of biological treatment. Advanced synthetic packing materials, such as polyurethane foam, can provide far higher surface-to-volume ratios than compost (about 600 m^2/m^3), and include much less "dead space" within the packing particles. They may also be of much lower bulk density, allowing construction of a lighter reactor structure.

Thus, as we learn how to choose the ideal packing pore size and pore-size distribution, as the problems of degrading packing materials become more apparent, and as their advantages seem less necessary, there is likely to be a trend to designer choice of inert packing materials.

17.5.4 Biomass Control

For biological systems receiving low loads of contaminants, it is apparently possible for the biofilm to reach a steady state. The activities of microbial predators and film-degrading organisms can reach levels such that they are consuming biofilm as rapidly as it is produced, and the system can theoretically operate indefinitely without clogging. This is commonly the case for odor control biofilters for which input concentrations are low. In many cases, however, the load may be high enough so that biomass accumulation becomes a problem. Pore radii are reduced, small pores are blocked, and biofilter performance suffers.

Some research (Ozis et al. 2003) also indicates that biofilms in biofilters rapidly become "too thick" – that is, they reach the point at which diffusion resistance prevents penetration of contaminant to the deeper portions of the biofilm, and it becomes inactive. Significant activity is likely to be restricted to the outer 100 μm or so of the film. Thus, once the biofilm is more than 100 μm thick, additional growth does not provide additional active biomass. Instead, the thickening film reduces the effectiveness of the system as pores shrink and clog, shrinking the surface area available and therefore the total volume of active biomass (equal to the surface area in contact with the air multiplied by the active depth).

The mean pore size chosen by system designers is typically governed by the need to avoid clogging. If a means were found to precisely control biofilm thickness during operation, biofilters and biotrickling filters could be made far more efficient. Pores could be much smaller, increasing the amount of surface and therefore active biomass per unit volume of biofilter, and allowing biofilters to be made much smaller.

Cox and Deshusses (1999) investigated biological control of biofilms in biotrickling filters – using protozoa to scavenge bacteria – and also did work on the use of chemical cleaners. Overall, however, the problem has not received much attention, despite its importance. It is therefore likely that research on biomass control will continue, and the new system designs will be developed with this objective in mind. Ultimately, because the size of bioreactors is commonly a limiting factor, the development of smaller reactors will broaden their applicability.

17.5.5 Inoculation

Biological treatment systems are only successful when they contain a vigorous culture of species capable of degrading the contaminant. Because the systems are open, new species are constantly being added, and it is presumed that there will eventually be a community that is dominated by those most capable of using the contaminant under the existing environmental conditions. However, there is inevitably a period of time between startup and achievement of full efficiency, and a vigorous inoculation procedure may shorten this period. Many researchers have studied this phenomenon through inoculation with selected microbial species, or by comparing alternate methods of inoculation. Many investigators have favored using a single, ideal species, known to vigorously degrade the compound of interest. However, there may also be some disadvantages to using "laboratory adjusted" microorganisms in a biological system that is constantly receiving wild strains from the environment – effective competitors or predators may overwhelm the inoculated species (Devinny et al. 1999).

Some studies have employed a recirculating inoculation method to provide an abundant inoculum and shorten the startup period (Kinney 2000). More sophisticated methods have also been attempted. Nicolay et al. (2003) trapped a selected consortium of microorganisms in a matrix that was used to coat a granular packing. Compared with classical bioaugmentation (spraying a suspended culture), inclusion in a synthetic matrix almost doubled the elimination capacity. This approach also provided control of the pressure drop for the airflow through the reactor. As designers more commonly choose inert instead of organic packing, improved techniques for inoculation will be needed.

17.5.6 Standards

There are currently no generally accepted standards for design and operation of biological treatment systems. A designer considering the characteristics of a particular discharge, and making decisions on packing type, pore size, empty bed detention time, or whether to use a biofilter or biotrickling filter has no manual to turn to. Indeed, no professional society has taken up the charge of developing standards of practice.

This is perhaps appropriate – biological air treatment is still new, and is developing rapidly. It is important that practitioners be given maximum latitude for innovation. But if biotechnology is to become a commonly employed option, it will eventually be necessary to develop and promulgate design standards.

17.5.7 Sensing and Automation

The environment within a bioreactor changes with the characteristics of the incoming air. Changes in the weather, including temperature and relative humidity, are compounded by changes in operating conditions. Plants shut down during the night and on weekends. Many discharges are episodic or irregular. Thus, the biological system is subject to changes in temperature, humidity, contaminant concentration, and airflow rate that are largely beyond the control of the operator.

These changes require changes in operating conditions. An increase in temperature, a decrease in humidity, or an increase in airflow rate will increase evaporation and require more water addition. A process upset may release hot air that should be prevented from entering the biofilter to avoid killing the microbial culture.

This suggests that improvements could be made in biological treatment system efficiency through the use of automated data collection and operation. Some biological systems have been made with load cells or other devices to continuously measure and readjust water content. The input airflow should be automatically shunted aside if temperature rises to the point of threatening the microorganisms (Kraakman 2003). For closed systems, input and output contaminant concentrations could be monitored to check on treatment success, and to provide data to demonstrate regulatory compliance.

17.5.8 Increasing Size

Larger installations will become common, and several examples of this trend can be cited. The City of Davenport, Iowa, owns an aerated static pile biosolids and yard waste composting facility with a capacity of 28 dry t per day. The 210,000 cfm of exhaust that is coming from the totally enclosed mixing and composting areas is treated through a biofiltration system. The system, which consists of inline humidification and eight independently operated cells, began operation in July 1995 (E&A Environmental Consultants 2003). In 1996, Western Lake Superior Sanitary District in Duluth, Minnesota, installed a biofilter to reduce emissions from the wastewater treatment plant. The system treats 50,000 cfm of exhaust gas through three individual cells (Mulder 2003).

A composting plant in Amsterdam operates two biofilters having a total discharge of $\sim$ 71,000 cfm of processed air (Mulder 2003). In the summer of 2000, two biofilters in Norway started operation. One has a capacity of 30,000 cfm and treats air from a municipal wastewater treatment plant, while the other, with a capacity of 11,770 cfm, cleans the air from sewer piping (BBK Bio Airclean 2003). As these examples suggest, in the future, as demand for waste recovery increases, larger systems will be built.

17.5.9 Wastewater Will Lead the Way

In the United States, installation at wastewater treatment plants will lead the development of the industry. Residential development will come ever closer to plants, increasing the likelihood of odor complaints. Hydrogen sulfide, the most common cause of odor at treatment plants, is readily treatable. The work of Gabriel and Deshusses (2003) in Orange County California showed biotrickling filters can efficiently remove hydrogen sulfide with an empty bed contact time of 1.6 s, making it possible to refit existing chemical scrubbers as biological systems. This approach will make installations much cheaper.

Finally, wastewater treatment plant operators are familiar with biological processes. This is an important issue in marketing. Vendors seeking to sell systems to facilities owners who are unfamiliar with biological systems have often encountered considerable skepticism. An owner accustomed to the simplicity of an incinerator is often reluctant to trust a complex and poorly understood ecosystem of invisibly small microorganisms. But the people who run wastewater treatment plants rely on biological systems every day and have no prejudices against them.

For all of these reasons, we have seen that the earliest and most widespread use of biological odor control systems is for wastewater odors. This is likely to continue, and will provide a base of successful experience from which the industry can expand.

17.5.10 Application to New Effluents

Biological air treatment is limited, by definition, to compounds that can be transformed to harmless products by the action of microorganisms. The first targets of biofilters and biotrickling filters have been readily degraded compounds like hydrocarbons and hydrogen sulfide. Researchers, however, are continually making the effort to utilize the advantages of biological treatment for new and more difficult compounds. Efforts are well under way, for example, to establish biological methods for treating ammonia (Choi et al. 2003).

The gasoline additive methyl tertiary butyl ether has become a common groundwater contaminant, partly because the tertiary structure makes it a difficult target for the soil microorganisms that tend to degrade other gasoline components. Several attempts have been made to develop biological air treatment systems that could be used in conjunction with air stripping on contaminated soils, but researchers have not yet developed systems to the level of practical application.

Researchers have also longed to apply biological treatment to chlorinated hydrocarbons, such as the trichloroethylene (TCE) that is such a widespread groundwater contaminant. Contaminated water could be brought to the surface and passed through an air stripper, and the resulting TCE vapors could then be destroyed by biological systems. TCE, however, is not readily degraded as a sole source of energy in aerobic systems. Investigators have instead sought to use cometabolites to induce the activity of enzymes that fortuitously degrade TCE. Lackey et al. (2003) devised a system for TCE degradation in which the biomass was periodically exposed to propane. This promoted synthesis of the active enzymes, which were then available to degrade TCE during the intermediate periods.

The future of biological air treatment will thus include a continued effort to apply systems to new kinds of pollutants, as new schemes for utilizing the powers of microbial ecosystems are developed.

17.5.11 Development of Green Manufacturing-Biosystem Combinations

In the past, the materials used for industrial processes have been chosen solely on their costs and suitability for the product being made. The principles of "green manufacturing" and "sustainability", however, are gaining new attention. Regulators try more and more to impose the costs of environmental protection on the process that generates the pollution. Facilities owners, in turn, are more commonly recognizing that they must be concerned with the total lifecycle cost of materials.

In the future, more of the materials used in manufacturing will be designed, where possible, to produce no toxic or odorous vapors. However, where this cannot be done, it may still be possible to make materials whose vapors are readily removed from exhaust streams by biological processes. Thus, a paint company may gain competitive advantage by formulating paints without chlorinated compounds or ethers, using instead only solvents that are readily degradable hydrocarbons or oxygenated hydrocarbons. They could then advertise that "Our paint – with a biofilter – will solve your air pollution regulatory problems".

17.6 Conclusions

The future prospects for biotechnology in odor control are bright. Certainly, the need for odor control will continue to expand. The advantages of biological systems, particularly for the low-concentration, high-volume flows typical of odor applications, will give them a substantial share of the market. While biotechnology is currently viable, there are also avenues for development of improved systems – and new technology will create further opportunities for use of biological treatment.

References

Accort J, Krause R, Rynk R (2001) Air handling holds key to odor management. BioCycle 42:54–59

Alix CM (1998) Retrofits curb biosolids composting odors. Biocycle Magazine, June

Anstine J (2003) Property values in a low populated area when dual noxious facilities are present. Growth Change 34:345–358

BBK Bio Airclean (2003) Odour removal at sewage treatment plants. http://www.bbk.dk/sewagethtm.html

Boyette RA, Bergstedt L (1998) Wastewater treatment plant odor control using a biofiltration system. In: Diluth MN (ed) Proc 9th Int Conf Cold Regions Engineering, American Society of Civil Engineers

Choi JH, Kim YH, Joo KJ, Choi SJ, Ha TW, Lee DH, Park H, Jeong YS (2003) Removal of ammonia by biofilters; a study with flow modified system and kinetics. J Air Waste Manage Assoc 53:92–103

Cox HHJ, Deshusses MA (1999) Biomass control in waste air biotrickling filters by protozoan predation. Biotechnol Bioeng 62:216–224

Devinny JS, Deshusses MA, Webster TS (1999) Biofiltration for air pollution control. Lewis, Boca Raton

E&A Environmental Consultants (2003) Biofiltration and odor control. http://members.aol.com/eaenviron/biofilter.html

Gabriel D, Deshusses MA (2003) Performance of a full-scale biotrickling filter treating H_2S at a gas contact time of 1.6 to 2.2 seconds. Environ Prog 22:111–118

Garner LG (2003) Overcoming marketing challenges for engineered biofiltration – a case study. In: Proc Air and Waste Management Association Natl Conf, 23–25 June 2003, San Diego, CA

Govind R (2002) Commercial potential for biofiltration. http://www.finishers-management.com/april2002/biofiltration.htm

Haines W, Brandt A (1999) Biosolids composting facility processing and odor control improvements, a case study. In: Proc WEF/AWWA Joint Residuals And Biosolids Management Conf

Kinney K (2000) Optimization of an innovative biofiltration system as a VOC control technology for aircraft painting facilities. http://www.serdp.org/research/CP/CP-1104.pdf

Kleemeier J, Dischler R, Dischler G, Carver B, Carver J (2002) Calhoun County pork producers sued for nuisance. Beazly Group, Inc. and Pork Innovation, Inc. http://www.iowapork.org/newsroom/nuisance.html

Kraakman NJR (2003) Robustness of a full-scale biological system treating industrial CS_2 emissions. Environ Prog 22:79–85

Lackey LW, Gamble JR, Boles JL (2003) Biofiltration of trichloroethylene-contaminated air: a pilot study. J Air Waste Manage Assoc 53:1248–1255

Luo J (2001) A pilot-scale study on biofilters for controlling animal rendering process odours. Water Sci Technol 44:277–285

Mann DD, DeBruyn JC, Zhang Q (2002) Design and operation of an open biofilter for treatment of odour from swine barns during sub-zero ambient temperatures. Can Biosyst Eng 44:6.21–6.26

Miner J, Brath R, Clyde L (1988) Controlling odors from swine buildings. Purdue University Cooperative Extension Service, WL, IN, Waste Management PIH-33

Moore GA (1991) Crossing the chasm. Harper Collins, New York

Mulder J (2003) PURVA composting plant, odour behaviour over 10 years. http://www.compostnetwork.info/downloads/odourworkshop/22_mulder_pruva_compostingplant.pdf

Nicolai R, Janni K, Schmidt D (2000) Biofilter design information. http://www.bae.umn.edu/extens/aeu/baeu18.html

Nicolay PT, Maesschalck ED, Simon JP (2003) Cell entrapment and coated-system immobilization (CECSI technology) for the biological treatment of industrial gaseous emissions. In: Proc Int Worksh Bioencapsulation. http://bioencapsulation.net/XI_IWB/index.html

Nukunya T, Tsotsis TT, Devinny JS (2002) Application of a pore network model to a lava rock biofilter treating ethanol. In: Reynolds FE (ed) Proc 2002 Conf Biofiltration (an Air Pollution Control Technology), 31 October–November 2002. The Reynolds Group, Tustin, California

Ozis F (2002) Biofiltration for air pollution control in composting facilities: an evaluation of regulatory and technical issues. USC White Paper, May 2002

Ozis F, Bina A, Yortsos Y, Devinny JS (2003a) A percolation-biofilm-growth model describing biomass clogging in biofilters. In: Proc Air and Waste Management Association 96th Annu Conf Exhibition, 22–26 June 2003, San Diego, California, Pap no 81866

Reynolds FE Jr, Grafton WD III (1999) Biofilter: an old technology comes of age. Environ Technol 9(4):51–52

Schiffman SS, Walker JM, Dalton P, Lorig TS, Raymer JH, Husterman D, Williams CM (2000) Potential health effects of odor from animal operations, wastewater treatment and recycling of byproducts. J Agromed 7:7–81

Schmidt D, Janni K, Nicolai R (2000) Biofilters for odor control. Department of Biosystems and Agricultural Engineering, University of Minnesota. http://manure.coafes.umn.edu/assets/biofilters.pdf

Song J, Kinney KA (2002) A model to predict long-term performance of vapor-phase bioreactors: a cellular automaton approach. Environ Sci Technol 36:2498–2507

Stewart WC, Barton TA, Thom RR (2000) High VOC loadings in biofilters – petroleum and industrial applications. In: Proc Air and Waste Management Association Conf, June 2000

Wade B (1999) The biology of odor control. American City County 1 Oct

Weida W, Hatz D (2002) Depopulation from rural areas linked to presence of factory farm operations. http://www.factoryfarm.org/docs/Weida.DepopCAFO.PrsRel.5021.doc

Wu L, Loo YY, Koe LCC (2001) A pilot study of a biotrickling filter for the treatment of odorous sewage air. Water Sci Technol 44:295–299

Subject Index

Made in the USA
Monee, IL
09 June 2026

52234753R00236